Elements of Quantum Mechanics with Chemical Applications

Note from the Publisher

This book has been set by monotype and printed by offset. The equations in the text were taken photographically from the original French volume. This unusual manufacturing method has been employed in order to bring students this advanced level translation at a low price in line with the objectives of the International Textbook Series.

Barnes & Noble International Textbook Series

Elements of Quantum Mechanics with Chemical Applications

Jean Barriol

Nancy, France

Translated by J. Warren Blaker, Vassar College

Barnes & Noble, Inc. New York

Publishers · Booksellers · Founded 1873

Library of Congress Catalogue Card No. 74-135117
ISBN 389 00450 2

Distributed:

In Canada by Methuen Publications, Toronto
In Australia and New Zealand by Hicks, Smith & Sons Pty. Ltd., Sydney and Wellington
In United Kingdom, Europe, and South Africa by Chapman & Hall Ltd., London

Printed in the United States of America

TRANSLATOR'S NOTE

In view of the significant contributions made to quantum chemistry by various European groups, it seems appropriate to have available in English an example of their approach in preparing their students. This text is a particularly good example of this approach, and I am grateful to have had the opportunity to prepare this translation.

A note on the references to the equations is in order. Equations within the chapter are referred to by the number alone. Those outside the chapter are referred to by the chapter number followed by the equation number.

I want to thank Professor J. Mucci for a number of helpful discussions and comments while I was doing the translation and for a complete reading of the final manuscript. I also want to thank Fannia Weingartner for her help and patience in copy editing the manuscript.

J. WARREN BLAKER

PREFACE

This book is the work of Professor Jean Barriol of the Faculty of Sciences at Nancy. It contains a modern development of the ideas of wave mechanics and of the formalism of quantum mechanics, along with numerous examples taken from the areas of atomic and molecular structure to which Professor Barriol has so heavily contributed. The author's purpose is to develop the fundamental principles of quantum chemistry.

The book is divided into three parts, the first of which is devoted to a thorough exposition of the fundamental ideas of wave mechanics and current quantum mechanical formalism, abundantly illustrated with examples. The author of this preface does not feel that the present formalism of quantum mechanics is sufficient to give us a complete physical representation of wave-particle duality, but it is nevertheless true that this powerful formalism permits the precise treatment of a number of important problems and is therefore of considerable value. In the first section, Professor Barriol gives a thorough, constructive treatment of the major applications of quantum mechanics to the single-particle case without considering spin. This section of the book, along with the appendices, presents an excellent general treatment of the material which should be introduced in a basic quantum mechanics text.

The second section of the book, after examining the structure of atoms and molecules, deals with the many-electron problem in an approximation which does not include the electron-electron interaction. Starting with the theory of the helium atom, the author discusses the problem of the covalent bond, the molecular orbital method, and the motion of an electron in a crystalline field.

In the final section of the book, the author treats the many-electron problem, including interactions, generally through the use of self-consistent fields. The author's profound and comprehensive understanding of this field has enabled him to make the presentation of these difficult problems—which are extremely important in the development of chemical theory—especially clear. This text will be useful to students beginning their study of wave mechanics and quantum mechanical formalism, and particularly useful to those interested in quantum chemistry. For both these groups this is an excellent guide.

LOUIS DE BROGLIE

INTRODUCTION

This work is addressed to students seeking an introduction to the methods of quantum mechanics for application in the field of chemistry. It corresponds to a course given for the third degree by the Faculty of Sciences at Nancy. For pedagogical reasons, the book is divided into three parts. The first part is devoted to general principles; the second to many-electron systems in a one electron approximation, and the final section to a treatment of many-electron systems with a more formal mathematical development.

Optical transitions and, in general, all problems requiring time-dependent perturbation theory are not treated. Various molecular applications are chosen from among those covered in the lectures on theoretical chemistry, but we have been unable to treat them all. Theoretical chemistry poses some difficult problems which can only be resolved by a comparison of experimental and theoretical results, and we have set these problems aside.

Even the most elementary exposition of theory often raises points which require the use of formal mathematical techniques. This is particularly true of quantum mechanics, especially in regard to its relationship with classical mechanics, and in the justification of the BWK approximation. We have not side-stepped these difficulties but have relegated them to the appendices in order to maintain the continuity of the text. Our treatment of spherical functions has also been left for the appendix so as to clearly separate the mathematical problem from its use in the theory.

We have not included a complete bibliography because of the vastness of the subject. In addition to several references within the text, we give a list of several fundamental treatises which have proven particularly useful to us.

Professor L. de Broglie has honored me by writing the Preface to this work and I hope that one can find here evidence of my gratitude to him and my admiration for him.

I am anxious to thank my colleague and friend, Professor A. Pacault, who read the draft, and whose comments I have found particularly useful.

A book which covers the material of a general course inevitably reflects the comments of students and thereby somewhat takes the form of a collective work. Thus, a number of the chapters have taken their final form only after long discussion. I thank in particular my former student and colleague, Professor M. Felden, and all my colleagues from the theoretical chemistry group who have helped me with the proofs.

J. BARRIOL

CONTENTS

(10) WAVE FUNCTIONS ON A TRANSLATIONAL LATTICE

PART THREE: GENERAL THEORY OF MANY-ELECTRON SYSTEMS

(11) DESCRIPTION OF THE STATES OF A SYSTEM OF MANY PARTICLES

(12) CLASSIFICATION OF THE STATES OF MANY-ELECTRON SYSTEMS

(13) THE COVALENT BOND

(14) SPECIAL PROBLEMS

TABLE OF SYMBOLS AND COMMONLY USED ABBREVIATIONS

[The number of the page on which these quantities are defined is given in parentheses.]

Algebraic functions and operators:

Δ	Laplacian (345).
$H(p_r, q_r)$	Hamiltonian (327).
$\mathscr{H}_r(\xi)$	Hermite polynomial (66).
$\mathscr{L}_{n,l}(\rho)$	Laguerre polynomial (82).
Λ	Laplacian angular operator (346).
P_l^m	Associated Legendre function (353).
$S(x_r, E)$	Maupertuis's action (328).
$W(x_r, t)$	Jacobi's action (321).

Monoelectric systems:

φ, θ, ψ	Wave functions.
$\alpha(\sigma), \beta(\sigma)$	Spin functions (95).
AO	Atomic orbitals (173).
MO	Molecular orbitals (159).

Magnitudes and the associated operator (in boldface print):

$\bar{a} = \langle a \rangle$	Mean value of a quantity *a* (56).
$\mathbf{H}$	Hamiltonian (46).
$\mathbf{I}$	Inversion operator (165).
J_r, $\mathbf{J}_r$	Component of the total angular momentum (124).
L_r, $\mathbf{L}_r$	Component of the orbital momentum (77).
P_r, $\mathbf{P}_r$	Component of the linear momentum (60).
S_r, $\mathbf{S}_r$	Component of the spin (93).

Nomenclature:

n, *l*, *m*	Atomic quantum numbers (83).
ns, *np*, *nd*	Atomic states (86).
k	Quantum number for a diatomic molecule (162).
σ, π, δ	States of the diatomic molecule (163).
g, *u*	Parity (165).
+, −	Reflection symmetry (163).

Many-electron systems:

Φ, Ψ, Θ	Antisymmetric wave functions
(Φ, Φ)	Normalization integral (285).

Variables and the associated operator:

$\mathscr{A}$	Antisymmetrization operator (240).
$\mathscr{H}$	Hamiltonian operator (238).
$(\Theta_r, \mathscr{H}\Theta_s)$	Hamiltonian matrix element (285).
$\mathscr{I}_r, \mathscr{T}_r$	Component of the total angular momentum (orbital and spin) (262).
$\mathscr{L}_r, \mathscr{L}_r$	Component of the total orbital angular momentum (253).
$S_r, \mathscr{S}_r$	Component of the total spin moment (253).

Nomenclature:

S, P, D	Atomic spectroscopic terms (255).
Σ, Π, Δ	Diatomic molecular spectroscopic terms (264).

FUNDAMENTAL CONSTANTS

Dielectric constant in vacuum (MKSA)	$\varepsilon_0 = 8.8543 \times 10^{-12}$
Vacuum permeability (MKSA)	$\mu_0 = 1.2566 \times 10^{-6}$
Speed of light in vacuum	$c = 2.9978 \times 10^8$ m-sec^{-1}
Electron charge	$e = 1.6021 \times 10^{-19}$ C
	$= 4.8029 \times 10^{-10}$ esu
Electron mass	$m_0 = 0.9109 \times 10^{-30}$ kg
Planck's constant	$h = 6.625 \times 10^{-34}$ J-sec
	$= 6.625 \times 10^{-27}$ erg-sec
	$h/2\pi = \hbar = 1.054 \times 10^{-34}$ J-sec
Bohr radius	$a_0 = 5.292 \times 10^{-11}$ m
Rydberg constant†	$E_0 = 21.796 \times 10^{-12}$ erg
	$= 13.605$ eV
	$= 109{,}737$ cm^{-1}

Energy units:

1 eV $= 1.6021 \times 10^{-12}$ erg $= 1.6021 \times 10^{-19}$ J $= 23.053$ kcal/mole

1 kcal/mole $= 6.946 \times 10^{-14}$ erg.

† This value of E_0 is calculated from equation (21), p. 83, under the assumption of a fixed nucleus. The experimental value gotten spectroscopically is slightly different (109,677.58cm^{-1}).

Part One

GENERAL THEORY AND PRINCIPLES OF QUANTUM MECHANICS

1 THE FOUNDATIONS OF WAVE MECHANICS

Historically, wave mechanics was developed by L. de Broglie via the theoretical analogy between certain forms of the laws of particle dynamics and wave propagation. These ideas are very fundamental but since they involve a lengthy mathematical development they will be found in Appendix I. This chapter is limited to an intuitive introduction developed through analogy with the wave phenomena associated with the propagation of light.

The Duality of Light 1

The inadequacies of the classical theory of light were demonstrated much sooner than were the failings of the classical laws of mechanics. Historically, two models were introduced to explain the laws of reflection and refraction. Newton advanced a particle theory while Huygens advanced a wave theory. The latter model was introduced because it was more suitable for the understanding of interference and diffraction phenomena. The wave theory was found even more satisfactory because of its analogy with Fresnel's theory of elasticity and Maxwell's electromagnetic theory. It is in the interpretation of the interaction of light and matter in the study of black bodies and the photoelectric effect that this wave theory breaks down. The photoelectric effect, discovered in 1887 by Hertz, involves the emission of electrons by a clean metal plate under a vacuum when exposed to radiation of sufficiently short wavelength. The peculiarity of this phenomenon rests on the fact that the velocity of the emitted electrons does not depend on the intensity of the illumination, as one would expect according to the principle of conservation of energy, but solely on the frequency, ν, of the light by the equation:

$$v = \mathrm{A}(\nu - \nu_0)^{1/2}$$

which we can rewrite as:

$$v^2 + \mathrm{A}^2 \nu_0 = \mathrm{A}^2 \nu,$$

with the constant, A, independent of the nature of the metal. The right-hand side of the equality is only dependent on the radiation while the left-hand side reflects the metal and the electron. The form of this equation led Einstein to assume that it expressed the conservation of energy for the electron emission process. To show this, he introduced the factor $m/2$ in the preceding equation to make evident the kinetic energy term corresponding to the emitted electron. The equation then becomes:

$$\frac{1}{2}mv^2 + W_0 = h\nu \qquad \text{with} \qquad W_0 = mA^2 \nu_0/2;\ h = mA^2/2.$$

The constant, h, which appears here† is the same constant that he had already introduced in his study of radiation and to which one can assign the value: $h = 6.623 \times 10^{-27}$ erg-sec or 6.623×10^{-34} J-sec.

In this form, W_0 represents the energy necessary to extract the electron from the metal, so that the left-hand side of the equation must represent the total energy furnished by the radiation to the metal. This energy is independent of the intensity of the illumination and depends only on the frequency ν. This then suggests that the interaction between light and matter takes place only in units of value $h\nu$, which suggests the concept of a discrete "particle of light" called a photon.

In developing this theory Einstein used the laws of special relativity as well as the laws of optics and mechanics. In particular, he used the result that a particle with rest-mass m_0 and velocity $\vec{v}$ has energy U and momentum $\vec{p}$ given by:

$$(1) \qquad U = \frac{m_0 c^2}{\sqrt{1-(v^2/c^2)}}; \qquad \vec{p} = \frac{m_0 \vec{v}}{\sqrt{1-(v^2/c^2)}}$$

where c is the velocity of light. These equations can be used for photons to demonstrate that these particles should travel with the speed of light. The first equation in (1) is not compatible with a finite energy of a photon $U = h\nu$ where the rest mass is zero. Also, the momentum as determined by (1), is indeterminate in the form 0/0. These difficulties can be resolved by forming the ratio p/U. One can see that by taking the modulus of $\vec{p}$:

$$(2) \qquad \frac{p}{U} = \frac{v}{c^2} = \frac{1}{c}, \qquad \text{so that} \qquad p = \frac{h\nu}{c} = \frac{h}{\lambda}$$

where $\lambda = c/\nu$ is the wavelength. One is therefore led to a fixed value of momentum for the photon, which can be used to understand various other effects—such as the Compton effect and radiation pressure—which involve classical collision processes governed by the conservation of momentum and energy. The following exercises illustrate the calculations involved.

---

† In order to simplify the equations we hereafter use either h or $\hbar = h/2\pi$ as required.

Exercise 1. The Compton effect involves the change in frequency of a scattered beam of X-rays as a function of the scattering angle θ (fig. 1).

This phenomenon is observed when X-rays are passed through various materials, such as paraffin or graphite, in which the electrons are only weakly bound to the atoms and can be thought to be at rest. We use the notation of figure 1.
(*a*) Write expressions for the conservation of energy and momentum and find the components of the momentum parallel to and normal to the direction of the incident photon.

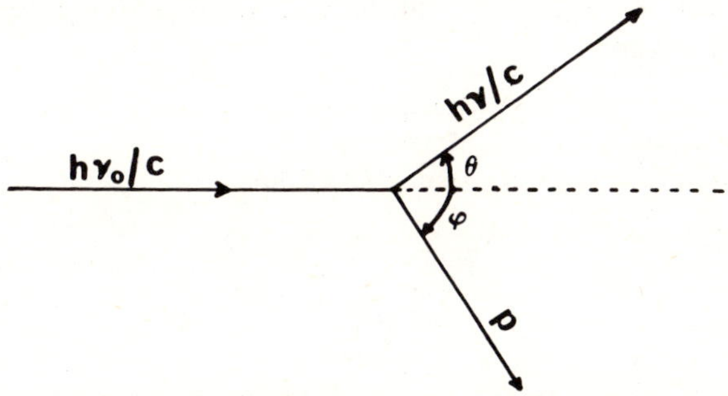

Figure 1. Diagram of the collision of a photon with momentum $h\nu_0/c$ and a stationary electron. Following the collision the scattered photon and the electron separate in the direction of the motion of the incident photon with deviations θ and φ.

(*b*) Elimination of φ is direct and gives:

$$\frac{h^2}{c^2}[\nu_0^2 - 2\nu_0\,\nu\cos\theta + \nu^2] = \frac{m_0^2\,v^2}{1-(v^2/c^2)}$$

which when combined with the expression for conservation of energy:

$$h[\nu_0 - \nu] = \frac{m_0\,c^2}{\sqrt{1-(v^2/c^2)}} - m_0\,c^2$$

gives:

$$\Delta\lambda = \lambda - \lambda_0 = \frac{h}{m_0 c}(1 - \cos\theta) = 0.0242(1 - \cos\theta), \text{ in Å}.$$

We note that the result is independent of the wavelength of the incident photon.

Exercise 2. Calculate the pressure which arises when radiation is incident on a perfectly reflecting surface.
(*a*) Consider the collision of a photon incident at angle θ. Find

the mean force acting over time τ in the same fashion as for a classical molecular collision with the surface. One finds:

$$\bar{f} = \frac{2}{\tau} p \cos \theta.$$

(*b*) Find the number dn of photons arriving during time τ on 1 cm^2 of the surface from an element of solid angle $d\Omega$ at angle θ; let N be the number of photons/cm^3 with random direction. We find:

$$dn = \mathrm{N}c\tau \cos \theta d\Omega.$$

(*c*) The radiation pressure can be calculated by integration over the angles to give:

$$\mathrm{P} = \mathrm{N}cp/3 = \mathrm{U}/3$$

where we have introduced the energy density:

$$\mathrm{U} = \mathrm{N}h\nu = \mathrm{N}cp.$$

Table I

Wave Model	Particle Model
Frequency: ν	Photon Energy: $E = h\nu$
Wavelength: λ	Momentum: $p = h/\lambda$
Energy Density: U	Photon Density: $N = U/h\nu$

In conclusion, light can be treated, depending on the case, as either a wave phenomenon or a particle phenomenon. A parallel beam of monochromatic light can equally well be treated as a beam of photons or a wave with the corresponding quantities given in Table I. We therefore have a dualistic model for light and are compelled by the nature of the particular experiment to use the appropriate model.

2 Material Particles

The classical laws of mechanics work quite well with macroscopic systems. However, the existence of stable, radiationless states of atoms and molecules seems to cast doubt upon the laws of mechanics and electromagnetism when applied to microscopic atomic and molecular systems. De Broglie suggested a solution to this problem by showing an analogy between particle momentum in classical mechanics and the velocity of optical rays in an isotropic, non-homogeneous medium. The analogy can be demonstrated by showing the equivalence of the variational form of dynamics to Fermat's principle in a medium of varying refractive index (Appendix I).

The failure of classical mechanical laws may be likened to the failure of geometric optics to explain diffraction and interference. The solution to the optical problem is well known and requires the use of optical waves,† which are treated either by Huygen's principle or by the wave equation. One can reason that the parallel solution may be made in the case of material particles according to the following scheme:

first approximation: classical mechanics $\xrightarrow{1}$ geometric optics

$\downarrow^{4}$ $\qquad\qquad$ ${}^{2}\downarrow$

second approximation: new mechanics $\xleftarrow{3}$ wave theory

We want to establish the correspondence (4) in the scheme by passing along (1), (2) and (3).

The rigorous equivalence of (1) is demonstrated in Appendix I, where it is shown that the trajectory of particles of equal energy in a field derived from a potential $V(x, y, z)$ is identical to the trajectory of rays of light in a medium where the refractive index is suitably chosen at every point. This can best be shown by a rigorous mathematical statement. However, we may take a special case where the correspondence is exact, namely, the case illustrated in the earlier section on light. If we use Table I in the left-to-right sense, we transform the model from a particle model, where we know the laws of motion, to the corresponding wave model, where the laws are *a priori* unknown.

We can then state: *a group of particles with mass m and energy* E *has momentum* $\vec{p} = m\vec{v}$ *and can be thought of as a beam of monochromatic radiation with frequency* ν *and wavelength* λ *as given by the two equations*:

(3) $\qquad E = h\nu \qquad$ Bohr

(4) $\qquad \lambda = h/p \qquad$ de Broglie.

---

† The essential point is the periodicity of the waves which requires the introduction of a wavelength. In the limit of *geometric optics* the periodicity vanishes.

For the time being we will not use the third relationship in Table I since it requires a more complete knowledge of the wave properties.

The transition (2) of the scheme is classical and consists in writing the equation of propagation of waves in a medium for which one knows at every point the properties which affect the phase velocity of the wave at that point.

We can write such an equation where φ is the wave variable:

$$\Delta\varphi - \frac{1}{u^2}\frac{\partial^2\varphi}{\partial t^2} = 0 \tag{5}$$

where Δ is the laplacian operator $\frac{\partial^2}{\partial x^2} + \frac{\partial^2}{\partial y^2} + \frac{\partial^2}{\partial z^2}$.

Equation (5) is the wave equation and is independent of the form of the wave being studied. One can derive it by assuming the existence of plane waves which propagate with arbitrary direction cosines α, β, and γ ($\alpha^2 + \beta^2 + \gamma^2 = 1$) and with constant phase velocity u. This suggests that the function φ ought to remain constant at time t for all points on the wavefront:

$$\alpha x + \beta y + \gamma z - ut = C,$$

where C is an arbitrary constant.

As a result φ may be thought of as a function of the single variable $\theta = \alpha x + \beta y + \gamma z - ut$ of the form

$$\varphi(x, y, z, t) = f(\theta) = f(\alpha x + \beta y + \gamma z - ut).$$

This can be verified by taking the required partial derivatives of φ giving:

$$\frac{\partial^2\varphi}{\partial x^2} = \alpha^2 f'', \quad \frac{\partial^2\varphi}{\partial y^2} = \beta^2 f'', \quad \frac{\partial^2\varphi}{\partial z^2} = \gamma^2 f'', \quad \frac{\partial^2\varphi}{\partial t^2} = u^2 f'',$$

from which $\Delta\varphi - \frac{1}{u^2}\frac{\partial^2\varphi}{\partial t^2} = (\alpha^2 + \beta^2 + \gamma^2 - 1)f'' = 0.$

The wave equation remains valid only when the velocity u varies sufficiently slowly with position and is independent of direction. This implies local isotropicity.

We can now treat step (3) of our scheme, in which we use equation

(5) to support the equivalence (3) and (4). To each particle with energy E, we assign a wave of frequency ν of the form†:

(6) $$\varphi(x, y, z, t) = \psi(x, y, z) \exp(-2\pi i \nu t).$$

Equation (5) then becomes:

(7) $$\Delta\varphi + \frac{4\pi^2\nu^2}{u^2}\varphi = 0$$

or:

$$\Delta\psi + \frac{4\pi^2\nu^2}{u^2}\psi = 0.$$

We can substitute the wavelength $\lambda = u/\nu$. Using the de Broglie relationship we get:

$$\frac{4\pi^2/\nu^2}{u^2} = \frac{4\pi^2}{\lambda^2} = \frac{mv^2}{\hbar^2}.$$

---

† Our choice of equation (6) is related to the form of the general wave equation:

$$\Delta\varphi - \frac{1}{u^2}\frac{\partial^2\varphi}{\partial t^2} = 0.$$

One can look for particular solutions to this equation by the method of separation of variables, taking:

$$\varphi(x, y, z, t) = \psi(x, y, z).\, \chi(t)$$

which, after substitution in (5) and division by φ yields:

$$\frac{\Delta\psi}{\psi} - \frac{1}{u^2}\frac{\ddot{\chi}}{\chi} = 0.$$

The term $\Delta\psi/\psi$ is a function of the coordinates and is equal to $\ddot{\chi}/u^2\chi$ which is only a function of time. Each of these two quantities varies independently and thus must be equal to a constant. The function χ thus satisfies $\ddot{\chi} - k\chi = 0$ where k is an arbitrary constant. For k negative ($k = -\omega^2$), the solution is $\chi = \exp(-i\omega t)$ to within an arbitrary multiplier and $\varphi^\star\varphi = \psi^\star\psi$. This represents an energy density ($\varphi^\star\varphi$) constant in time at every point and equal to the square of the modulus of the complex amplitude ψ which characterizes the stationary states. Note: we can omit the complex notation and use only real functions to represent physical magnitudes. To do this we postulate that:

$$\psi(x, y, z) = A(x, y, z) \exp[i\alpha(x, y, z)]$$

from which:

$$\varphi(x, y, z, t) = A \exp[i(\alpha - \omega t)]$$
$$= A\cos(\omega t - \alpha) - iA\sin(\omega t - \alpha).$$

The vibration can be characterized by the real part $\varphi_1 = A\cos(\omega t - \alpha)$ but this expression is inconvenient since φ_1^2 is not time independent and cannot represent the energy density. Thus we see the significance of the complex notation.

The speed of the particle is fixed at each point by a force field derived from the potential energy $V(x,y,z)$. The energy, E, can be written explicitly as:

$$E = \frac{1}{2}mv^2 + V(x,y,z)$$

from which, on rearrangement, we get:

$$\frac{m^2 v^2}{\hbar^2} = \frac{2m}{\hbar^2}(E - V).$$

This substituted into (7) gives:

$$\Delta\varphi + \frac{2m}{\hbar^2}(E - V)\varphi = 0 \tag{8}$$

which may be written for a monochromatic wave function φ where the amplitude ψ differs from φ only by a factor independent of the spatial coordinates. Equation (8) is the Schrödinger equation of the first kind, called the time-independent Schrödinger equation and is related to stationary states. We will return to it later.

A more general equation, which does not specify constant energy, can be obtained by using Bohr's equation (3). When this is substituted into equation (6) we get:

$$\varphi(x, y, z, t), = \psi(x, y, z) \exp(-iEt/\hbar) \tag{9}$$

which has as a time derivative:

$$\frac{\partial\varphi}{\partial t} = -i\frac{E}{\hbar}\varphi. \tag{10}$$

The elimination of E from this equation simultaneously with the Schrödinger equation (8) gives the Schrödinger equation of the second kind, the time-dependent Schrödinger equation which is usually written in the form:

$$-\frac{\hbar^2}{2m}\Delta\varphi + V\varphi = -\frac{\hbar}{i}\frac{\partial\varphi}{\partial t}. \tag{11}$$

Through the analogy suggested in our scheme we have been able to derive the most general equation of the new mechanics. We have yet to establish the meaning of φ and ψ and we will deal with this point next.

The Significance of the Wave Function 3

We begin here with the equivalence developed in the study of light as illustrated in Table I. We saw that the density of photons, N, is proportional to the energy density U, and, from the study of vibrational motion, one knows that U is proportional to the amplitude of the oscillation. The amplitude is given here by $|\psi^2| = |\varphi^2|$ so that we can say that, in the case of light:

$$N = \varkappa \, | \varphi^2 |. \tag{12}$$

We shall show that this result remains valid in the new mechanics and that the density of particles is proportional to the square of the modulus of the wave function. The physical significance of the wave function is the correspondence between $|\psi^2|$ and the particle density.

The above interpretation may be a little too weak, with too much dependence on optical phenomena. The argument developed in Appendix I is stronger. It shows that the tie between the mechanical interpretation and the wave interpretation is found by relating the speed of the particle to the group velocity of the waves. The group velocity of a wave is the velocity of propagation of wave energy. It is thus legitimate, regardless of the significance of the waves, to retain the correspondence between the particle distribution and the energy as previously postulated.

Another direct justification may be obtained by showing that the interpretation leading to (12) is compatible with the time-dependent Schrödinger equation (11). Equation (11) involves the complex number i which suggests that the solutions φ may be non real. This result presents no difficulty since we will need only the square of the amplitude $|\psi^2|$.†

We write the time-dependent Schrödinger equation again to illustrate the formation of the conjugate equation:

$$(13) \qquad \begin{aligned} -\frac{\hbar^2}{2m}\Delta\varphi + V(x, y, z)\varphi &= -\frac{\hbar}{i}\frac{\partial\varphi}{\partial t} \qquad (a) \\ -\frac{\hbar^2}{2m}\Delta\varphi^\star + V(x, y, z)\varphi^\star &= \frac{\hbar}{i}\frac{\partial\varphi^\star}{\partial t}. \qquad (b) \end{aligned}$$

Multiplying (*a*) by $\varphi^\star$ and (*b*) by φ and subtracting we get:

$$\varphi^\star\frac{\partial\varphi}{\partial t} + \varphi\frac{\partial\varphi^\star}{\partial t} = \frac{\partial}{\partial t}(\varphi^\star\varphi) = \frac{i\hbar}{2m}(\varphi^\star\Delta\varphi - \varphi\Delta\varphi^\star).$$

† By $|\psi^2|$ we mean $\psi^\star\psi$ where $\psi^\star$ is the complex conjugate of ψ (ed.).

The right-hand side of this equation may be transformed by use of the identity:

$$\varphi^* \frac{\partial^2 \varphi}{\partial x^2} - \varphi \frac{\partial^2 \varphi^*}{\partial x^2} = \frac{\partial}{\partial x}\left(\varphi^* \frac{\partial \varphi}{\partial x} - \varphi \frac{\partial \varphi^*}{\partial x}\right)$$

which permits us to write the equation in the form:

$$\frac{\partial}{\partial t}(\varphi^* \varphi) + \sum_{x,y,z} \frac{\partial}{\partial x}\left[\frac{i\hbar}{2m}\left(\varphi \frac{\partial \varphi^*}{\partial x} - \varphi^* \frac{\partial \varphi}{\partial x}\right)\right] = 0. \tag{14}$$

This equation has the form of a classical conservation equation (the equation of continuity) written:

$$\frac{\partial \rho}{\partial t} + \operatorname{div} \vec{J} = 0 \tag{15}$$

where ρ represents a fluid density and $\vec{J}$ the corresponding current density.

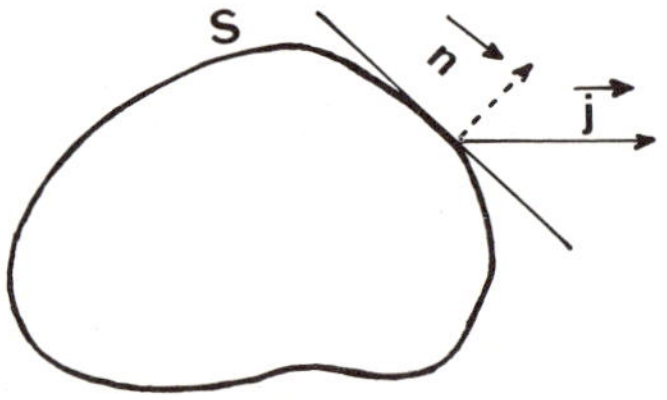

Figure 2. The arrangement of the unit normal $\vec{n}$ and the current density $\vec{J}$ for application of Gauss's theorem to the volume enclosed by surface S.

Equation (15) expresses the variation of the quantity of fluid inside a closed, continuous surface S (fig. 2), where this variation is due only to exchange with the exterior. The mass Q of the fluid can be calculated by means of the volume integral:

$$Q = \int \rho d\tau.$$

The quantity of fluid crossing the surface per unit time is given by the flux:

$$\int \vec{J}.\vec{n} dS,$$

where $\vec{n}$ is a unit outgoing normal vector. The conservation of matter over time dt within the volume is given by:

$$dQ + dt.\int \vec{J}.\vec{n} dS = 0,$$

which becomes:

$$\int \left(\frac{\partial \rho}{\partial t} + \operatorname{div} \vec{J}\right) d\tau = 0$$

on the application of Gauss's theorem. The necessary and sufficient condition for this equation to be true is that the surface S is closed and continuous at every point.

As a consequence of this comparison we can state that the total number of particles is conserved, where:

$$\rho = \varphi^* \varphi \tag{16}$$

represents the number of particles per unit volume (the particle density) and:

$$\vec{J} = \frac{i\hbar}{2m} (\varphi \, \overrightarrow{\text{grad}} \, \varphi^* - \varphi^* \, \overrightarrow{\text{grad}} \, \varphi) \tag{17}$$

represents the current density of particles.

If φ tends to zero sufficiently rapidly at ∞, then we can say:

$$\int \vec{J} . \vec{n} dS \rightarrow 0, \tag{18}$$

for a surface S which encloses all space. As a consequence:

$$\frac{\partial}{\partial t} \int \rho d\tau \rightarrow 0 \quad \text{so that} \quad \int \rho d\tau = \text{const.}$$

Thus if (18) is true, the function φ represents the behavior of a finite number of particles. We note that equation (11) is linear and since we have defined φ only to within a multiplicative constant factor, it is always possible to impose the condition:

$$\int \varphi^* \varphi d\tau = N \tag{19}$$

where N represents an arbitrary number of particles.

The arbitrary number N is related to the statistical aspect of the interpretation. In effect, equation (13) relates to a single particle, and (19) represents N identical examples of this single particle system which may be treated independently in the sense of the canonical ensemble in statistical mechanics. One gets the concept of the probability interpretation by taking the number of particles per unit volume at some point, which, using (16), is given by:

$$n = \rho d\tau = \varphi^* \varphi d\tau.$$

The N particles are independent. The probability dp of finding one of them in the volume element $d\tau$ is:

$$dp = \frac{n}{N} = \frac{1}{N} \varphi^* \varphi d\tau = \varphi'^* \varphi' d\tau \tag{20}$$

where one defines:

$$\varphi' = \frac{1}{\sqrt{N}}\varphi.$$

The function φ' thus defined describes the behavior of a single particle and should satisfy the normalization condition set in (19):

$$\int \varphi'^{*}\varphi' d\tau = 1. \tag{20'}$$

An important practical case is that of monochromatic waves for which one has:

$$\varphi^{*}\varphi = \psi^{*}\psi$$

which allows us to use the time-independent form of the Schrödinger equation with monochromatic waves.

4 Uniform Motion in One Dimension

We will now examine the solution of the Schrödinger equation for a particle moving in one dimension with no applied force. We want to use this special case to study the relationship between the wave model and the particle model.

(*a*) *The time-independent equation.* In this case the potential energy is zero and E is the total kinetic energy which is always positive. We can write the Schrödinger equation (8) in one dimension (taken as the x-direction) with wave amplitude ψ as:

$$-\frac{\hbar^2}{2m}\frac{d^2\psi}{dx^2} = E\psi, \quad \text{so that} \quad \frac{d^2\psi}{dx^2} + \frac{2mE}{\hbar^2}\psi = 0.$$

The general solution of this equation is:

$$\psi = c_1 \exp(i\alpha x) + c_2 \exp(-i\alpha x) \tag{21}$$

with

$$\alpha^2 = \frac{2mE}{\hbar^2} \geqslant 0.$$

To interpret this equation we consider the special case where one of the two constants c_1 or c_2 is zero.

If we assume that $c_1 = c$ and $c_2 = 0$, then $\psi = c\exp(i\alpha x)$. The energy is all kinetic and we can write:

$$E = \frac{1}{2}mv^2 \qquad \text{from which} \qquad \alpha = \frac{mv}{\hbar} = \frac{2\pi}{\lambda}.$$

Thus we again get the de Broglie wavelength, which is to be expected since the derivation of the wave equation is based on the de Broglie relationship.

The physical significance of this solution can be examined by looking at the particle density ρ, and the current density $\vec{J}$, as previously defined. This case is somewhat different from that in the previous section in that the integration is limited to the single variable x. Thus:

$$\rho = \psi^{\star}\psi = c^{\star}c$$
$$J = \frac{i\hbar}{2m}\left(\psi\frac{d\psi^{\star}}{dx} - \psi^{\star}\frac{d\psi}{dx}\right) = vc^{\star}c = \rho v.$$

The wave function may be thought of as representing an infinite number of particles distributed along the axis with a uniform density ρ, and moving with a constant velocity v in the positive direction.†

If we take the other particular solution corresponding to $c_1 = 0$ and $c_2 = c$, then we have $\psi = c\exp(-i\alpha x)$, and we again find a uniform particle distribution, this time with a uniform velocity in the negative direction.

We will now take the general solution (21) and try to interpret the charge density and the particle current. The calculation leads to the results:

$$\rho = \psi^{\star}\psi = c_1^{\star}c_1 + c_2^{\star}c_2 + c_1c_2^{\star}\exp(2i\alpha x) + c_1^{\star}c_2\exp(-2i\alpha x)$$

which can always be put in the form:

$$\rho = c_1^{\star}c_1 + c_2^{\star}c_2 + 2\,|\,c_1c_2\,|\,\cos(2\alpha x + \theta)$$

where we have introduced the argument θ of the complex number $c_1c_2^{\star}$. For the current density we get:

$$J = v(c_1^{\star}c_1 - c_2^{\star}c_2). \tag{22}$$

This expression for the current suggests that it is the resultant of two particle currents, one moving in the positive direction with uniform density $c_1^{\star}c_1$, the other moving in the negative direction with density $c_2^{\star}c_2$. The surprising point is, that ρ is not given by

† The current density in this one-dimensional case is defined as the number of particles which pass a fixed point in unit time. Since the particles move with constant velocity v, the particles passing the arbitrary point in a unit time are those which lie on the segment of length v adjoining the fixed point. The current density will then be ρv.

the value $c_1^\star c_1 + c_2^\star c_2$, which one would intuitively expect. However, if we consider the mean density on an arbitrary segment of length L, we find:

$$\bar{\rho} = \frac{1}{L}\int_0^L \rho dx \to c_1^* c_1 + c_2^* c_2. \tag{22'}$$

Equations (22) and (22′), giving the current density and the mean distribution, suggest an interpretation of the state represented by the general solution ψ. The particles on an arbitrary segment of length L can be divided into two groups, with densities $c_1^\star c_1$ and $c_2^\star c_2$, respectively. The form of (22) for J implies that the first family has velocity v and the second family $-v$. Thus, if one particle is picked at random on the segment, the probability of finding velocity v is $c_1^\star c_1/(c_1^\star c_1 + c_2^\star c_2)$, and that of finding $-v$ is $c_2^\star c_2/(c_1^\star c_1 + c_2^\star c_2)$. Thus we can see that the general solution ψ is a superposition state of the two particular solutions corresponding to the real physical situation.

This interpretation can be used to stress the difference between this new description and the classical view by examining two uniform currents of particles moving in opposite directions. With a classical model the density truly remains constant at every point. It is the representation of a state by means of a wave function which introduces the variation of the density along the axis. Precisely the same situation arises in optics when one treats the interference of two rays of light falling on a screen (Young's experiment). The total intensity of the light remains the same but the distribution is altered to give bright and dark fringes.

(*b*) *Solution of the time-dependent equation.* The solutions which we previously examined corresponded to states with known velocity but whose position was determined only to the extent that the particle distribution was uniform. We cannot use this description with a single particle and to get at the single particle case it is necessary to look at the solutions of the more general time-dependent equation. For $V = 0$ in one dimension, equation (11) reads:

$$\frac{\partial^2 \varphi}{\partial x^2} + \frac{2im}{\hbar}\frac{\partial \varphi}{\partial t} = 0. \tag{23}$$

This is the equation of heat conduction—a comparatively simple form for a partial differential equation. We note to begin with that (23) has solutions of the form:

$$\varphi = c \exp\left(i\alpha x - \frac{iEt}{\hbar}\right) = c \exp\left[i\left(\alpha x - \frac{\hbar\alpha^2 t}{2m}\right)\right] \tag{24}$$

which is natural, since it is from such waves that one derives the equation. The absence of the parameters α and E in the differential equation suggests that all linear combinations of solutions in the form of (24) are solutions of (23). One may have as a solution of this equation a function which depends on an arbitrary function $c(\alpha)$. This is, in fact, the way one usually treats a partial differential equation, i.e., by writing the solution as an integral relative to a parameter α, or, in other words, as a superposition of an infinity of particular solutions of the type (24).

Thus we can write:

$$\varphi = \int_{-\infty}^{+\infty} c(\alpha) \exp\left[i\left(\alpha x - \frac{\hbar\alpha^2 t}{2m}\right)\right] d\alpha.$$

For ease in interpretation we limit ourselves here to the superposition of two simple waves of the type (24) characterized by α and α'. We then replace the integral by a sum which we write at time $t = 0$ as:

$$\varphi = c \exp(i\alpha x) + c' \exp(i\alpha' x).$$

The physical significance of the state appears when we calculate ρ and J. We find for ρ, assuming c and c' real:

$$\rho = [c \exp(i\alpha x) + c' \exp(i\alpha' x)]\,[c \exp(-i\alpha x) + c' \exp(-i\alpha' x)]$$
$$= c^2 + c'^2 + 2cc' \cos(\alpha - \alpha')x.$$

The distribution is not uniform along the axis, but it has as mean value:

$$\bar{\rho} = \frac{1}{L}\int_0^L \rho\, dx \to c^2 + c'^2, \qquad \text{as} \qquad L \to \infty.$$

The current density can be calculated without difficulty and we get:

$$J = \frac{\hbar}{m}[\alpha c^2 + \alpha' c'^2 + (\alpha + \alpha')cc' \cos(\alpha - \alpha')x].$$

J is not constant along the axis, but it is possible to define a mean current density $\bar{J}$, given by:

$$\bar{J} = \frac{1}{L}\int_0^L J\, dx \to \frac{\hbar}{m}(\alpha c^2 + \alpha' c'^2), \qquad \text{as} \qquad L \to \infty.$$

With a view to interpretation we show, as in the case of (*a*), the simple solution:

(25) $$\alpha = \frac{mv}{\hbar}; \qquad \alpha' = \frac{mv'}{\hbar}$$

so that the mean current can be written as:

$$\bar{J} = vc^2 + v'c'^2 = \overline{v\rho}$$

and thus for the mean velocity we have

$$\bar{v} = \frac{vc^2 + v'c'^2}{c^2 + c'^2}.$$

The mean velocity which we calculate is the same as one would get by taking two independent beams of particles† with velocities v and v', and distributions c^2 and c'^2, respectively. In taking a

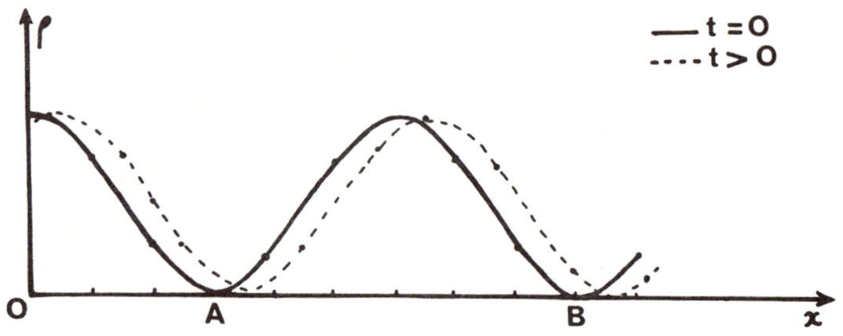

Figure 3. Diagram of the probability density ρ of finding a particle along the x-axis for a state which is the superposition of two states of velocity v and v'. The curve is displaced by a translation during the course of time.

particle at random, the probability of picking one or the other varies as c^2 or c'^2, respectively.

We now consider the special case $c = c'$. The expressions for ρ and J simplify. For ρ we get:

$$\rho = 2c^2[1 + \cos(\alpha - \alpha')x] = 4c^2 \cos^2 \frac{(\alpha - \alpha')x}{2}.$$

The variation of ρ as a function of x is given by the curve in figure 3 with the zeros determined by:

$$\text{(26)} \qquad \frac{\alpha - \alpha'}{2} x = (2k + 1)\frac{\pi}{2}; \qquad \text{or} \qquad x = \frac{(2k + 1)\pi}{\alpha - \alpha'}.$$

The current density can also be seen to be zero at these points.‡

In conclusion, a particle situated between A for $k = 0$ and B for $k = 1$ at $t = 0$ may not escape, since $J = 0$ at the limits of the region.

† Exactly reiterating the interpretation of 22 and 22′ developed more fully in part (*a*).

‡ This is a consequence of (17) which implies $J = 0$ when $\varphi = 0$.

We must examine the evolution of this state with time. The calculation is done in exactly the same way as for $t = 0$. Thus we get:

$$\rho = 4c^2 \cos^2 \left[\frac{\alpha - \alpha'}{2} x - \frac{\hbar}{4m}(\alpha^2 - \alpha'^2)t\right]$$

$$J = 4c^2(\alpha + \alpha') \cos^2 \left[\frac{\alpha - \alpha'}{2} x - \frac{\hbar}{4m}(\alpha^2 - \alpha'^2)t\right].$$

The points corresponding to $\rho = 0$ and $J = 0$ are given by:

$$\frac{\alpha - \alpha'}{2} x - \frac{\hbar}{4m}(\alpha^2 - \alpha'^2)t = (2k + 1)\frac{\pi}{2}$$

and they can be seen to travel with a uniform velocity:

$$\frac{\hbar}{2m}(\alpha + \alpha') = \frac{v + v'}{2} = \bar{v}.$$

Thus we contend that if a particle is initially found localized on an interval A to B, it remains over time confined to this fixed length segment which moves with velocity $\bar{v}$ defined above. In particular, if v and v' are nearly equal, we get $\bar{v} \rightarrow v$ and we can imagine a localized particle which is displaced with a given speed. We can make the following observations:

(*a*) The localization of a particle on a straight line segment of finite length Δx requires the use of two waves with different velocities v and v'. The hypothesis $c^2 = c'^2$ corresponds to an equal probability that the particle has one or the other of these two values, and from this an uncertainty in the speed of magnitude $\Delta v = |v' - v|$ results.

(*b*) The length Δx which characterizes the region of confinement of the particle can be gotten from (26) by taking two consecutive values of k which gives:

$$|\Delta x| = \frac{2\pi}{|\alpha - \alpha'|} = \frac{2\pi\hbar}{m|\Delta v|} = \frac{h}{m|\Delta v|}$$

which can in turn be written

(27) $$|\Delta x| \cdot m|\Delta v| = |\Delta x| \cdot |\Delta p| = h.$$

This relationship, called the Heisenberg uncertainty principle, implies that the two quantities—x the particle position, and p the particle momentum along the same axis—cannot be simultaneously determined with arbitrarily fine precision. The general nature of this relationship is much broader than the simple case treated here.

(*c*) If we take a fixed point M on the axis, the fact that the particle is confined on a segment Δx leads to an uncertainty in the time at which the particle actually passes M. The time uncertainty Δt is given by:

$$\Delta t = \frac{\Delta x}{v} = \frac{h}{mv\Delta v}.$$

Now $mv \cdot \Delta v = \Delta(mv^2/2)$ represents the uncertainty ΔE in the energy of the particle. Thus we get a second uncertainty relationship:

$$|\Delta t| . |\Delta \mathrm{E}| = h. \qquad (27')$$

Comparison of the two uncertainty relationships again illustrates the parallel between x and t on the one hand, and p and E on the other, imposed by classical analytical mechanics.

The Concept of Group Velocity

We conclude this discussion by stressing the line of reasoning developed for passing from the idea of a monochromatic wave as representative of a beam of non-localized particles to the case of an isolated particle. An analogous point of view may be developed for the general case of wave propagation. A monochromatic wave which propagates in a homogeneous medium is characterized by a phase velocity u, which is used in the definition of the wavelength, and which governs the interference and diffraction effects. Another fundamental concept is that of v_g, the velocity of energy transmission or group velocity, where the definition requires the truncation of an infinite monochromatic wave-train into pulses which can be observed passing between two points in space (Fizeau experiment). This measurement becomes more precise as the pulse is made shorter. The creation of pulses destroys the monochromatic nature of the wave, and it can now be represented by a superposition of waves of different frequencies, the spectrum growing wider as the pulse shortens. In acoustics, for example, we know that a sound loses its pitch if it is sounded too briefly because it may then excite many different frequencies to resonance.

Inversely, the superposition of monochromatic waves of slightly different frequency, produces, by interference, a series of pulses regularly spaced and propagating with a uniform velocity. We apply the notation used in the previous calculation to the superposition of two waves of equal amplitude. We use here real variables and get:

$$y = \mathrm{A} \cos 2\pi\left(\nu t - \frac{x}{\lambda}\right) + \mathrm{A} \cos 2\pi\left(\nu' t - \frac{x}{\lambda'}\right)$$

which can be written

$$y = 2A\cos\pi\left\{(\nu' - \nu)t - \left(\frac{1}{\lambda'} - \frac{1}{\lambda}\right)x\right\}\cdot\cos\pi\left\{(\nu' + \nu)t - \left(\frac{1}{\lambda'} - \frac{1}{\lambda}\right)x\right\}.$$

When ν' tends to ν, λ' tends to λ, and one can see that the expression for y has a factor $\cos 2\pi(\nu t - x/\lambda)$ characteristic of the propagation of a monochromatic wave with an amplitude which varies slowly in time, given as:

$$A\cos\pi\left\{(\nu' - \nu)t - \left(\frac{1}{\lambda'} - \frac{1}{\lambda}\right)x\right\}.$$

This amplitude goes to zero at points M_k defined by:

$$(\nu' - \nu)t - \left(\frac{1}{\lambda'} - \frac{1}{\lambda}\right)x = k + \frac{1}{2} \text{ with } k \text{ integral.}$$

Two consecutive points M_k and M_{k+1} define a wave packet whose energy remains constant in time since it cannot exchange energy with adjacent packets. The relationship defining M_k leads to an equation of uniform motion of the packet where the velocity is given by:

$$v_g = (\nu' - \nu)\Big/\left(\frac{1}{\lambda'} - \frac{1}{\lambda}\right),$$

which defines that velocity of propagation usually called the group velocity. At the limit as $\nu' \to \nu$, we get the Rayleigh relationship:

$$v_g = \frac{d\nu}{d(1/\lambda)}. \tag{28}$$

If we reason in a more general fashion, we still find that, in the general case, the Rayleigh relationship remains valid regardless of the form of the monochromatic waves used to construct the wave packets. We note that in the absence of dispersion $u(\nu) = \text{const.}$:

$$\frac{1}{\lambda} = \frac{\nu}{u} \quad \text{and} \quad v_g = u,$$

and the group velocity is equal to the phase velocity.

Diffraction of Particles 5

The de Broglie hypothesis relating the behavior of a particle and the associated wave has been experimentally verified by Davisson and Germer, who studied the scattering of a beam of electrons by a crystal surface (a single crystal of nickel in the original experiment). Figure 4 illustrates the experimental scheme. The beam was produced by an electron gun and the distribution of the scattered electrons was measured with a Faraday cup and electrometer. The experiment shows the same scattering maxima that one finds with a conventional beam of X-rays of wavelength λ. That is to say, the angles θ of the scattering maxima (see fig. 4) obey the Bragg law:

$$k\lambda = 2d \sin\frac{\theta}{2}$$

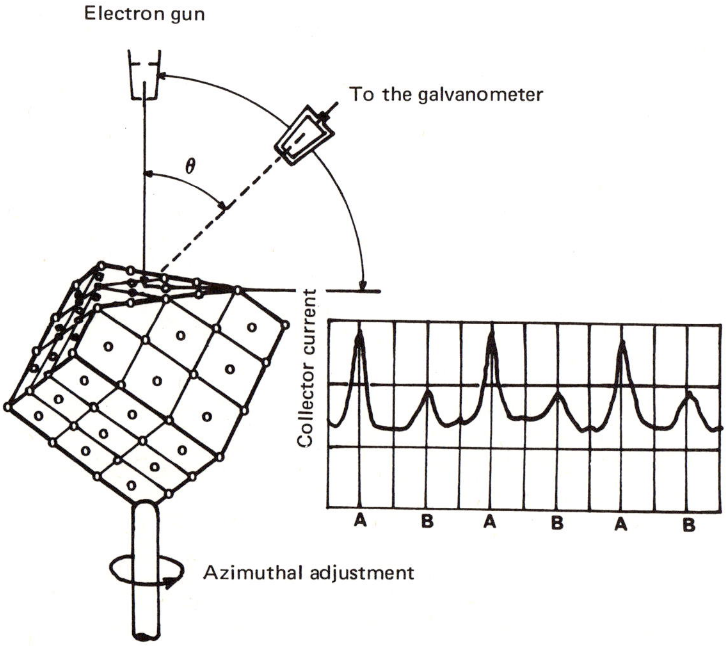

Figure 4. Diagram of the Davisson and Germer apparatus and the curve obtained when the crystal is rotated.

where k is an integer and d is the separation of the crystal planes on which the reflection occurs. For a given acceleration voltage V, λ can be picked so as to verify the de Broglie relation which, taking into account $mv^2/2 = e\mathrm{V}$, gives:

$$\lambda = \frac{h}{mv} \sim \frac{1}{\sqrt{\mathrm{V}}}$$

and the distribution of the maxima should satisfy:

$$\sqrt{\mathrm{V}} \sin\frac{\theta}{2} = \text{const.}$$

This scattering can also be observed using powder targets (Ponte experiment). Similar experiments have been done with a helium beam (Stern and Estermann), with hydrogen molecules, and with neutrons. We note that light materials must be used so that the wavelength does not become too small.

Compatible and Incompatible Variables 6

The equivalence between a beam of particles and a monochromatic wave introduced in section 2, and directly verified by the Davisson-Germer experiment, requires an important modification in our ideas of the physical magnitudes of these particles. To examine

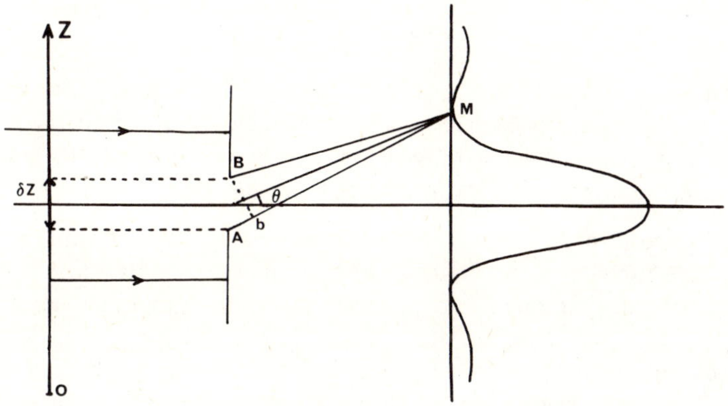

Figure 5. Diagram of the electron density obtained on a screen when a monochromatic beam of electrons is diffracted by slit $\overline{AB}$.

this question we may begin with the simplest diffraction experiments. In observations of the action of the three-dimensional grating of a real crystal the analysis becomes complex, so we will here consider the case of diffraction by a thin slit. Here also, the behavior of a beam of monoenergetic particles must be deduced from the corresponding plane monochromatic waves. The diffraction pattern is independent of the specific wave phenomenon and its analysis for the case of light is well known (see fig. 5). We recall that an intensity distribution of the light falling on a screen normal to the ray is seen to have a diffraction pattern. The same pattern is found for the distribution of particles on a screen. It is interesting to examine this experiment as a means of determining precisely the mechanical state of particles whose speed is known, but whose position is totally unknown at any given instant. In particular, the z displacement on an axis normal to the beam may be examined. The insertion of a slit of width z represents a technique which can select those particles which have the same position on the z-axis (at least to within δz). This would be the case in the absence of diffraction, but diffraction resulting from passage through the slit could leave the v_z component with any value. The mean value of v_z is, of course, zero, corresponding to the propagation as originally set along the direction of the beam. The uncertainty of v_z may be calculated in the usual way by the position of the minimum in the diffraction pattern corresponding to the angle θ (fig. 5) given by:

$$\overline{Ab} = \Delta z \cdot \theta = \lambda, \qquad \text{from which} \qquad \Delta v_z = v \cdot \theta = \frac{\lambda v}{\Delta z}.$$

By introducing the de Broglie relation we get:

$$m \Delta v_z \cdot \Delta z = h.$$

Thus we again get (27), one of the Heisenberg uncertainty relations, but here from a physically realistic experiment, at least in principle, even if the physical meaning of the two waves which leads us to (27) in the original derivation is not apparent.

The uncertainty relations imply the impossibility of finding any particle for which the value of one of the coordinates z and the corresponding momentum $p_z = mv_z$ are simultaneously known with arbitrarily fine precision. It is a fundamental aspect of wave mechanics that the state of a material particle cannot be defined with as much precision as in classical mechanics. The statement: let us take a particle with coordinates x, y, z, and speed v_x, v_y, v_z can no longer have any significance. For the same reason, the idea of a definite trajectory for a particle can no longer have significance.

It is often said of the previous results that the two variables, coordinates z and the conjugate momentum p_z, are incompatible variables. Though the idea of incompatibility between certain variables could be regarded as a characteristic of the new mechanics, it does not lead to results which differ from those obtained by the application of classical mechanics when applied to compatible variables.

One might be tempted to use the preceding results to examine the behavior of an isolated particle "by a thought experiment," however, this does not correspond to a real physical situation and we must be very careful in drawing conclusions. To begin with, it is natural to assume that the particles do not interact with one another within the beam and the probability distribution on the screen represents the impact probability density for a single particle (see fig. 5). One gets the idea that the behavior of an isolated particle is indeterminate, in spite of the fact that intuition suggests that we should be able to follow the particle. We would be unwise to insist on this reasoning, since localizing a particle in the beam is an operation devoid of physical meaning, and every effort to construct an apparatus to determine the location (by inserting a thin slit for example) will necessarily lead to diffraction. A particularly clear example of the difficulties which arise is seen in

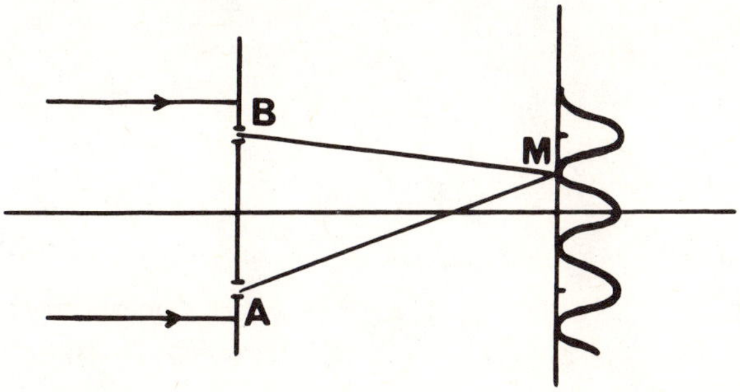

Figure 6. Arrangement of Young's slits and the diffraction fringes observed on the screen.

Young's experiment (fig. 6). At point M, corresponding to a dark fringe, we do not find particles. But if one considers the behavior of one particle, one must surely insist that it passes through one of the slits, A, for example. The fact that it cannot get to M implies that the behavior of the particle must be influenced by the existence of the other slit, B, through which it does not pass. This again leads to the basic problem of incompatible variables.

Note on the Calculation of J. The current density of particles is defined with the wave function φ by equation (17). For a monochromatic wave, the result is independent of time and may be calculated exactly from the amplitude ψ, as one does for the probability distribution. This ceases to be true when the wave function is the superposition of many waves corresponding to different energies.

2 THE FREE ELECTRON

The Step Potential 7

The problem which we will now treat is less important for direct applications than for illustration of the differences between classical and quantum mechanics. The experiment will correspond in principle to charged particles entering a Faraday cup, but the potential step which one gets in that case will never be sufficiently abrupt to allow the observation of the quantum effects.

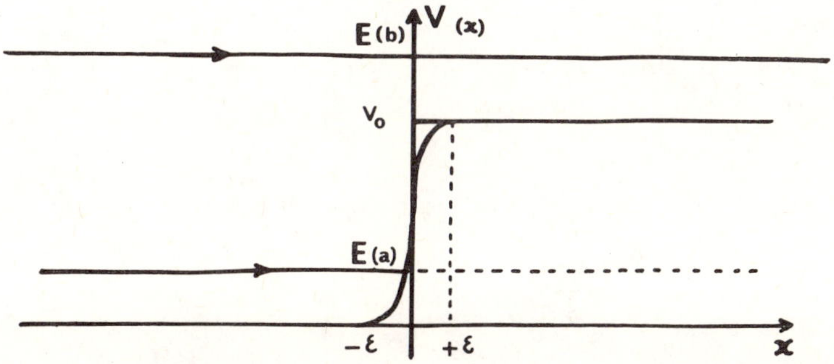

Figure 7. Energy diagram of a particle incident on a potential step of height V_0 for the two cases (*a*), $E < V_0$ and (*b*), $E > V_0$. This step can be considered as the limit of a continuous curve in the interval $-\varepsilon$ to $+\varepsilon$.

As with all these studies, we take as the origin O of the coordinates the position of the potential jump V_0 (fig. 7). The energy of the free particle will be measured from the origin.

We will first consider the classical behavior of a particle arriving from $-\infty$ onto a potential step. Two cases are possible:
(*a*) The kinetic energy of the particle is less than V_0 so that $mv^2/2 < V_0$. The particle cannot pass the origin and is reflected with conservation of energy ($v' = -v$).
(*b*) The kinetic energy of the particle exceeds V_0 and the particle enters the positive region with a velocity v'. The quantity v' may be found by applying conservation of energy:

$$\frac{1}{2}mv^2 = \frac{1}{2}mv'^2 + V_0.$$

The wave mechanical behavior will be quite different and we will treat the two cases individually.

(*a*) $E < V_0$. The system is governed by the one-dimensional Schrödinger equation:

$$\text{(1)} \qquad \frac{d^2\psi}{dx^2} + \frac{2m}{\hbar^2}(E - V(x))\psi = 0.$$

The function $V(x)$ is discontinuous at $x = 0$, so that equation (1) takes the following forms in each region:

$$\text{(1a)} \quad x < 0: \quad \frac{d^2\psi_1}{dx^2} + \frac{2m}{\hbar^2}E\psi_1 = 0 \qquad \text{with} \qquad E > 0$$

$$\text{(1b)} \quad x > 0: \quad \frac{d^2\psi_2}{dx^2} + \frac{2m}{\hbar^2}(E - V_0)\,\psi_2 = 0 \quad \text{with} \quad E - V_0 < 0.$$

If we take into account the respective signs of E and $E - V_0$ the general solutions of these linear second-order equations can be written:

$$\text{(2a)} \qquad \psi_1 = c_1 \exp(i\alpha x) + c_1' \exp(-i\alpha x)$$

$$\text{(2b)} \qquad \psi_2 = c_2 \exp(\beta x) + c_2' \exp(-\beta x)$$

with α and β defined by:

$$\text{(3a)} \qquad -\alpha^2 + \frac{2m}{\hbar^2}E = 0.$$

$$\text{(3b)} \qquad \beta^2 + \frac{2m}{\hbar^2}(E - V_0) = 0.$$

The coefficients c_1, c_1', c_2', and c_2 should satisfy the boundary conditions, which we will define.

To begin with, the fact that the equations (1*a*) and (1*b*) are derived from a single equation (1) implies that the continuity conditions may be gotten by examination of the potential discontinuity as a limit of a function $V(x)$ varying rapidly from 0 to V_0 in the arbitrarily small interval $-\epsilon$ to ϵ. Under these conditions, by multiplying (1) by the differential element dx and integrating from $-\epsilon$ to ϵ:

$$\text{(4)} \qquad \left(\frac{d\psi}{dx}\right)_{+\varepsilon} - \left(\frac{d\psi}{dx}\right)_{-\varepsilon} + \frac{2m}{\hbar^2}\int_{-\varepsilon}^{+\varepsilon}(E - V)\,\psi dx = 0.$$

The physical significance of ψ which allows us to calculate the particle density implies that ψ is finite, so that the integral in (4) tends to zero as $\epsilon \to 0$. Thus:

$$\text{(5)} \qquad \left(\frac{d\psi_1}{dx}\right)_0 = \left(\frac{d\psi_2}{dx}\right)_0$$

$(d\psi/dx)$ is therefore a continuous function at a discontinuity in the potential. One may also form the integral:

$$\psi_{+\varepsilon} - \psi_{-\varepsilon} = \int_{-\varepsilon}^{+\varepsilon} \left(\frac{d\psi}{dx}\right) dx.$$

This integral also goes to zero as $\epsilon \to 0$ and this implies the continuity of ψ:

(6) $$(\psi_1)_0 = (\psi_2)_0.$$

We note that the two continuity conditions (5) and (6) also insure the continuity of J defined by (1-17). The coefficients which define ψ_1 and ψ_2 in (2*a*) and (2*b*) should satisfy (5) and (6), so that:

(7) $$i\alpha(c_1 - c_1') = \beta(c_2 - c_2').$$
$$c_1 + c_1' = c_2 + c_2'.$$

These equations are not sufficient to define the coefficients completely and one of the coefficients must be fixed arbitrarily. In making this choice we note that the particles are distributed on the positive axis with density:

$$|\psi_2|^2 = \overset{*}{c_2}c_2 \exp(2\beta x) + c_2'^{*}c_2' \exp(-2\beta x) + \overset{*}{c_2}c_2' + c_2c_2'^{*}$$

which would grow indefinitely with x. Thus we take $c_2 = 0$, making $J = 0$, as one would expect from (1-17). A non-zero value for particle current density in this case would imply an infinite number of particles which we cannot have in any physical sense. We can therefore resolve equation (7) by making $c_2 = 0$ which gives:

(8) $$c_1' = \frac{i\alpha + \beta}{i\alpha - \beta} c_1; \qquad c_2' = \frac{2i\alpha}{i\alpha - \beta} c_1.$$

The parameters α and β are real, thus we have†

$$|(i\alpha + \beta)/(i\alpha - \beta)| = 1,$$

from which $c_1'^2 = c_1^2$. The state described by ψ_1 may therefore be thought of as the superposition of two beams; one incident on, and the other reflected from the potential step, with equal density corresponding to the same absolute value of the speed v related to α by (3*a*), which, when taken with $E = mv^2/2$, leads to:

(9) $$v = \frac{\hbar\alpha}{m}.$$

† The two complex numbers $i\alpha + \beta$ and $i\alpha - \beta$ have the same modulus, so that their ratio has modulus unity.

This solution is similar to the classical solution in the region $x < 0$. The difference between the wave mechanical and classical picture occurs in the region $x > 0$. In this region the existence of a solution $\psi_2 = c_2' \exp(-\beta x)$ implies a finite probability for the particle to be found here, the probability decreasing exponentially with distance. The "penetration" of the particle into the high potential region† is of the order of $1/\beta$ so that, following (3*b*), and letting $V_0 \gg E$:

$$\frac{1}{\beta} = \frac{\hbar}{\sqrt{2mV_0}}. \tag{10}$$

If V_0 is very large, the penetration becomes infinitely small and we can assume:

$$c_1' = -c_1; \qquad c_2' = 0. \tag{11}$$

Exercise 1. By assuming: $E = mv^2/2$ and by letting $V_0/E = k$, compare the depth of penetration to the length of the de Broglie wave associated with the particle. This will illustrate why the effect is not observable in the usual macroscopic case.

(*b*) $E > V_0$. The problem is parallel to that of section 7(*a*), but the two equations corresponding to (1*a*) and (1*b*) are now of similar form and we get a general solution:

$$\begin{aligned} \psi_1 &= c_1 \exp(i\alpha_1 x) + c_1' \exp(-i\alpha_1 x) \\ \psi_2 &= c_2 \exp(i\alpha_2 x) + c_2' \exp(-i\alpha_2 x) \end{aligned} \tag{12}$$

with:

$$\begin{aligned} -\alpha_1^2 + \frac{2m}{\hbar^2} E &= 0 \\ -\alpha_2^2 + \frac{2m}{\hbar^2}(E - V_0) &= 0. \end{aligned} \tag{13}$$

The continuity conditions (5) and (6) give:

$$\begin{aligned} c_1 + c_1' &= c_2 + c_2' \\ \alpha_1(c_1 - c_1') &= \alpha_2(c_2 - c_2'). \end{aligned}$$

Applying the results of section 4, we see that ψ_1 represents a superposition of states in which the particle has velocities v_1 and $-v_1$, respectively, with $v_1 = \hbar\alpha_1/m$, and distributed with the probability c_1^2 and $c_1'^2$. The state ψ_2 which has the same analytic form can be interpreted in the same way. As in (*a*), one of the constants must be chosen arbitrarily, and we let $c_2' = 0$, so that the only

† This penetration is defined by the distance in which the wave amplitude falls to $1/e$ of its original value.

particles arriving at the barrier are moving in a positive direction. Thus:

$$c_1' = \frac{\alpha_1 - \alpha_2}{\alpha_1 + \alpha_2} c_1; \qquad c_2 = \frac{2\alpha_1}{\alpha_1 + \alpha_2} c_1. \tag{14}$$

The solution which results corresponds to an incident beam of particles, distributed with density $|c_1^2|$, which gives rise to a transmitted beam (density $|c_2^2|$) and a reflected beam (density $|c_1'^2|$). Conservation of the number of particles is assured by the continuity of J:

$$\frac{\hbar}{m}(\alpha_1 \mid c_1^2 \mid - \alpha_1 \mid c_1'^2 \mid) = \frac{\hbar}{m} \alpha_2 \mid c_2^2 \mid$$

which is a direct consequence of the continuity conditions as verified in (14). This solution is clearly different from that expected classically, but it does have direct analogy to the behavior of a wave incident on the surface separating regions with two different indices of refraction. This point of view can be illustrated by examining the value of the reflection coefficient: $r = |c_1'^2|/|c_1^2|$ calculated from (14):

$$r = \left(\frac{\alpha_1 - \alpha_2}{\alpha_1 + \alpha_2}\right)^2 = \left(\frac{\lambda_2 - \lambda_1}{\lambda_2 + \lambda_1}\right)^2$$

where $\lambda_1 = h/mv_1 = 2\pi/\alpha_1$ and $\lambda_2 = h/mv_2 = 2\pi/\alpha_2$ designate the de Broglie wavelengths in the two regions. The ratio $n = \lambda_1/\lambda_2$ represents the index of refraction of the second region with respect to the first, so that by division we get Fresnel's classical expression for the reflection coefficient for normal incidence:

$$r = \left(\frac{n-1}{n+1}\right)^2.$$

The problem we have just gone through gives us a chance to return to the question introduced at the end of section 6, which we dealt with when we looked at the behavior of an isolated particle. Here also, if we follow an incident particle "by a thought experiment," we should be able to find it either in the transmitted beam or in the reflected beam. This of course leads to an "uncertainty" in its behavior. However, this thought experiment is questionable, since if we isolate a single particle in the incident beam, its velocity must be either $+v$ or $-v$, so that the problem is not one of a lack of knowledge of the behavior of the particle but rather one of incomplete knowledge—of uncertainty—about the state in which we find the particle initially.

8 The Tunnel Effect

The case of the potential barrier (fig. 8) is treated in exactly the same way as that of the potential step. We will limit ourselves to the case where the kinetic energy of the beam is too small to permit crossing of the barrier in the classic sense. The wave mechanical treatment differs from that in the previous section only by the fact that we have three regions to treat rather than two. The wave function in each of these regions takes the form:

$$x < 0: \qquad \psi_1 = c_1 \exp(i\alpha_1 x) + c_1' \exp(-i\alpha_1 x)$$

$$0 < x < l: \qquad \psi_2 = c_2 \exp(\beta_2 x) + c_2' \exp(-\beta_2 x)$$

$$l < x: \qquad \psi_3 = c_3 \exp(i\alpha_3 x) + c_3' \exp(-i\alpha_3 x).$$

The outer regions are at the same potential so that $\alpha_3 = \alpha_1 = \alpha$; the values of $\alpha_1 = \alpha$ and $\beta_2 = \beta$ are given by equations (3*a*) and

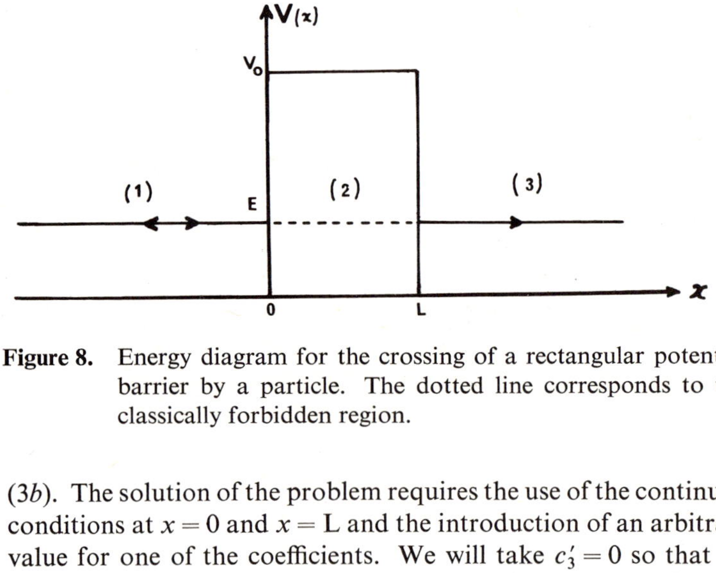

Figure 8. Energy diagram for the crossing of a rectangular potential barrier by a particle. The dotted line corresponds to the classically forbidden region.

(3*b*). The solution of the problem requires the use of the continuity conditions at $x = 0$ and $x = \mathrm{L}$ and the introduction of an arbitrary value for one of the coefficients. We will take $c_3' = 0$ so that we are only evaluating the behavior of the particles approaching the barrier from one side. The continuity conditions may be written:

$$(15) \qquad \begin{aligned} x = 0: \quad & c_1 + c_1' = c_2 + c_2' \\ & i\alpha(c_1 - c_1') = \beta(c_2 - c_2') \\ x = l: \quad & c_2 \exp(\beta l) + c_2' \exp(-\beta l) = c_3 \exp(i\alpha l) \\ & \beta[c_2 \exp(\beta l) - c_2' \exp(-\beta l)] = i\alpha c_3 \exp(i\alpha l). \end{aligned}$$

The "transparency" of the barrier is defined by c_3. We have:

$$c_2 = \frac{i\alpha + \beta}{2\beta} c_3 \exp (i\alpha - \beta)l$$

$$c_2' = \frac{-i\alpha + \beta}{2\beta} c_3 \exp (i\alpha + \beta)l.$$

We can get c_1 in terms of c_3 from (15):

$$c_1 = \frac{c_3}{4i\alpha\beta} [(i\alpha + \beta)^2 \exp (i\alpha - \beta)l - (i\alpha - \beta)^2 \exp (i\alpha + \beta)l].$$

The "transmission coefficient," τ, of the barrier is defined as the ratio of the incident and transmitted currents, and since the velocity is the same in each region:

$$\tau = \frac{J_3}{J_1} = \frac{|c_3^2|}{|c_1^2|}. \tag{16}$$

Using our expression for c_1 in terms of c_3 we get:

$$\frac{1}{\tau} = \frac{1}{16\alpha^2\beta^2} \{ (i\alpha-\beta)^2(i\alpha+\beta)^2 [\exp (-2\beta l) + \exp (2\beta l)] - (i\alpha-\beta)^4 - (i\alpha+\beta)^4 \}. \tag{17}$$

As the width of the barrier goes to zero, the transmission coefficient tends to unity. As this occurs, one exponential term becomes predominant and as an approximation for τ we get:

$$\tau \simeq \frac{16\alpha^2\beta^2}{(\alpha^2 + \beta^2)^2} \exp (- 2\beta l) \tag{18}$$

to make this more explicit, using (3) we have:

$$\chi = \frac{\beta}{\alpha} = \sqrt{\frac{V_0 - E}{E}}:$$

$$\tau \simeq \frac{16\chi^2}{(1+\chi^2)^2} \exp\left(-\frac{2}{\hbar} \sqrt{2m(V_0 - E)}\, l\right) \tag{19}$$

$$= \frac{16E(V_0 - E)}{V_0^2} \exp\left(-\frac{2l}{\hbar} \sqrt{2m(V_0 - E)}\right).$$

As one would expect, the transmission coefficient decreases as the width l of the barrier increases and also as the relative height $V_0 - E$ increases. Finally, the coefficient will be zero if we take Planck's constant to be zero and this illustrates the basic quantum nature of the process. The possibility of a particle crossing a potential which is too high for classical crossing is known as the tunnel effect. Merely to state that this is a quantum effect does not constitute an explanation and a paradox remains since, according to the classical picture, the particle should have a negative kinetic energy during the crossing of the barrier. In fact, during the crossing, the particle's position is fixed within the width of the barrier and this leads to an uncertainty in momentum by virtue of the Heisenberg relations. The barrier crossing demands classically a minimum value of kinetic energy:

$$\frac{1}{2}mv'^2 = \frac{1}{2m}p'^2 > V_0 - E \qquad \text{so that} \qquad p' > \sqrt{2m(V_0 - E)}.$$

We can suppose that such large values are obtained if one has:

$$p'l \sim h \quad \text{so that} \quad \sqrt{2m(V_0 - E)}\, l/h \sim 1.$$

The above condition emphasizes the importance of the factor which appears in (19) for the barrier transmission.

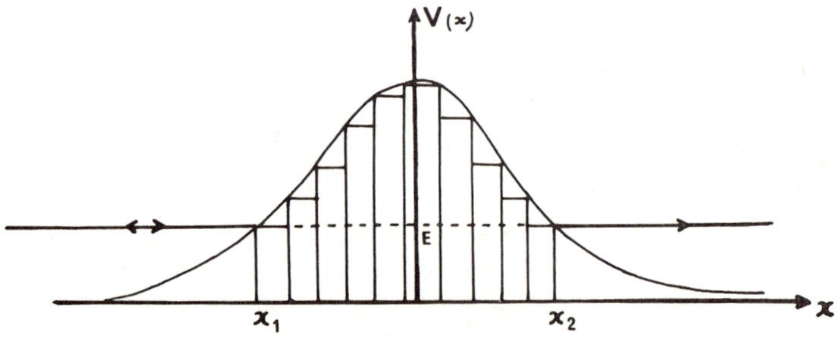

Figure 9. Decomposition of a potential barrier of general form into a succession of adjacent rectangular barriers.

Finally, we note that (19) is the product of two factors, one depending only on the width of the barrier if we take V_0 as fixed. It is tempting to use this result to compute the transmission coefficient for a barrier of arbitrary form (fig. 9) by taking a succession of elementary barriers, the coefficient then being the product of the coefficients for the elementary barriers. We might assume, as a result of having an exponential factor in (19), that the exponent for an arbitrary barrier would be equal to the sum of the exponents for the elementary barriers, which would permit us to use a relationship of the form:

$$\tau = K \exp\left[-\frac{2}{\hbar}\int_{x_1}^{x_2}\sqrt{2m(V(x) - E)}\,dx\right]. \tag{20}$$

The complete argument† effectively leads to an expression for τ with $K \sim 1$, but the reasoning as we have just sketched it cannot

† E. S. Kemble, *Fundamental Principles of Quantum Mechanics* (New York: McGraw-Hill, 1937), Section 21j.

easily be treated in an exact way. We must note that the expression for τ requires an approximation for the elementary barrier in going from (17) to (18) which takes $2\beta l \gg 1$ and this is not valid for an infinitely narrow barrier.

Equation (20) allows us to treat quantitatively a certain number of problems in which the tunnel effect occurs, including field-assisted electron emission from metals, and radioactive α-decay.

The Potential Well 9

The same equations also allow us to treat the problem of a particle localized on a segment of length L (fig. 10). Confinement is assured classically if the potential energy relative to the free particle

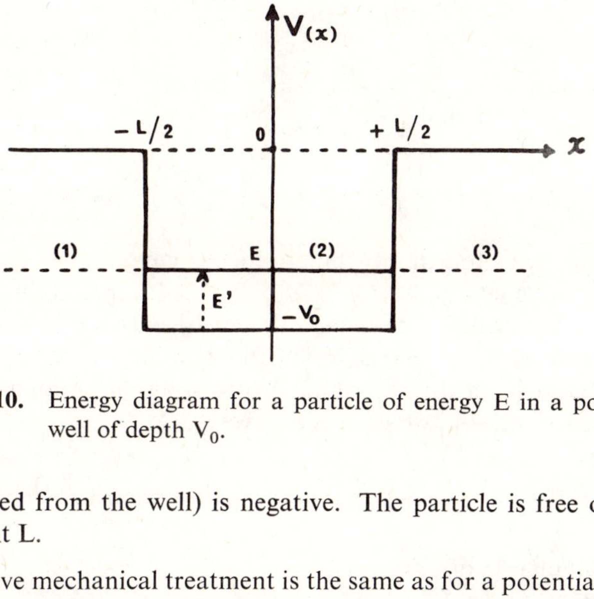

Figure 10. Energy diagram for a particle of energy E in a potential well of depth V_0.

(removed from the well) is negative. The particle is free on the segment L.

The wave mechanical treatment is the same as for a potential step, but with three regions instead of two. The Schrödinger equation takes different forms depending on the region under consideration. The origin is placed at the center of the well to exploit the symmetry:

$$x < -\frac{L}{2}: \qquad \psi_1 = c_1 \exp(\beta_1 x) + c_1' \exp(-\beta_1 x)$$

$$-\frac{L}{2} < x < \frac{L}{2}: \qquad \psi_2 = c_2 \exp(i\alpha_2 x) + c_2' \exp(-i\alpha_2 x)$$

$$x > \frac{L}{2} \qquad \psi_3 = c_3 \exp(\beta_3 x) + c_3' \exp(-\beta_3 x).$$

These functions satisfy the Schrödinger equation in each region. Regions 1 and 3 are the same and this reduces the problem to two regions. We write $\beta_1 = \beta_3 = \beta$ and $\alpha_2 = \alpha$:

$$\begin{aligned} &\beta^2 + \frac{2m}{\hbar^2} \mathrm{E} = 0 \\ &- \alpha^2 + \frac{2m}{\hbar^2} (\mathrm{E} + \mathrm{V}_0) = 0. \end{aligned} \tag{21}$$

The argument developed for the case of the potential step can be used here to show $c_1 = 0$, so that the density of particles goes to zero as $x \to \infty$. This choice restricts us to the problem of particles localized in the vicinity of the well.†

Using the continuity conditions (5) and (6) for L/2 and –L/2 we can now write:

$$\begin{aligned} c_1 \exp\left(-\frac{\beta \mathrm{L}}{2}\right) &= c_2 \exp\left(-\frac{i\alpha \mathrm{L}}{2}\right) + c_2' \exp\left(\frac{i\alpha \mathrm{L}}{2}\right) \\ \beta c_1 \exp\left(-\frac{\beta \mathrm{L}}{2}\right) &= i\alpha\left[c_2 \exp\left(-\frac{i\alpha \mathrm{L}}{2}\right) - c_2' \exp\left(\frac{i\alpha \mathrm{L}}{2}\right)\right] \\ c_3' \exp\left(-\frac{\beta \mathrm{L}}{2}\right) &= c_2 \exp\left(\frac{i\alpha \mathrm{L}}{2}\right) + c_2' \exp\left(-\frac{i\alpha \mathrm{L}}{2}\right) \\ -\beta c_3' \exp\left(-\frac{\beta \mathrm{L}}{2}\right) &= i\alpha\left[c_2 \exp\left(-\frac{i\alpha \mathrm{L}}{2}\right) - c_2' \exp\left(\frac{i\alpha \mathrm{L}}{2}\right)\right]. \end{aligned} \tag{22}$$

This system of four linear, homogeneous equations in c_1, c_2, c_2', and c_3, will give a non-zero result only if the compatibility conditions are satisfied. These conditions may be treated in the most elementary way by the vanishing of the determinant of the system of equations and then by simultaneous solution. Elimination of c_1 between the first two equations gives:

$$c_2(i\alpha - \beta) \exp\left(-\frac{i\alpha \mathrm{L}}{2}\right) = c_2'(i\alpha + \beta) \exp\left(\frac{i\alpha \mathrm{L}}{2}\right)$$

and elimination of c_3' gives:

$$c_2(i\alpha + \beta) \exp\left(\frac{i\alpha \mathrm{L}}{2}\right) = c_2'(i\alpha - \beta) \exp\left(-\frac{i\alpha \mathrm{L}}{2}\right).$$

Multiplying these two results together yields:

$$(\alpha^2 + \beta^2)c_2^2 = (\alpha^2 + \beta^2)c_2'^2.$$

† We will show, in general for stationary states (Chapter 3), that this choice imposes the requirement of a finite value on the normalization integral.

It appears as if there are two classes of solution:

Symmetric Solution: $c_2' = c_2$

This gives immediately $c_3' = c_1$ and the condition of continuity can be written:

$$\tan\frac{\alpha L}{2} = \frac{\beta}{\alpha}\,. \tag{23a}$$

Antisymmetric Solution: $c_2' = -c_2$

From this $c_3' = -c_1$ and

$$\tan\frac{\alpha L}{2} = -\frac{\alpha}{\beta}\,. \tag{23b}$$

In either case these equations may be combined with (21) to eliminate E and then:

$$\alpha^2 + \beta^2 = \frac{2mV_0}{\hbar^2}\,. \tag{24}$$

The solutions for α and β cannot be gotten by analytic methods. The solution may be obtained graphically and it is thus possible to obtain the successive values of E. The E values are quantized. For given V_0 and L, the wave functions can be calculated and the first two have the form shown in figure 11.

The equations are greatly simplified when the kinetic energy of the particle $E' = T = E + V_0$ is much less than V_0. This occurs frequently in practice. We get $\alpha \ll \beta$, and (23*a*) and (23*b*) take the simple form:

Symmetric Solution:

$$\tan\frac{\alpha L}{2} = \infty \qquad \text{so that} \qquad \alpha L = (2k+1)\pi\,. \tag{25a}$$

Antisymmetric Solution:

$$\tan\frac{\alpha L}{2} = 0 \qquad \text{so that} \qquad \alpha L = 2k\pi\,. \tag{25b}$$

The two families combine in the general form:

$$\alpha L = n\pi \qquad \text{with} \qquad n = 1, 2, 3 \ldots$$

The even values ($n = 2k$) correspond to antisymmetric solutions, and the odd values ($n = 2k + 1$) to symmetric solutions. The possible values of the kinetic energy can then be written

$$E' = E + V_0 = \frac{n^2\hbar^2\pi^2}{2mL^2} = \frac{n^2h^2}{8mL^2} \tag{26}$$

and they form a discrete-valued series. We then say that the energy is quantized and n, which serves to identify and fix the different levels, is called a quantum number. The energy data is sufficient to characterize the state and we term such energy values simple or non-degenerate.

We can also say that the quantization of the energy will not be observed unless the quantity $h^2/8m\text{L}^2$ in equation (26) is within

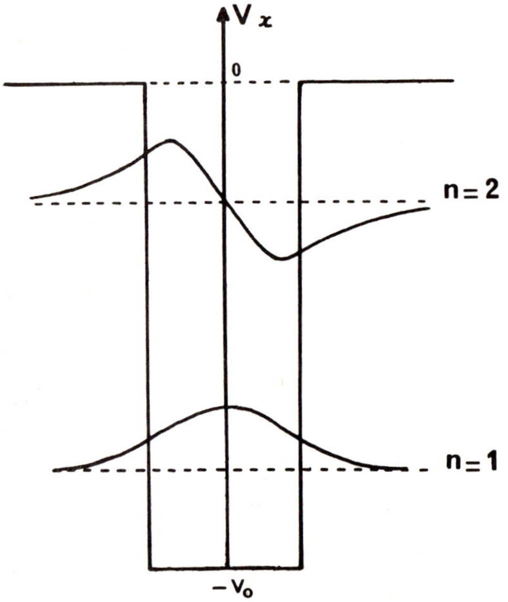

Figure 11. Representation of the first two wave functions for the potential well, the abscissa in each case representing the corresponding energy level.

detectable limits, most likely when L and m are small. For example, this is the case for an electron confined in a region of atomic dimensions (1 Å). With $\text{L} = 1$ Å $= 10^{-8}$ cm and $m = 0.9 \times 10^{-7}$ g we get:

$$h^2/8m\text{L}^2 = 6.1 \times 10^{-11} \text{ erg.}$$

The significance of this is best illustrated by expressing the solution in electron-volts or kcal/mole. We get 38.2 eV or 880 kcal/mole.

We have considered the limiting case $\alpha \ll \beta$ which allowed us to replace (23a, b) by (25a, b). A little better approximation can easily be gotten by introducing a new parameter θ. Setting $\rho^2 = 2m\text{V}_0/\hbar^2$, (24) can be written:

$$\alpha = \rho \sin \theta; \qquad \beta = \rho \cos \theta.$$

(23*a*) and (23*b*) then take the form:

$$\tan\frac{\alpha L}{2} = \cot\theta \qquad \text{and} \qquad \tan\frac{\alpha L}{2} = -\tan\theta$$

and

$$\alpha L = n\pi - 2\theta = n\pi - 2 \text{ Arc} \sin\frac{\alpha}{\rho}.$$

When V_0 is sufficiently large that one can pass to the limit of the arc sine series expansion:

$$\alpha\left(L + \frac{2}{\rho}\right) = n\pi \tag{27}$$

which gives the solution for the case of V_0 infinite, provided that we introduce a "width correction." This correction, independent of L, measures the penetration as previously defined in the step potential. As V_0 becomes infinite the correction goes to zero.

Normalization

We have seen that the function ψ may represent the behavior of a single particle if the normalization condition (1-20) is satisfied. In the one-dimensional case this condition takes the form:

$$\int_{-\infty}^{+\infty} |\psi^2|\,dx = \int_{-\infty}^{-\frac{L}{2}} |\psi_1^2|\,dx + \int_{-\frac{L}{2}}^{+\frac{L}{2}} |\psi_2^2|\,dx + \int_{\frac{L}{2}}^{+\infty} |\psi_3^2|\,dx = 1.$$

We will examine the case where V_0 can be considered infinite which lets us write $c_1 = c_3' = 0$. We treat the symmetric case first:

$$\psi_2 = c\cos\alpha x, \qquad \text{with:} \qquad c = 2c_2, \qquad \alpha = \frac{(2k+1)\pi}{L}.$$

ψ_1 and ψ_3 are identically zero and the normalization fixes c as:

$$\int_{-\frac{L}{2}}^{+\frac{L}{2}} |\psi_2^2|\,dx = |c^2| \int_{-\frac{L}{2}}^{+\frac{L}{2}} \cos^2\frac{(2k+1)\pi x}{L}\,dx = \frac{1}{2}|c^2|\,L = 1. \tag{28}$$

Only the modulus of c (generally complex!) is determined, and not its argument. This remains undetermined because the time-independent wave function has the form, following (1-9):

$$\varphi(x, t) = \psi(x)\exp\left(-\frac{iEt}{\hbar}\right)$$

with arbitrary origin in time, which modifies the argument of c.

The antisymmetric solution can be treated in the same way and one gets:

$$\psi_2 = c \sin \alpha x, \qquad \text{with: } \quad c = 2ic_2, \qquad \alpha = \frac{2k\pi}{L}$$

with the same normalization coefficient as in the symmetric case.

10 The Particle in a Box

The preceding results may be used to study the behavior of a particle inside a three-dimensional box, which we will take as a parallelepiped with sides *a*, *b*, and *c*. The Schrödinger equation must take the complete form:

$$\Delta\psi + \frac{2m}{\hbar^2}(E - V_0)\psi = 0. \tag{29}$$

We will treat this case with a slightly different notation, the energy V will be zero inside the box and infinite outside. The boundary conditions are particularly simple in this case; $\psi(x, y, z)$ disappears at the surface of the box. Under these conditions equation (29) yields:

$$\Delta\psi + \frac{2m}{\hbar^2} E\psi = 0 \tag{30}$$

on the interior.

We will look for solutions of (30) in the form:

$$\psi(x, y, z) = X(x) \cdot Y(y) \cdot Z(z).$$

Substituting into (30) we get:

$$\frac{1}{X}\frac{d^2X}{dx^2} + \frac{1}{Y}\frac{d^2Y}{dy^2} + \frac{1}{Z}\frac{d^2Z}{dz^2} + \frac{2m}{\hbar^2} E = 0.$$

The term $\frac{1}{X}\frac{d^2X}{dx^2}$ depends only on *x*, and the other two analogous terms depend only on *y* or *z*, respectively. Their sum is independent of any variable so that each of them separately is equal to a constant and we have:

$$\frac{1}{X}\frac{d^2X}{dx^2} + \frac{2m}{\hbar^2} E_1 = 0; \quad \frac{1}{Y}\frac{d^2Y}{dy^2} + \frac{2m}{\hbar^2} E_2 = 0; \quad \frac{1}{Z}\frac{d^2Z}{dz^2} + \frac{2m}{\hbar^2} E_3 = 0$$

with

$$E = E_1 + E_2 + E_3.$$

For each coordinate we have reduced the problem to that treated in the previous section. Thus we get solutions of the form:†

$$\psi = c \sin\left(\frac{k_1\pi x}{a}\right) \sin\left(\frac{k_2\pi y}{b}\right) \sin\left(\frac{k_3\pi z}{c}\right) \tag{31}$$

with:

$$E = \frac{h^2}{8m}\left(\frac{k_1^2}{a^2} + \frac{k_2^2}{b^2} + \frac{k_3^2}{c^2}\right) \quad k_1, k_2, k_3 \text{ positive integers.} \tag{32}$$

The definition of a state now requires the knowledge of three quantum numbers, k_1, k_2, and k_3.

We will now show that the general solution of (30) can be formed as a linear combination of the particular solutions (31).

The reasoning we use to demonstrate this principle can be generalized to a number of other cases when one has an ensemble of particular solutions which form a complete orthogonal system (see Appendix II). We assume that we know a solution $\psi(x,y,z)$ of (30) satisfying the boundary conditions (ψ goes to zero on the surface of the box). In particular, $\psi(o,y,z) = \psi(a,y,z)$. Moreover, the function satisfying (30) has at least first- and second-order derivatives, so that the conditions of regularity of the solution are fulfilled. We can, therefore, take ψ as a function of x, given by the expansion:

$$\psi(x, y, z) = \sum_{k_1=1}^{\infty} c_{k_1}(y, z) \sin\left(\frac{k_1\pi x}{a}\right). \tag{33}$$

This expansion can be made because the ensemble of functions $\sin(\pi kx/a)$ forms a complete set of orthogonal functions (Appendix II) with domain (o,z) and has the same boundary conditions, $f(o) = f(a) = 0$, as the original function. This is the same expansion used in the problem of a vibrating string with fixed ends. Taking (33) as the x-solution and resubstituting into (30) gives:

$$\sum_{k_1=1}^{\infty}\left[\frac{\partial^2 c_{k_1}}{\partial y^2} + \frac{\partial^2 c_{k_1}}{\partial z^2} + \left(\frac{2mE}{\hbar^2} - \frac{\pi^2 k_1^2}{a^2}\right)c_{k_1}\right] \sin\left(\frac{\pi k_1 x}{a}\right) = 0.$$

The orthogonal nature of the function $\sin(\pi kx/a)$ implies that the sum cannot be zero unless the coefficients are all zero, so that:

$$\frac{\partial^2 c_{k_1}}{\partial y^2} + \frac{\partial^2 c_{k_1}}{\partial z^2} + \left(\frac{2mE}{\hbar^2} - \frac{\pi^2 k_1^2}{a^2}\right)c_{k_1} = 0. \tag{34}$$

† The origin of coordinates is taken here at one of the corners of the box, which means that we sacrifice the symmetry of solutions, but this leads to a single analytic form for all the wave functions, viz., $\sin\alpha x \sin\alpha' y \sin\alpha'' z$.

We now have a problem analogous to the original problem but with one less variable. We seek solutions for $c_{k_1}(y,z)$, which are zero on the surface of the box. Using the series representation as above:

$$c_{k_1}(y, z) = \sum_{k_2=1}^{\infty} c_{k_1,k_2}(z).\sin\left(\frac{\pi k_2 y}{b}\right).$$

By the same procedure as above we get a new differential equation:

$$\frac{d^2 c_{k_1,k_2}}{dz^2} + \left(\frac{2mE}{\hbar^2} - \frac{\pi^2 k_1^2}{a^2} - \frac{\pi^2 k_2^2}{b^2}\right) c_{k_1,k_2} = 0.$$

Again applying the boundary conditions, $c_{k_1 k_2}(z)$ is of the form:

$$c_{k_1,k_2}(z) = \sum_{k_3=1}^{\infty} c_{k_1,k_2,k_3} \sin\left(\frac{\pi k_3 z}{c}\right)$$

with

$$E = \frac{h^2}{8m}\left(\frac{k_1^2}{a^2} + \frac{k_2^2}{b^2} + \frac{k_3^2}{c^2}\right).$$

Thus, ψ can be given as a linear combination of functions of the type (31) which we have gotten as particular solutions of the Schrödinger equation with k_1, k_2, and k_3 satisfying (32).

If two of the three sides a, b, or c of the box do not have a rational ratio, there are no two systems of quantum numbers k_1, k_2, k_3, which give the same value of energy E. E then forms a simple level. If two of the dimensions of the box form a rational ratio then the situation is different. In particular, in the case of a cubic box, there are multiple systems of the values k_1, k_2, k_3, which lead to the same energy, for example, 2, 1, 1; 1, 2, 1; 1, 1, 2. The energy values are not sufficient to uniquely characterize the state and are thus called degenerate levels. The general solution of the Schrödinger equation, corresponding to this value of energy, is then a linear combination of different particular solutions, as we have just demonstrated. For example, in the case taken above of a level relative to a cubic box with:

$$E = \frac{h^2}{8ma^2}\left(k_1^2 + k_2^2 + k_3^2\right) = \frac{3h^2}{4ma^2}$$

(35)

$$\psi = c_1\psi_{2,1,1} + c_2\psi_{1,2,1} + c_3\psi_{1,1,2}$$

where $\psi_{2,1,1}$ and the two analogous functions are of the form (31). The coefficient c is selected to satisfy the normalization condition. The normalization integral can be easily evaluated using the one-dimensional result and gives $c = (2/a)^{3/2}$. Since the three functions

$\psi_{2,1,1}$, $\psi_{1,2,1}$, $\psi_{1,1,2}$ are on the one hand normalized, and on the other hand orthogonal, that is to say, they satisfy in pairs:

$$\int \psi^*_{2,1,1}\psi_{1,2,1}d\tau = 0$$

and the normalization condition:

$$\int |\psi^2| d\tau = 1$$

for ψ to be normalized it is necessary that c_1, c_2, and c_3, satisfy:

$$c_1^2 + c_2^2 + c_3^2 = 1,$$

since the ψ_{k_1,k_2,k_3}'s are orthogonal.

Therefore, there is an infinity of those solutions ψ, corresponding to the arbitrary nature of the choice of the constants c_1, c_2, c_3. One cannot assume that each of these functions ψ represents a distinct state. They are, rather, a superposition of the three base states given by $\psi_{2,1,1}$, $\psi_{1,2,1}$, $\psi_{1,1,2}$, which is a generalization of the interpretation given in section 4. Thus, all of the physical quantities associated with a state ψ of the particle can be calculated as the mean values of each of the three basis states with weighting $|c_1^2|$, $|c_2^2|$, $|c_3^2|$. The basis states are so named because they are sufficient to describe the system in state ψ. We should note the very important analogy of the superposition state with the representation of a vector in an n-dimensional system as a geometric sum of the components of some basis set of reference axes.

Finally, we recall that the quantities $h^2/8ma^2$, $h^2/8mb^2$, and $h^2/8mc^2$, which occur in the energy E, will become very small when we consider a box of macroscopic dimensions. Thus there is a very large number $n(\mathrm{E})$ of states with energy less than E, if E is large. The function $n(\mathrm{E})$ which is required in many statistical studies can be calculated readily by examining the representation of states by the points in a system of coordinates k_1, k_2, k_3 (fig. 12). Each cubic elementary volume of the system with side unity represents a state so that the number of states with energy less than the given E is equal to the number of elementary cubic volumes in the total volume defined by:

$$\frac{k_1^2}{a^2} + \frac{k_2^2}{b^2} + \frac{k_3^2}{c^2} \leqslant \frac{8m\mathrm{E}}{h^2} = \rho^2$$

limited to the positive coordinate values. We are then lead to calculate one-eighth of the volume of an ellipsoid with half axes $a\rho$, $b\rho$, $c\rho$, and

$$n(\mathrm{E}) = \frac{1}{8} \cdot \frac{4\pi}{3} abc\rho^3 = \frac{\pi}{6} abc \left(\frac{8m\mathrm{E}}{h^2}\right)^{3/2} = \frac{\pi}{6} \mathrm{V} \left(\frac{8m\mathrm{E}}{h^2}\right)^{3/2}$$

where V is the volume of the box. The energy E is purely kinetic and can be expressed as a function of the momentum p:

$$E = \frac{1}{2}mv^2 = \frac{p^2}{2m}$$

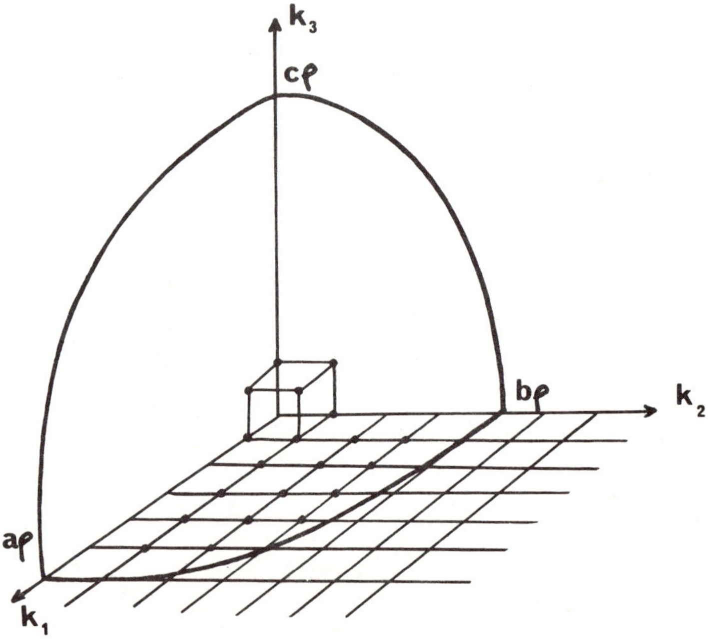

Figure 12. Representation of the states of a particle in a three-dimensional box by means of the nodes of a lattice on the k_1, k_2, k_3 diagram. The contour of the constant energy ellipsoid in this representation.

which leads to:

$$n(p) = \frac{4\pi}{3} V \frac{p^3}{h^3}$$

where:

$$dn = \frac{4\pi V}{h^3} p^2 dp$$

represents the number of states where the momentum of the particle is between p and $p + dp$.

3 ONE-PARTICLE NON-DEGENERATE SYSTEMS

Properties of the Schrödinger Equation 11

The first examples of wave mechanics that we discussed, dealt with currents of particles. Only in the case of the potential well or the particle in a box did we treat systems where we could describe the behavior of single particles. For the study of atomic and molecular systems it is of utmost importance for us to be able to describe the properties of an individual particle. Before proceeding to an examination of the atomic and molecular systems, it is worthwhile to develop some of the general properties of the Schrödinger equation. We begin by summarizing the results previously obtained in Chapter 1.

(a) Status of the Problem

The behavior of a particle in a field derived from a potential $V(x,y,z)$ can be determined by starting with the Schrödinger equation:

(1) $$\Delta\psi + \frac{2m}{\hbar^2}[E - V(x, y, z)]\psi = 0.$$

From this we get the energy E and the wave function $\psi(x,y,z)$. However, all of the solutions we might get from this equation are not physically acceptable. For a solution to represent the behavior of a single particle, the wave function must satisfy the normalization condition:

(2) $$\int \psi^* \psi d\tau = 1.$$

If E is fixed arbitrarily, the corresponding solutions for $\psi(x,y,z)$ will in general not satisfy (2). The values of the energy which give a normalizable wave function generally form a discrete set of values. The quantum nature of the acceptable solutions then results not only from the solution of the Schrödinger equation, but also from the requirement of normalization.

(b) The Hamiltonian Operator

The Schrödinger equation can be written in a different form by transposition of the total energy (E) term:

$$-\frac{\hbar^2}{2m}\Delta\psi + V\psi = E\psi.$$

The left-hand side represents a function ψ' derived from $\psi(x,y,z)$ by a series of operations (derivatives, addition after multiplication by $V(x,y,z)$), which can be worked out for any continuous function and not only for the solutions of the Schrödinger equation. A convenient notation may be used for this new function ψ' ,which we get from ψ by the action of an operator traditionally written **H** and called the hamiltonian operator.† Thus, by definition:

$$\psi' = \mathbf{H}\psi = -\frac{\hbar^2}{2m}\Delta\psi + V\psi. \quad (4)$$

The operator **H**, as defined, has two fundamental properties which are basic to all wave mechanical calculations.
H *is linear, that is to say*:

$$\mathbf{H}(c\psi) = c\mathbf{H}\psi,$$

where c is an arbitrary complex constant and:

$$\mathbf{H}(\psi_1 + \psi_2) = \mathbf{H}\psi_1 + \mathbf{H}\psi_2,$$

where ψ_1 and ψ_2 are two arbitrary functions.

These two relationships together permit us to write for the case of a general linear combination of functions ψ_1 and ψ_2 with arbitrary constant coefficients c_1 and c_2:

$$\mathbf{H}(c_1\psi_1 + c_2\psi_2) = c_1\mathbf{H}\psi_1 + c_2\mathbf{H}\psi_2. \quad (5)$$

H *is a hermitian operator*. To describe this new property it is necessary to introduce an integral which naturally appears in calculations involving the Schrödinger equation. We can rewrite (3) using the notation for **H** introduced in (4) as:

$$\mathbf{H}\psi = E\psi. \quad (6)$$

If ψ is normalized, that is to say, if ψ satisfies (2), E can be calculated from the integral:

$$\int \psi^*(\mathbf{H}\psi)d\tau = E\int \psi^*\psi d\tau = E. \quad (7)$$

Even if the linear character of **H** places no restriction on the functions to which the operator may be applied, we limit ourselves here to those functions ψ_i which satisfy (2) and (7), and which have

† We will use boldface characters to represent those operators which are associated with physical variables.

an infinite domain of integration. If ψ_1 and ψ_2 are two such functions, and we examine the mixed integral:†

(8)
$$\int \psi_1^*(\mathbf{H}\psi_2)d\tau$$

which appeared first in section 3 and in which the ordering of ψ_1 and ψ_2 is immaterial. Thus, as a result of definition (4) of **H**:

$$\int \psi_1^*(\mathbf{H}\psi_2)d\tau - \int (\mathbf{H}\psi_1^*)\psi_2 d\tau = -\frac{\hbar^2}{2m}\int (\psi_1^*\Delta\psi_2 - (\Delta\psi_1^*)\psi_2)d\tau.$$

The integral on the right-hand side has three terms of the form:

$$\int\left(\psi_1^*\frac{\partial^2\psi_2}{\partial x^2} - \psi_2\frac{\partial^2\psi_1^*}{\partial x^2}\right)d\tau = \int\frac{\partial}{\partial x}\left(\psi_1^*\frac{\partial\psi_2}{\partial x} - \psi_2\frac{\partial\psi_1^*}{\partial x}\right)d\tau$$

so that this integral may, as in section 3, be transformed into a flux integral:‡

$$-\frac{\hbar^2}{2m}\int (\psi_1^*\ \overrightarrow{\text{grad}}\ \psi_2 - \psi_2\ \overrightarrow{\text{grad}}\ \psi_1^*)_n dS$$

extended over the surface which encloses the region in which the particle can be found. This generally tends to infinity. The finite character of the normalization integrals of $\psi_1(M)$ and $\psi_2(M)$ implies a sufficiently rapid decrease in the functions at $M(x,y,z)$ so that the integral goes to zero as M goes to infinity. Thus we find that the two integrals are equal when evaluated over all space:

(9)
$$\int \psi_1^*(\mathbf{H}\psi_2)d\tau = \int (\mathbf{H}\psi_1^*)\psi_2 d\tau$$

and this is what is meant by the hermitian character of the operator **H**.

(c) Application to Solutions of the Schrödinger Equation

The two important properties of **H**, namely, normalizability and hermiticity have important consequences.

We begin by introducing the vocabulary associated with this operator. The Schrödinger equation written as in (6) shows that solutions of this equation are transformed into proportional functions by **H**. We call these functions the eigenfunctions of this operator and the coefficients of proportionality are the eigenvalues.

† Integrals of this type appear in normalization integrals when we have ψ as a linear combination of some other ψ_i's.

‡ The index n signifies that we are taking the component of the vector normal to the surface of integration.

We also say that a solution ψ_r of equation (6) represents the eigenfunction of **H** belonging to the eigenvalue E_r. Having established these definitions, we can go on to examine the consequences of the hermitian character of **H**.

The eigenvalues E_r *of* **H** *are real.* If ψ_r belongs to E_r we can write:

$$\mathbf{H}\psi_r = E_r\psi_r$$

and integrating over all space following multiplication by the conjugate $\psi_r^\star$:

$$\int \psi_r^*(\mathbf{H}\psi_r)d\tau = E_r \int \psi_r^* \psi_r d\tau. \tag{10}$$

In addition, if two complex variables are equal, so is the conjugate and we can write:

$$(\mathbf{H}\psi_r)^* = E_r^* \psi_r^* \quad \text{or} \quad \mathbf{H}\psi_r^* = E_r^* \psi_r^*.$$

This is true because **H** does not have any term with the imaginary *i*. Thus, by analogy:

$$\int \psi_r(\mathbf{H}\psi_r^*)d\tau = E_r^* \int \psi_r^* \psi_r d\tau. \tag{11}$$

The two integrals have the same value as a result of the hermiticity property, and the right-hand sides of (10) and (11) are therefore equal. The normalization integral: $\int \psi_r^\star \psi_r d\tau$ is always positive unless ψ is identically zero, a trivial result which we exclude, and thus we have $E_r = E_r^\star$. This is necessary, of course, so that the energies will have physical significance, but it is apparent here that this is a fundamental property of the hermitian character of the operator.

The eigenfunctions ψ_1 *and* ψ_2 *belonging to different eigenvalues* E_1 *and* E_2 *are orthogonal.* By hypothesis we have:

$$\mathbf{H}\psi_1 = E_1\psi_1; \qquad \mathbf{H}\psi_2 = E_2\psi_2.$$

With the first of these we take the integral over all space:

$$\int \psi_2^* \mathbf{H}\psi_1 d\tau = E_1 \int \psi_2^* \psi_1 d\tau$$

and similarly with the second (remembering $E_2^\star = E_2$):

$$\int (\mathbf{H}\psi_2^*)\psi_1 d\tau = E_2 \int \psi_2^* \psi_1 d\tau.$$

The hermiticity of **H** gives us equality of the left-hand sides of these two equations and on subtraction:

$$(E_2 - E_1)\int \psi_2^* \psi_1 d\tau = 0.$$

Since $(E_2 - E_1) \neq 0$ by the original hypothesis:

$$\int \psi_2^* \psi_1 d\tau = 0 \tag{12}$$

and the two functions are orthogonal (also see Appendix II).

(d) Completeness of a System of Eigenfunctions

We have limited ourselves here to systems for which all of the eigenvalues of energy are non-degenerate, that is to say, if E_r is an eigenvalue of $\mathbf{H}$, the group of functions ψ_r which satisfy:

$$\mathbf{H}\psi_r = E_r\psi_r$$

differ only by a coefficient of proportionality. Assuming that we can always represent an arbitrary function $f(x,y,z)$ by the series:

$$f(x, y, z) = \sum_r c_r\psi_r(x, y, z). \tag{13}$$

The coefficients c_r can be calculated using the orthogonality of ψ_r. These functions are also normalized, so that, following the method outlined in Appendix II:

$$c_r = \int \psi_r^* f d\tau. \tag{14}$$

We know that the series (13) with coefficients set by (14) only converges to $f(x,y,z)$ if certain conditions on ψ_r are maintained. When this is the case for an arbitrary regular function $f(x,y,z)$, it satisfies the same conditions and limits which are imposed on ψ_r, and we say that the set of orthogonal functions ψ_r forms a complete orthogonal set.† This property exists for each of the physical systems we study and we are going to see this expansion used frequently to represent general states.

Superposition 12

The arguments previously advanced have essentially involved particular solutions of the Schrödinger equation of the form (1-9):

$$\varphi_r(x, y, z, t) = \psi_r(x, y, z) . \exp\left(-\frac{iE_r t}{\hbar}\right) \tag{15}$$

† Thus, in the case of the three-dimensional potential well with perfectly reflecting surface (§10) the function $f(x,y,z)$ can be zero on the surface, quite like the set ψ_r, and (13) then leads to a three-dimensional Fourier expansion.

where the amplitudes ψ_r are solutions of the time-independent equation corresponding to the energy E_r. The form of this wave function implies that the probability density calculated by means of $|\varphi_r^2| = |\psi_r^2|$, and used to locate the particle, is time independent or *stationary*. To study these *stationary states* determined from ψ_r it is not necessary to use the complete wave function φ_r.

In addition to these solutions we want to introduce the idea of the *most general state* of the system, which is here a single particle state. We could have gotten this state through an arbitrary function† $\theta(x,y,z)$, describing the particle distribution at the initial time $(t=0)$, where this function has the required second derivative and satisfies the normalization condition:

$$\int \theta^* \theta d\tau = 1.$$

The function $\theta(x,y,z)$, defining the state at the initial time, is going to evolve in the course of time and we call $\theta(x,y,z,t)$ its value at time t and position x, y, z. An important mathematical property of this function $\theta(x,y,z,t)$ is that it can be represented by a series on the base functions φ_r written with time-independent coefficients.‡ Current usage consists in saying that the various states considered may be described as a superposition of the base states, implying a physical interpretation which we will now examine.

We begin with a mathematical property which is easily demonstrated by observing that if the function $\theta(x,y,z,t)$ is known, it can be represented by an expansion of the form (13) on the complete orthogonal system of $\psi_r(x,y,z)$. The coefficients c_r of this expansion are functions of time and are found as in (14):

$$c_r(t) = \int \psi_r^*(x, y, z)\, \theta(x, y, z, t) d\tau. \tag{16}$$

The expansion (13) can thus be written:

$$\theta(x, y, z, t) = \sum_r c_r(t)\, \psi_r(x, y, z). \tag{17}$$

† This introduction of the most general state by means of a wave function $\theta(x,y,z)$, known as the initial state, is convenient for discussion but does not correspond to any real physical problem. Moreover, by admitting that we are capable of effecting a state of known probability density at a given time, we then know $|\theta^2|$, from which we can get the modulus $|\theta|$, though not its argument which governs the evolution of the state with time. Such a definition of a state is incomplete.

‡ This approach reflects the most general way of introducing the time dependence.

The functions $c_r(t)$ are not arbitrary, since the evolution of a wave function with time is necessarily governed by the time-dependent Schrödinger equation (1-11), which we write in the abbreviated form using $\mathbf{H}$:

$$\mathbf{H}\theta = -\frac{\hbar}{i}\frac{\partial\theta}{\partial t}. \tag{18}$$

Taking into account the linear character of $\mathbf{H}$, substitution of the expansion (17) allows us to write:

$$\sum_r c_r \mathbf{H}\psi_r = -\frac{\hbar}{i}\sum_r \dot{c}_r\psi_r, \qquad \text{with:} \quad \dot{c}_r = \frac{dc_r}{dt}.$$

By hypothesis, ψ_r is the eigenfunction of $\mathbf{H}$ belonging to the eigenvalue E_r, and by rearranging terms we get:

$$\sum_r \left(c_r E_r + \frac{\hbar}{i}\dot{c}_r\right)\psi_r = 0.$$

The orthogonal nature of the system of functions ψ_r implies that the coefficients of the expansion should all be zero (Appendix II). This condition may be immediately integrated and:

$$c_r(t) = c_r^{(0)} \exp\left(-\frac{iE_r t}{\hbar}\right). \tag{19}$$

As a consequence, the expansion (17) takes the form:

$$\theta(x, y, z, t) = \sum_r c_r^{(0)}\psi_r \exp\left(-\frac{iE_r t}{\hbar}\right) = \sum_r c_r^{(0)}\varphi_r \tag{20}$$

which yields the desired result.

The second part of this treatment concerns the interpretation of the expansion (20) and necessitates the introduction of a new idea. The comparison with the case of a vibrational motion allows a better understanding of the problem. We began with an example taken from the field of acoustics which seems particularly suitable. We know that the profile of a vibrating cord over time is a superposition of the normal modes of vibration which are harmonics of the fundamental, and can be derived from an equation closely analogous to (20). Again the mathematical property of interest corresponds to a well-known physical situation, where the measurement of the frequency of an emitted sound by means of a resonator can be used to analyze this sound and allows the isolation of each of the normal modes, as well as measurement of their relative intensity—important in the representation of a complex vibration.

The example is inverted in the case of the energy of the state represented by a wave function $\theta(x,y,z,t)$. We know that the measurement of energy may only lead to eigenvalues E_r of the hamiltonian of the system, which is the same behavior as one finds with the eigenfrequencies of an acoustical system. The probability of measuring a value E_r will be proportional to the modulus of the coefficient $c_r^{(0)}$ of φ_r in the expansion (20) of θ.

This comparison is designed to illustrate the status of the problem, namely, that the measurement of the energy in an arbitrary state will not lead to a definite result, but rather to different values from the energy spectrum of the eigenvalues E_r, where the probabilities are determined by the particular state. This interpretation can be strengthened, as we will show. We begin with the integral of the form (7), but this time calculated on any state function $\theta(x,y,z,t)$, so that:

$$\text{(21)} \qquad W(t) = \int \theta^*(x, y, z, t)\, \mathbf{H}\theta\,(x, y, z, t) d\tau.$$

We treat this integral by using the expansion (20) of θ:

$$W(t) = \int \Big(\sum_r c_r^{(0)*} \varphi_r^*\Big)\Big(\mathbf{H} \sum_s c_s^{(0)} \varphi_s\Big) d\tau.$$

The linear nature of $\mathbf{H}$ allows us to write the integral as:

$$W(t) = \int \theta^* \mathbf{H}\theta d\tau = \sum_{r,s} c_r^{(0)*} c_s^{(0)} \int \varphi_r^* \mathbf{H} \varphi_s d\tau$$

which upon integration gives:

$$\int \varphi_r^* \mathbf{H} \varphi_s d\tau = \exp\left[\frac{i}{\hbar}(E_r - E_s)t\right] . \int \psi_r^* \mathbf{H} \psi_s d\tau.$$

The functions ψ_r and ψ_s are eigenfunctions of $\mathbf{H}$ so that:

$$\int \psi_r^* \mathbf{H} \psi_s d\tau = E_s \int \psi_r^* \psi_s d\tau = 0 \quad \text{if} \quad r \neq s \quad \text{and} \quad E_r \quad \text{if} \quad r = s.$$

The cross terms vanish and $W(t)$ takes the form:

$$\text{(22)} \qquad W(t) = \int \theta^* \mathbf{H}\theta d\tau = \sum_r c_r^{(0)*} c_r^{(0)} E_r.$$

The integral $W(t)$ is *time independent*. Its value is equal to E_r when all of the coefficients c_s^0 are zero, with the exception of $c_r^0 = 1$, which corresponds to $\theta = \varphi_r$. As a consequence, we have defined a quantity W which has the dimensions of energy, is equal to the energy in all cases where the latter is well defined (state φ_r), and remains constant with time when calculated for any state. It is reasonable to say that this quantity W defines the energy for the state represented by the wave function $\theta(x,y,z,t)$.

We can examine the second point in the interpretation by normalizing the wave function θ using the expansion (20):

$$\int \theta^{\star}\theta d\tau = \sum_{r,s} c_r^{(0)\star} c_s^{(0)} \exp\left[\frac{i}{\hbar}(\mathrm{E}_r - \mathrm{E}_s)t\right] . \int \psi_r^{\star}\psi_s d\tau.$$

The orthogonality of the eigenfunctions ψ_r and ψ_s leads to:

$$\int \theta^{\star}\theta d\tau = \sum_{r} c_r^{(0)\star} c_r^{(0)} = 1. \tag{23}$$

Thus, using a different approach, we have obtained a result from Chapter 1 (§3), namely, that the normalization remains fixed over time. The two equations (22) and (23) form the basis of the interpretation. If we note that the elements $p_r = c_r^{(0)\star} \cdot c_r^{(0)} = |c_r^{(0)2}|$ of the sum (23) are all positive with sum unity, we can rewrite (22) and (23) as:

$$\mathrm{W} = \sum_r p_r \mathrm{E}_r \, ; \qquad \sum_r p_r = 1 \, .$$

These are exactly the equations we would write if told that the measurement of energy in a state defined by θ leads to a value E_r with probability p_r. Thus, W represents the mean energy corresponding to a large number of measurements, or, alternatively, the result of the simultaneous measurement on a large number of identical independent systems. This is the experimental significance of the coefficients $c_r^{(0)}$ of the expansion of the wave function and the result of the superposition principle.†

Finally, we ask ourselves whether—with the general state defined by the wave function θ—two distinct systems of coefficients $c_r^{(0)}$ and $\gamma_r^{(0)}$ can define the same state. A first condition is, evidently, that the probability of finding E_r in an energy measurement must be the same calculated from either system, which necessitates:

$$|\gamma_r^{(0)}| = |c_r^{(0)}| \quad \text{so that} \quad \gamma_r^{(0)} = c_r^{(0)} . \exp(i\lambda_r)$$

λ_r being any real number. The condition is not sufficient, for if we consider the probability of the particle being in a volume element $d\tau$, we get:

$$\theta^{\star}\theta d\tau = \sum_{r,s} c_r^{(0)\star} c_s^{(0)} \exp\left[\frac{i}{\hbar}(\mathrm{E}_r - \mathrm{E}_s)t\right] \psi_r^{\star}\psi_s d\tau$$

---

† A particular case of this principle was introduced in section 4 for the problem of the localization of a free particle moving on an infinite line.

and this expression is generally different when calculated with the coefficients $\gamma_r^{(0)}$ or $c_r^{(0)}$, which means that we have for all pairs of indices r and s:

$$c_r^{(0)*}c_s^{(0)} = \gamma_r^{(0)*}\gamma_s^{(0)} = c_r^{(0)*}c_s^{(0)} \exp\left[i(\lambda_r - \lambda_s)\right]$$

which implies the equality of all the λ_r's. This allows us to state that if θ and θ' represent the same state, we must have $\theta' = \theta e^{i\gamma}$ with λ real.

Variational Calculation of the Eigenvalues of **H**

The preceding discussion allows us to introduce an approximation method often used in chemistry for calculating the eigenvalues of **H**. To do this, we take the function defined by the expansion (20), representative of an arbitrary state, and we let $|c_r^{(0)}| = \rho_r$. We then look for each of those functions which correspond to an extremum of energy (22) in this state, written:

$$W = \sum_r \rho_r^2 E_r$$

taking into account the normalization condition, we have:

$$\sum_r \rho_r^2 = 1.$$

The condition for an extremum can be written:

$$\delta W = 2\sum_r E_r\rho_r\delta\rho_r = 0.$$

The variations $\delta\rho_r$ cannot be regarded as independent since the normalization condition leads to the relation:

$$2\sum \rho_r\delta\rho_r = 0.$$

The method of Lagrangian multipliers (Appendix III) lets us put these two conditions in a more convenient form by introducing a parameter λ, such that:

$$E_r\rho_r = \lambda\rho_r, \quad \text{or} \quad (E_r - \lambda)\rho_r = 0 \text{ for all values of } r.$$

If we exclude the solution where all the values of ρ_r are zero, which corresponds to a wave function which is identically zero and which is without physical meaning, we see that this condition necessitates that the parameter λ be equal to one of the eigenvalues of **H**, E_i for example. To prove this, it is necessary to take $\rho_r = 0$ for all values of r different from i, with the result that the extremum condition must lead to the eigenfunctions and eigenvalues of **H**.

In practice, the value of W is independent of time and we can make the variational calculation on the value of the wave function at the initial time $\theta(x,y,z)$. We have thus shown that the eigenstate of **H** corresponds to the extremum of the integral:

$$W = \int \theta^*(\mathbf{H}\theta)d\tau,$$

the function θ being constrained by the normalization condition:

$$\int \theta^*\theta d\tau = 1.$$

As we will see, this method is frequently used in the study of molecules. Here we will limit ourselves to a single example.

Exercise 1. Take the Schrödinger equation for the linear harmonic oscillator, which we can write as:

$$\mathbf{H}\psi = -\frac{\hbar^2}{2m}\frac{d^2\psi}{dx^2} + \frac{1}{2}kx^2\psi = E\psi.$$

We want to use the preceding method of approximation to calculate the energy of the ground state level from the family of functions:

$$\psi = C \cos \alpha x,$$

defined between:

$$\alpha x_1 = -\frac{\pi}{2} \quad \text{and} \quad \alpha x_2 = \frac{\pi}{2},$$

and zero outside this interval.

Find the relation between the parameters C and α imposed by the normalization conditions. Calculate the integral $W(\alpha)$ and find the minimum. This gives:

$$\frac{\pi C^2}{2\alpha} = 1\,; \qquad W(\alpha) = \frac{\alpha^2\hbar^2}{2m} + \frac{k}{4\alpha^2}\left(\frac{\pi^2}{6} - 1\right); \qquad W_{min} = 0.57\ h\nu_0,$$

where ν_0 is the mechanical frequency of the oscillator.

Measurement of a Variable in Quantum Mechanics 13

In the preceding argument we discussed the general idea of the evolution of the state of a single-particle system, fixed at some instant by a function $\theta(x,y,z)$ whose development with time we know. The problem we now face is to determine what the result of measuring some variable associated with the state given by $\theta(x,y,z)$ will be.

The interpretation of the wave function θ, which allows us to get the probability of finding the particle in a volume element surrounding a given point by $\theta^\star \theta d\tau$, gives an immediate answer to this question for one of the coordinates of the particle, x, for example. The only quantity which has physical meaning, given the basis of the distribution, is the definition of the mean value of x by means of:

$$\bar{x} = \int x\theta^\star \theta d\tau.$$

We will write this expression in the following form, convenient for the general case, and will use the notation $\langle x \rangle$ to represent the mean value of x in the state under consideration.

$$\bar{x} = \langle x \rangle = \int \theta^\star x \theta d\tau. \tag{24}$$

We will now treat velocity. We start with the fact that the function θ evolves with time,† this evolution being governed by the time-dependent Schrödinger equation (18), and as a result the mean value defined above is a function of time. We can compute the velocity by differentiation under the integral. We get:

$$\bar{v}_x = \frac{d\bar{x}}{dt} = \int \left(\frac{\partial \theta^\star}{\partial t} x\theta + \theta^\star x \frac{\partial \theta}{\partial t} \right) d\tau$$

where by substituting for $\partial\theta/\partial t$ and $\partial\theta^\star/\partial t$ from (18):

$$\bar{v}_x = \frac{i}{\hbar} \int (x\theta \mathbf{H} \theta^\star - x\theta^\star \mathbf{H} \theta) d\tau = \frac{\hbar}{2mi} \int (x\theta . \Delta\theta^\star - x\theta^\star . \Delta\theta) d\tau.$$

The integral can be broken down into three terms of which the first is:

$$\frac{\hbar}{2mi} \int \left(x\theta \frac{\partial^2 \theta^\star}{\partial x^2} - x\theta^\star \frac{\partial^2 \theta}{\partial x^2} \right) d\tau = \frac{\hbar}{2mi} \int x \frac{\partial}{\partial x} \left(\theta \frac{\partial \theta^\star}{\partial x} - \theta^\star \frac{\partial \theta}{\partial x} \right) d\tau.$$

On integration by parts:

$$\int x \frac{\partial}{\partial x} \left(\theta \frac{\partial \theta^\star}{\partial x} - \theta^\star \frac{\partial \theta}{\partial x} \right) d\tau$$

$$= \int \frac{\partial}{\partial x} \left[x \left(\theta \frac{\partial \theta^\star}{\partial x} - \theta^\star \frac{\partial \theta}{\partial x} \right) \right] d\tau - \int \left(\theta \frac{\partial \theta^\star}{\partial x} - \theta^\star \frac{\partial \theta}{\partial x} \right) d\tau.$$

† A normalized function $\theta(x, y, z)$ defines a state at a given time, which we take as the origin of time, but this state is, in general, not stationary and evolves in the course of time. We are forced to write $\theta(x, y, z, t)$ to designate the evolution of the wave function of the state defined initially as $\theta(x, y, z)$.

The first of the two integrals on the right-hand side can be transformed into a surface integral over the domain of integration and this, as we saw earlier, goes to zero. The second integral can be transformed using the same integration by parts which leads to the zero-valued surface integral. We get:

$$\frac{\hbar}{2mi}\int\left(\theta^*\frac{\partial\theta}{\partial x}-\theta\frac{\partial\theta^*}{\partial x}\right)d\tau$$

$$=\frac{\hbar}{2mi}\left[-\int\frac{\partial}{\partial x}(\theta^*\theta)d\tau+2\int\theta^*\frac{\partial\theta}{\partial x}d\tau\right]=\frac{\hbar}{mi}\int\theta^*\frac{\partial\theta}{\partial x}d\tau.$$

The other two terms of the expression for $\bar{v}_x$ are treated the same way and from these we get only surface integrals which vanish. The final result is:

$$\bar{v}_x=\frac{\hbar}{mi}\int\theta^*\frac{\partial\theta}{\partial x}d\tau$$

where by introducing the momentum $\vec{p}=m\vec{v}$:

$$\bar{p}_x=\langle p_x\rangle=\int\theta^*\frac{\hbar}{i}\frac{\partial\theta}{\partial x}d\tau. \tag{25}$$

We can thus see that the expression giving the mean value of the momentum p_x in the state θ, has the same form as that giving the mean value of the coordinate x, but with the introduction of an operator $\mathbf{P}_x=(\hbar/i)(\partial/\partial x)$, which is said to be associated with the variable p_x. In classical physics all variables associated with a particle are functions of the coordinates and velocity so that we must expect that the mean value of such a variable in a state θ will have the same form. We have an excellent example of this in (22), the mean energy expression, which we can write:

$$\langle E\rangle=W=\int\theta^*\mathbf{H}\theta d\tau.$$

The hamiltonian (4) is the sum of two terms, which gives:

$$\langle E\rangle=\int\theta^*\left(-\frac{\hbar^2}{2m}\Delta\theta\right)d\tau+\int\theta^*V\theta d\tau.$$

The integral on the right can immediately be interpreted as representing the mean value of the potential energy $V(x,y,z)$. The first integral represents the mean value of the kinetic energy. Thus we get three expressions for mean values in analogous forms:

$$\langle E\rangle=\int\theta^*\mathbf{H}\theta d\tau;\qquad\langle T\rangle=\int\theta^*\left(-\frac{\hbar^2}{2m}\Delta\right)\theta d\tau;\qquad\langle V\rangle=\int\theta^*V\theta d\tau. \tag{26}$$

We are now lead to state generally that the mean value $\langle a \rangle$ of all variables a, defined in a state represented by θ, is given by means of the integral using the operator $\mathbf{A}$ associated with a as:

$$\langle a \rangle = \int \theta^* \mathbf{A} \theta d\tau. \tag{27}$$

This result can be taken as the point of departure for an axiomatic development of quantum mechanics. We will use the term wave mechanics to designate the form of the new mechanics which we have previously treated on the basis of the analogy between dynamics and optics. Quantum mechanics is the study of the construction and properties of the operators associated with variables.

The axiomatic approach is necessary for the complete study of wave mechanics, principally because we will deal with generalizations of the theory, such as the spin of particles and the many particle systems. This point of view is systematically adopted in a number of books and we cannot treat it here in comparable detail. We will confine ourselves to dealing with some of the aspects of the generalizations of the properties just treated.

The Fundamental Postulate of Quantum Mechanics

All variables a are associated with a linear hermitian operator $\mathbf{A}$ acting on the state function θ. The measurement of the variable a in state θ yields a mean value given by:

$$\langle a \rangle = \int \theta^* \mathbf{A} \theta d\tau.$$

Consequently we find that:

First: The states where a is rigorously known are defined by the eigenfunctions θ_r of $\mathbf{A}$ which satisfy:

$$\mathbf{A}\theta_r = a_r \theta_r.$$

The preceding equation shows that in a state θ_r, the mean value $\langle a \rangle$ of a is equal to the eigenvalue a_r.

Second: The eigenfunctions θ_r of the hermitian operator $\mathbf{A}$ form an orthogonal system† which is also complete, and which allows the general expansion:

$$\theta = \sum_r c_r \theta_r. \tag{28}$$

† The proof is identical with that given earlier for $\mathbf{H}$.

The use of (27) leads to the following expression for $\langle a \rangle$ in the state represented by θ, as:

$$\langle a \rangle = \sum_{r,s} c_r^* c_s \int \theta_r^* \mathbf{A} \theta_s d\tau = \sum_{r,s} c_r^* c_s \, a_s \int \theta_r^* \theta_s d\tau$$

where

$$\langle a \rangle = \sum_r c_r^* c_r \, a_r, \quad \text{with} \quad \sum_r c_r^* c_r = \int \theta^* \theta d\tau = 1.$$

This expression for $\langle a \rangle$ is in agreement with the physical interpretation of the series (28), which involves showing that the probability p_r of finding the value a_r when measuring a in the state θ is calculated from the coefficients of the series (28) by:

(29) $$p_r = c_r^* c_r.$$

Third: If two variables a and b are compatible, the operators $\mathbf{A}$ and $\mathbf{B}$ associated with them commute, that is to say they satisfy:

$$\mathbf{AB}\theta = \mathbf{BA}\theta$$

regardless of the state θ.

We have thus listed without proof the properties which we will use later.

In the case of the operators associated with compatible variables, the above property is evident in the particular instance where the eigenvalues a_r of one of the operators, $\mathbf{A}$, for example, are all non-degenerate. We start in effect with eigenfunctions θ_r defined by:

$$\mathbf{A}\theta_r = a_r \theta_r$$

and apply the operator $\mathbf{B}$ to two identical functions $\mathbf{A}\theta_r$ and $a_r \theta_r$. We get:

$$\mathbf{BA}\theta_r = \mathbf{B}(a_r \theta_r)$$

assuming $\mathbf{B}$ commutes with $\mathbf{A}$, we can write, using the linear property of the operator:

$$\mathbf{AB}\theta_r = a_r \mathbf{B}\theta_r$$

which shows that the function $\mathbf{B}\theta_r$ is also an eigenfunction of $\mathbf{A}$ with the same eigenvalue a_r as θ_r. Since this value is non-degenerate the functions $\mathbf{B}\theta_r$ and θ_r are proportional and:

$$\mathbf{B}\theta_r = b_r \theta_r.$$

Thus, θ_r is also an eigenfunction of $\mathbf{B}$ and the variable b has an exactly defined value in all states where a is known.

Another fundamental question which may be simply treated in this particular case is the relationship of operators and their variables.

A rigorous study leads to the following result: it is always possible to define the sum c of two variables a and b whether or not they are compatible, and the associated operator $\mathbf{A} + \mathbf{B}$ is given by:

$$\mathbf{C}\theta = \mathbf{A}\theta + \mathbf{B}\theta.$$

One has, in effect, in a state given by θ:

$$\langle c \rangle = \langle a \rangle + \langle b \rangle.$$

On the other hand, we can only define a product of two variables, a and b, when these variables are compatible, and the operators commute ($\mathbf{AB} = \mathbf{BA}$). Thus we see that the relationship:

$$\mathbf{AB}\theta_r = a_r \mathbf{B}\theta_r = a_r b_r \theta_r$$

shows that θ_r is an eigenstate of $\mathbf{AB}$ with eigenvalue $a_r b_r$, which allows us to associate the operator $\mathbf{C} = \mathbf{AB} = \mathbf{BA}$ with ab.

These properties can be illustrated by considering the position operators $\mathbf{X} = x$ and:†

$$\mathbf{P}_x = \frac{\hbar}{i} \frac{\partial}{\partial x}$$

the momentum operator in the x-direction.
(*a*) These variables are not compatible and the operators should not commute. One has in fact:

$$\mathbf{P}_x(x\theta) - x(\mathbf{P}_x\theta) = \frac{\hbar}{i}\left[\frac{\partial}{\partial x}(x\theta) - x\frac{\partial \theta}{\partial x}\right] = \frac{\hbar}{i}\theta.$$

This identity, independent of θ, may be written as an operator equation:

(30) $$\mathbf{P}_x x - x\mathbf{P}_x = \frac{\hbar}{i}.$$

(*b*) The variable p_x is compatible with itself defining p_x^2. The associated operator $\mathbf{P}_x^2$ is:

$$\mathbf{P}_x^2\theta = \mathbf{P}_x(\mathbf{P}_x\theta) = -\hbar^2 \frac{\partial}{\partial x}\left(\frac{\partial \theta}{\partial x}\right) = -\hbar^2 \frac{\partial^2 \theta}{\partial x^2}.$$

The addition of variables causes no particular problem and we can construct the operator $\mathbf{T}$ associated with the kinetic energy $(p_x^2 + p_y^2 + p_z^2)/2m$ which gives:

$$\mathbf{T}\theta = -\frac{\hbar^2}{2m}\left(\frac{\partial^2\theta}{\partial x^2} + \frac{\partial^2\theta}{\partial y^2} + \frac{\partial^2\theta}{\partial z^2}\right) = -\frac{\hbar^2}{2m}\Delta\theta$$

the expression previously introduced in (26).

† The identity: $\mathbf{X}\theta = x\theta$. It is unnecessary to write the operand when we use boldface to distinguish the fact that $\mathbf{X}$ is an operator.

The Matrix Associated with a Variable

We postulate a one-particle system in which the eigenfunctions of some operator are known, for example ψ_r of the operator **H** are particularly important because they define stationary states. Consider now any variable a and **A** its associated operator. In the general case a is not compatible with the energy, so that **A** and **H** do not commute. As a result, the application of **A** to ψ_r is a new function, which may be written as a series expansion on the orthonormal set ψ_r:

$$\mathbf{A}\psi_r = \sum_s a_{sr}\psi_s \tag{31}$$

the coefficients a_{sr} are found as in (14):

$$a_{sr} = \int \psi_s^* \mathbf{A}\psi_r d\tau \tag{32}$$

and the notation adopted leads to the same order of the indices s and r in the matrix element and in the integral.

The table of coefficients a_{ij} thus forms the matrix associated with the variable a. This matrix may eventually be of infinite order. Knowledge of it completely defines the variable a because it allows the calculation of $\langle a \rangle$ of a in any state θ. In effect, the function θ may be represented by expansion on the orthogonal system ψ_r:

$$\theta = \sum_r c_r\psi_r$$

and thus:

$$\langle a \rangle = \int \theta^* \mathbf{A}\theta d\tau = \sum_{r,s} c_r^* c_s \int \psi_r^* \mathbf{A}\psi_s d\tau = \sum_{r,s} c_r^* a_{rs} c_s. \tag{33}$$

A physical variable a may be defined either by the operator **A** or by the matrix which represents it in an orthogonal system of eigenfunctions (not necessarily the eigenfunctions of **H**). The relationship between the operator and the associated matrix is similar to that which occurs in the study of vectors between intrinsic representation by a single letter $\vec{v}$ and definition of that vector by its components in a cartesian system.

Finally, we have limited ourselves to the case of variables whose eigenvalues form a discrete set. A second case can occur corresponding to a continuous set of eigenvalues. The treatment of continuous eigenvalues merely requires the substitution of integrals for the sums in the preceding treatment.

Exercise 2. Find the eigenfunctions of $\mathbf{P}_x$ in the case of a free electron on an infinite line. A difficulty arises with the normalization which may be made on a segment (–L/2 to L/2) and then L is allowed to go to infinity. Another method permits passage from a set of discrete eigenvalues to a set of continuous values by means of the Born cyclic boundary conditions. To do this, we divide the axis into an infinite number of adjacent segments. Any period L in this system for finite L admits to the expansion in a Fourier series and then simply requires taking the limit when L is infinitely augmented.

(*a*) Find the eigenvalues and the eigenfunctions of $\mathbf{P}_x$ using Born conditions on the segment –L/2 to L/2. One gets $p = kh/\mathrm{L}$ with k an integer and $\theta_k = \mathrm{C}\exp(2\pi kix/\mathrm{L})$. The normalization coefficient will be found to be $\mathrm{C} = \mathrm{L}^{-1/2}$.

(*b*) Consider the function $\theta = \mathrm{C}'\exp(-x^2/2\sigma^2)$ which represents a particle in the neighborhood of the origin. The coefficient C′ may be found using:

$$\int_{-\infty}^{+\infty} \exp\,(-\,\xi^2)d\xi = \sqrt{\pi},$$

and assuming $\sigma \ll \mathrm{L}$.

(*c*) Represent θ by a series using the eigenfunctions θ_k of $\mathbf{P}_x$. The calculation of the coefficients is easily made by taking $\sigma \ll \mathrm{L}$ to permit the evaluation of the integral between $-\infty$ and $+\infty$. The following relationship will be needed:

$$\int_{-\infty}^{+\infty} \exp\,[-\,(\xi + i\eta)^2]d\xi = \int_{-\infty}^{+\infty} \exp\,(-\,\xi^2)d\xi = \sqrt{\pi}$$

which is a direct result of the integration of $\exp(-z^2)$ in the complex plane.

(*d*) Show that the series obtained may be put in the form of an integral, and note that for large L, the number Δk of terms which should be found in dp is such that $dp = (\Delta k)h/\mathrm{L}$. In completing the integral we must get the function θ from which we started.

14 The Linear Harmonic Oscillator

We will next study the system made up of a particle of mass m moving along a line about a center O constrained by a restoring force proportional to the displacement ($\mathrm{F} = -kx$). The classical equation of motion is given by:

$$m\ddot{x} + kx = 0.$$

Integration of this equation leads to a sinusoidal function with two arbitrary constants:

$$x = a \cos (2\pi\nu_0 t + \chi) \quad \text{with} \quad \nu_0 = \frac{1}{2\pi}\sqrt{\frac{k}{m}}.$$

This model corresponds to a potential energy of the particle having the form $V = kx^2/2$ and this potential is what appears when we approximate to the first order a potential energy curve which has a minimum. For example, we can treat a diatomic molecule near equilibrium in this fashion (fig. 13).

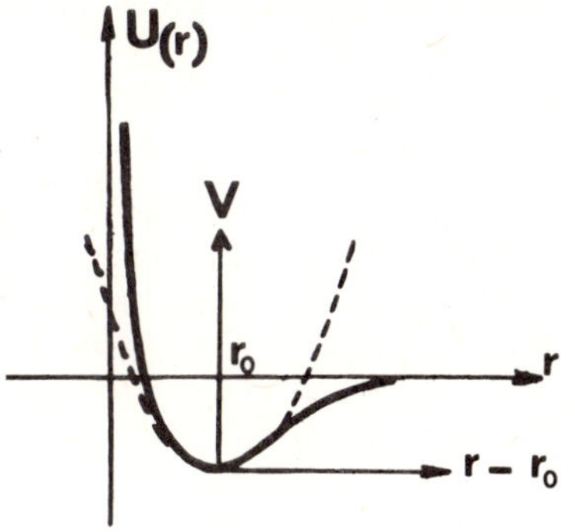

Figure 13. The potential energy of a diatomic molecule as a function of the internuclear distance. The dotted curve is the parabola of oscillation near the equilibrium point.

To treat this problem in wave mechanics, we write the Schrödinger equation for a single (x) variable:

$$-\frac{\hbar^2}{2m}\frac{d^2\psi}{dx^2} + \frac{kx^2}{2}\psi = E\psi.$$

Dividing by $h\nu_0$ (energy) the form can be reduced and k can be replaced by an expression in terms of ν_0, the classical oscillator frequency, giving:

$$-\frac{h}{8\pi^2 m\nu_0}\frac{d^2\psi}{dx^2} + \frac{2\pi^2 m\nu_0}{h}x^2\psi = \frac{E}{h\nu_0}\psi.$$

This form suggests a change in variables:

$$2\pi\sqrt{\frac{m\nu_0}{h}}\,x = \xi; \qquad \frac{2E}{h\nu_0} = \lambda$$

which will lead to a reduced form for the equation, thus:

(34) $$\mathbf{H}'\psi = -\frac{d^2\psi}{d\xi^2} + \xi^2\psi = \lambda\psi, \quad \text{hence} \quad \frac{d^2\psi}{d\xi^2} + (\lambda - \xi^2)\psi = 0.$$

From the study of differential equations with constant coefficients we are lead to predict a solution:

$$\psi = c \exp (f(\xi))$$

which upon substitution yields:

$$f'' + f'^2 + \lambda - \xi^2 = 0.$$

This equation has as particular solutions:

$$f'' = \pm 1; \qquad f' = \pm \xi; \qquad f = \pm \frac{\xi^2}{2}, \qquad \text{for} \quad \lambda = \mp 1.$$

The normalization condition on ψ requires that ψ vanish at infinity, so that we must discard one of the solutions and take only $f = -\xi^2/2$, the solution for $\lambda = 1$. This is the immediate solution of the equation, but we must note that for $|\xi| \gg \lambda$ the equation is approximately equivalent to:

$$f'' + f'^2 - \xi^2 = 0$$

which is independent of λ, with the result that all of the solutions show the same asymptotic behavior. In other words, the expression $f = -\xi^2/2$, which is an exact solution for $\lambda = 1$, represents for any λ the asymptotic behavior of the wave function and we then must seek solutions of the form:

$$\psi = \mathcal{H}(\xi) . \exp\left(-\frac{\xi^2}{2}\right). \tag{35}$$

Substitution into (34) shows that $\mathcal{H}(\xi)$ satisfies the equation:

$$\mathcal{H}'' - 2\xi\mathcal{H}' + (\lambda - 1)\mathcal{H} = 0. \tag{36}$$

The general theory of linear differential equations shows that when the coefficients of $\mathcal{H}''$, $\mathcal{H}'$ and $\mathcal{H}$ are polynomials on ξ, the solutions can be given as a series of the form:

$$\mathcal{H} = \sum_{r=0}^{\infty} a_r \xi^r, \tag{37}$$

which satisfies (36). The coefficients of ξ_r must be identically zero when (37) is substituted into (36), since each term of the substituted expansion is independent and must go to zero. We then get:

$$(r + 2)(r + 1)a_{r+2} - (2r + 1 - \lambda)a_r = 0. \tag{38}$$

The coefficient of a_{r+2} is always non-zero, so that for arbitrary a_0, it is possible to calculate a_2 and, one by one, all of the coefficients of the even set. Those of the odd set can be gotten in the same way from a_1, so that we get for $\mathcal{H}$ a series of the form:

$$\mathcal{H} = a_0\left(1 + \frac{1 - \lambda}{2}\xi^2 + \dots\right) + a_1\left(\xi + \frac{3 - \lambda}{6}\xi^3 + \dots\right).$$

The solution has two arbitrary constants, as we anticipate for the solution of a second order differential equation. It remains to be seen whether this solution can lead to a normalizable function ψ, which requires that $\mathscr{H}$ go to zero at infinity. To examine the asymptotic behavior of $\mathscr{H}$ we note that for arbitrarily large values of r we can write:

$$\frac{a_{r+2}}{a_r} \sim \frac{2}{r}.$$

Starting from an arbitrary r, the general term of (37) is:

$$a_{r+2n}\xi^{r+2n} \sim a_r \frac{2^n}{r(r+2)\ldots(r+2n-2)} \xi^{r+2n}$$

which we can also write (taking for example the case of the even coefficients $r = 2k$):

$$a_{2k} \cdot k! \frac{(\xi^2)^{k+n}}{(k+n-1)!}.$$

Thus for large values of ξ, $\mathscr{H}$ behaves like $\exp(\xi^2)$, so that ψ given by (35) cannot be normalized. Physically acceptable solutions can only be obtained if, beginning from a certain value r, all the coefficients of (37) vanish. The existence of a recurrence relation brings this about when λ takes the form:

$$\lambda = 2r + 1$$

with the integer r zero or positive.

The coefficient a_{r+2} is then zero, and term by term all the coefficients of the same parity become zero, so that the expansion (37) reduces to a polynomial of degree r.† Thus we see that the Schrödinger equation only admits to normalizable solution for a discrete series of values of λ, from which arise the quantum energy levels which can only have the following values, fixed by the quantum number r:

$$\text{(39)} \qquad E_r = \lambda \frac{h\nu_0}{2} = \left(r + \frac{1}{2}\right) h\nu_0, \qquad \text{with} \qquad r \geqslant 0.$$

These energy values $h\nu_0/2$, $3h\nu_0/2$, $5h\nu_0/2$, ..., form an arithmetic progression on $h\nu_0$. The lowest of these values is $h\nu_0/2$, and not zero corresponding to the particle at rest on the center of force as predicted by classical mechanics. The impossibility of such a state from the standpoint of quantum mechanics can be related to the uncertainty principle which does not allow $x = 0$ and $p = 0$ at the same time.

† If we take for example, $\lambda = 3$, the expansion $\mathscr{H}$ reduces to $a_1 \xi$ provided $a_0 = 0$. For $\lambda = 1$ we have $\mathscr{H} = a_0$ assuming $a_1 = 0$.

Wave Function of the Harmonic Oscillator

The preceding calculation gives to each value of $\lambda = 2r + 1$ a polynomial and a normalization constant, and therefore a unique wave function, making the energies simple (non-degenerate). These polynomials, known as Hermite polynomials, have been studied in great detail, but we will not discuss them here. A more rapid method of getting these results begins with a technique frequently used in problems of this type, that is, by attempting to obtain the linearization of $\mathbf{H}'$ (34) by trying to decompose it into factors analogous to those which appear in the difference of two squares. The second derivative is similar to a square. We therefore introduce two linear operators:

$$\text{(40)} \qquad \mathbf{C} = \xi - \frac{d}{d\xi}; \qquad \mathbf{D} = \xi + \frac{d}{d\xi}.$$

Of course, this is only an analogy. In fact, $\mathbf{H}'\psi$ is not identical with $\mathbf{CD}\psi$. Substitution leads to the result:

$$\mathbf{CD}\psi = -\frac{d^2\psi}{d\xi^2} + \xi^2\psi - \psi = \mathbf{H}'\psi - \psi.$$

Thus the equation $\mathbf{H}'\psi = \lambda\psi$ can be written in terms of $\mathbf{C}$ and $\mathbf{D}$ as:

$$\text{(41)} \qquad \mathbf{CD}\psi - (\lambda - 1)\psi = 0.$$

In this calculation, we can transpose $\mathbf{C}$ and $\mathbf{D}$ and we get:

$$\text{(42)} \qquad \mathbf{DC}\psi - (\lambda + 1)\psi = 0.$$

Having established these basic relationships, we consider the function ψ_r with eigenvalues $\lambda = 2r + 1$, and check the two equations. Let

$$\psi'_r = \mathbf{C}\psi_r,$$

be the function resulting from ψ_r by the action of $\mathbf{C}$. Putting this function into (41), we see:

$$\mathbf{CD}\psi'_r - (\lambda - 1)\psi'_r = \mathbf{CDC}\psi_r - (\lambda - 1)\mathbf{C}\psi_r$$

ψ_r satisfies (42) so that:

$$\mathbf{CDC}\psi_r = \mathbf{C}(\mathbf{DC}\psi_r) = (\lambda + 1)\mathbf{C}\psi_r = (\lambda + 1)\psi'_r$$

with the result that we get:

$$\mathbf{CD}\psi'_r - (\lambda + 2 - 1)\psi'_r = 0$$

which shows that $\psi'_r = \mathbf{C}\psi_r$ is a function related to $(\lambda + 2)$, thus corresponding to a quantum number $r + 1$. It is therefore possible to get all of the wave functions, beginning with the first corresponding to $r = 0$, which we already know [$\exp(-\xi^2/2)$]. For the moment putting aside the determination of the normalization coefficient, with the result that the functions ψ_r are written to within a constant factor, and starting with:

$$\psi_0 = \exp\left(-\frac{\xi^2}{2}\right)$$

it is possible to deduce the set of wave functions by successive applications of $\mathbf{C}$:

$$\psi_r = \mathbf{C}\psi_{r-1} = \mathbf{C}^r\psi_0.$$

The calculation is easily made by noting that:

$$\mathbf{C}\psi = \left(\xi - \frac{d}{d\xi}\right)\psi = -\exp\left(\frac{\xi^2}{2}\right)\cdot\frac{d}{d\xi}\left[\psi\exp\left(-\frac{\xi^2}{2}\right)\right]$$

and thus:

$$\psi_1 = \mathbf{C}\psi_0 = -\exp\left(\frac{\xi^2}{2}\right)\cdot\frac{d}{d\xi}\exp(-\xi^2)$$

$$\psi_2 = \mathbf{C}\psi_1 = -\exp\left(\frac{\xi^2}{2}\right)\cdot\frac{d}{d\xi}\left[\psi_1\exp\left(-\frac{\xi^2}{2}\right)\right]$$

$$= \exp\left(\frac{\xi^2}{2}\right)\cdot\frac{d^2}{d\xi^2}\exp(-\xi^2)$$

and term by term:

$$\psi_r = \mathbf{C}\psi_{r-1} = (-1)^r\exp\left(\frac{\xi^2}{2}\right)\cdot\frac{d^r}{d\xi^r}\exp(-\xi^2).$$

Normalization Constants and the Matrix Elements of *x* and *p*

We will now assume the ψ_r's normalized. Under this condition, in going from ψ_r to ψ_{r+1} by means of $\mathbf{C}$, we introduce a numerical coefficient, thus:

$$\mathbf{C}\psi_r = c_{r+1,r}\psi_{r+1}.$$

By an argument parallel to the one we have already presented for $\mathbf{C}$, we can show that $\mathbf{D}\psi_r$ satisfies the Schrödinger equation for $(r-1)$, and we can write:

$$\mathbf{D}\psi_r = d_{r-1,r}\psi_{r-1}.$$

The two series of coefficients are not independent. In effect, if θ_1 and θ_2 are any two normalizable functions:

$$\int_{-\infty}^{+\infty}\theta_1^*\mathbf{C}\theta_2 d\xi = \int_{-\infty}^{+\infty}\theta_1^*\left(\xi - \frac{d}{d\xi}\right)\theta_2 d\xi = \int_{-\infty}^{+\infty}\theta_1^*\xi\theta_2 d\xi - \int_{-\infty}^{+\infty}\theta_1^* d\theta_2.$$

The integration by parts leads to an exact differential whose integral between $\pm\infty$ is zero, since θ_1 and θ_2 are normalizable by hypothesis and thus vanish at infinity, and:

$$\int_{-\infty}^{+\infty} \theta_1^* \mathbf{C} \theta_2 d\xi = \int_{-\infty}^{+\infty} \left(\xi\theta_1^* + \frac{d\theta_1^*}{d\xi}\right)\theta_2 d\xi = \int_{-\infty}^{+\infty} (\mathbf{D}\theta_1^*)\theta_2 d\xi.$$

We apply this equation to the eigenfunctions ψ_r and:

$$c_{r+1,r} = \int_{-\infty}^{+\infty} \psi_{r+1}^* \mathbf{C} \psi_r d\xi = \int_{-\infty}^{+\infty} (\mathbf{D}\psi_{r+1}^*)\psi_r d\xi = d_{r,r+1}^*.$$

The functions ψ_r are all real, so that we need only consider real coefficients and:

$$d_{r,r+1} = c_{r+1,r}.$$

The value of the coefficients can thus be determined where ψ_r is the solution of the Schrödinger equation as in (41) with $\lambda = 2r + 1$. The successive action of $\mathbf{D}$ followed by $\mathbf{C}$ gives:

$$\mathbf{CD}\psi_r = d_{r-1,r}\mathbf{C}\psi_{r-1} = c_{r,r-1} \cdot d_{r-1,r} \cdot \psi_r = 2r\psi_r$$

and

$$c_{r,r-1} = d_{r-1,r} = \sqrt{2r}$$

or

(43) $$c_{r+1,r} = d_{r,r+1} = \sqrt{2(r+1)}.$$

Thus for the normalized function:

$$\mathbf{C}^r\psi_0 = c_{r,r-1} \ldots c_{2,1}\, c_{1,0}\psi_r = (2^r r!)^{1/2}\psi_r.$$

All of the normalization coefficients can be derived from ψ_0. To normalize ψ_0 we must solve the following integral:

(44) $$\int_{-\infty}^{+\infty} \psi_0^2 d\xi = c_0^2 \int_{-\infty}^{+\infty} e^{-\xi^2} d\xi = 1,$$

so that:

$$c_0 = \pi^{-1/4}$$

and using the previously determined wave function:

(45) $$\int_{-\infty}^{+\infty} \psi_0^2 dx = c_0'^2 \int_{-\infty}^{+\infty} \exp\left(-\frac{x^2}{x_1^2}\right) dx = 1$$

with

$$x_1 = x/\xi = (h/4\pi^2 m\nu_0)^{1/2}$$

we see that:

$$c_0' = c_0/x_1^{1/2} = (h/4\pi^2 m\nu_0)^{-1/4}\pi^{-1/4}.$$

With the normalization (44) and using reduced coordinates ξ, the first state functions are written:

$$\psi_0 = \pi^{-1/4} \exp\left(-\frac{\xi^2}{2}\right)$$

(46) $$\psi_1 = 2^{-1/2}\pi^{-1/4}\cdot 2\xi \exp\left(-\frac{\xi^2}{2}\right)$$

$$\psi_2 = 2^{-3/2}\pi^{-1/4}(4\xi^2 - 2) \exp\left(-\frac{\xi^2}{2}\right)$$

and, in general:

$$\psi_r = (-1)^r(2^r \cdot r\,!)^{-1/2}\pi^{-1/4} \exp\left(\frac{\xi^2}{2}\right)\cdot \frac{d^r}{d\xi^r} \exp(-\xi^2).$$

Functions which satisfy (45) can be gotten from these functions by multiplying by:

$$x_1^{-1/2} = (4\pi^2 m\nu_0/h)^{1/4}.$$

The functions ψ_r are orthogonal and by virtue of the general property of eigenfunctions lead to distinct energy values. We can write as a general theorem:†

$$\int_{-\infty}^{+\infty} \psi_r^* \psi_{r'} d\xi = \delta_{rr'}, \quad \text{so that } \begin{array}{ll} 0, & \text{if } r \neq r' \\ 1, & \text{if } r = r'. \end{array}$$

All the matrix elements of the operators **C** and **D** are necessarily zero, with the exception of those set by equation (43). We get for example:

$$c_{r,r'} = \int_{-\infty}^{+\infty} \psi_r^* \mathbf{C} \psi_{r'} d\xi = c_{r'+1,r'} \int_{-\infty}^{+\infty} \psi_r^* \psi_{r'+1} d\xi = 0, \quad \text{if } r \neq r' + 1.$$

The matrix elements of x and p_x can be found simply. We begin by using the reduced variables. We then have:

$$\xi\psi = \frac{1}{2}(\mathbf{C}\psi + \mathbf{D}\psi) \qquad \frac{d\psi}{d\xi} = \frac{1}{2}(\mathbf{D}\psi - \mathbf{C}\psi).$$

The matrix elements of ξ are then defined by:

$$\xi_{r,s} = \int_{-\infty}^{+\infty} \psi_r^* \xi \psi_s d\xi = \frac{1}{2}(c_{r,s} + d_{r,s}).$$

These elements are generally zero with the exception of:

$$\xi_{r,r-1} = \frac{1}{2} c_{r,r-1} = \sqrt{\frac{r}{2}}$$

$$\xi_{r,r+1} = \frac{1}{2} d_{r,r+1} = \sqrt{\frac{r+1}{2}}.$$

† We use the classical Kronecker delta $\delta_{rr'}$, defined by: $\delta_{rr'} = 0$ if $r \neq r'$, and 1 if $r = r'$.

The matrix associated with ξ is of infinite order and has the following structure:

$$\frac{1}{\sqrt{2}}\begin{vmatrix} 0 & \sqrt{1} & 0 & 0 & 0 & \cdots \\ \sqrt{1} & 0 & \sqrt{2} & 0 & 0 & \cdots \\ 0 & \sqrt{2} & 0 & \sqrt{3} & 0 & \cdots \\ 0 & 0 & \sqrt{3} & 0 & \sqrt{4} & \cdots \\ \cdot & \cdot & \cdot & \cdot & \cdot & \\ \cdot & \cdot & \cdot & \cdot & \cdot & \\ \cdot & \cdot & \cdot & \cdot & \cdot & \end{vmatrix}$$

and the matrix associated with x can be gotten from this by multiplication (term by term) by x_1 as in (45).

We can associate with the operator $d/d\xi$ the matrix with elements:

$$\int_{-\infty}^{+\infty} \psi_r^* \frac{d}{d\xi} \psi_s d\xi = \frac{1}{2}(d_{r,s} - c_{r,s})$$

and thus we get the matrix for momentum which is given by the operator:

$$\mathbf{P} = \frac{\hbar}{i}\frac{d}{dx} = \frac{1}{i}\sqrt{mh\nu_0}\,\frac{d}{d\xi}\,.$$

Comparison to the Classical Oscillator

The distribution of the particle in the eigenstate ψ_r can be represented by plotting the probability density $dp/dx = |\psi_r^2|$. Figure 14 shows the values of the reduced coordinate ξ. It is important to see the relationship between the states of the wave mechanical oscillator and the classical sinusoidal motion. First of all, we remember that the idea of a stationary state characterized by a fixed energy value implies a description independent of time, and this is essentially different from the classical case. The comparison can be illustrated by examining the probability density of finding the particle at various points on the axis. If we write the equation for the classical motion in the form:

$$x = a \cos(2\pi\nu_0 t + \chi), \quad \text{so that} \quad \frac{dx}{dt} = -2\pi\nu_0 a \sin(2\pi\nu_0 t + \chi)$$

we can see that the probability of finding the particle on the element dx is proportional to the time, dt, that the particle spends in the element, the proportionality constant being the time necessary for going from $-a$ to $+a$ or a half-period T/2. Thus:

$$dp = \frac{2dt}{T} = \frac{|dx|}{\pi\sqrt{a^2 - x^2}}, \quad \text{so that} \quad \left|\frac{dp}{dx}\right| = \frac{1}{\pi\sqrt{a^2 - x^2}}$$

The curve representing $|dp/dx|$ can be compared to that obtained in wave mechanics for a state ψ_r, where the classical and wave

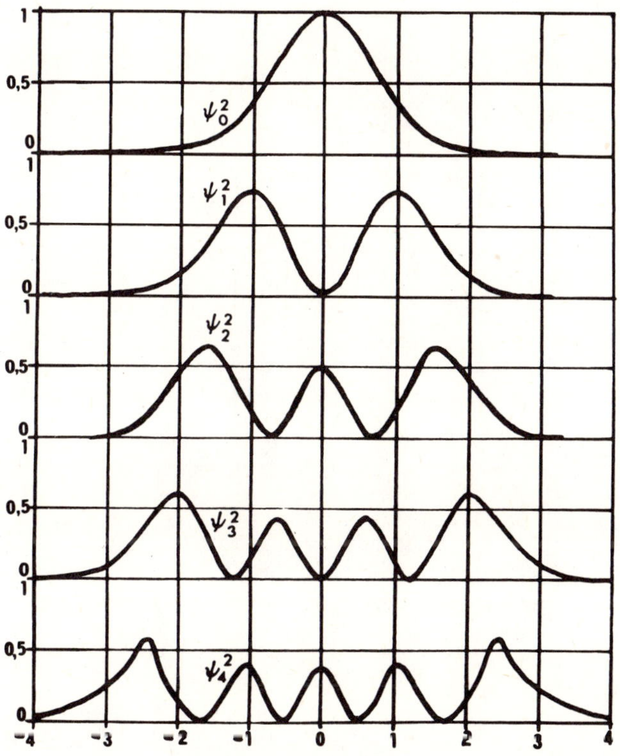

Figure 14. Diagram representing the probability density distribution for a particle in the case of the first five states of the harmonic oscillator, where the position of the particle is defined using the reduced coordinate ξ.

mechanical energies are given the same value. We find that in the wave mechanical case, the mean distribution tends toward the classical as the index r increases, as is shown in figure 15 for $r = 10$.

The analogy cannot be extended to the case where the quantum number is small, but the argument does hold for all cases where the interval $h\nu_0$ is relatively small compared to the energy (in other words, in the case of high quantum numbers). Under those conditions, it is possible to form a wave packet by combining neighboring energy states ψ_r and to localize the particle completely while maintaining approximately the defined energy value.

L. Schiff† has devised a particularly interesting example which we will treat as an exercise by slightly modifying the method of computation, and by using the generating functions of the Hermite polynomials.

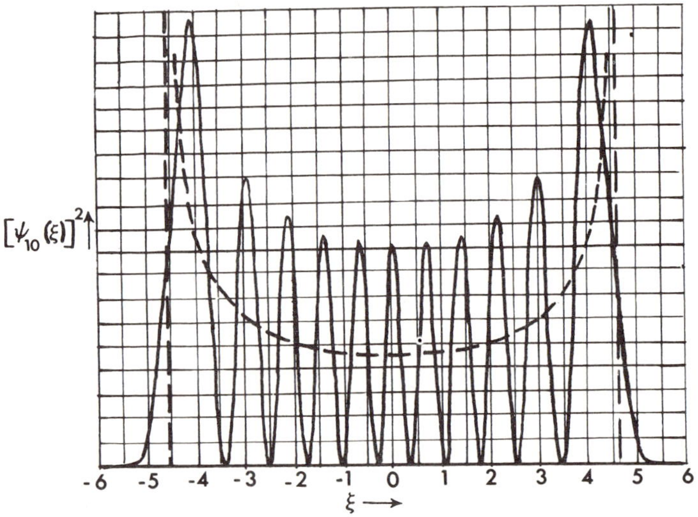

Figure 15. Probability density for a particle in the level $r = 10$ of the harmonic oscillator. The dotted curve represents the probability density for a classical oscillator with the same energy (after Pauling).

Exercise 3. The general form (46) of the wave functions of the harmonic oscillator leads to the expression of the general term by an expansion in a Maclaurin series, so that by introducing the auxiliary variable z:

$$\sum_{n=0}^{\infty} \frac{z^n}{n!} \cdot \frac{d^n}{d\xi^n} (\exp(-\xi^2)) = \exp[-(\xi - z)^2]$$

where the expansion on the powers z of the function:

$$\exp\left[\frac{\xi^2}{2} - (\xi - z)^2\right]$$

† L. I. Schiff, *Quantum Mechanics* (New York: McGraw-Hill, 1968), Chapter IV, Section 13.

gives, to within a constant, the wave function (46). In the same way, expansion of:

$$\exp\left[\xi^2 - (\xi - z)^2\right] = \exp\left(2\xi z - z^2\right)$$

on z gives the Hermite polynomials with a constant multiplier which should be determined. This function of ξ and z is called the generating function of the Hermite polynomials.

Exercise 4. The calculations can be treated first using the reduced coordinate ξ. Taking, at $t = 0$, a state represented by the wave function $\exp -(\xi - \xi_0)^2/2$, which localizes the particle in the neighborhood of $\xi = \xi_0$, calculate the normalization coefficient.

We can represent this wave function ψ as an expansion on ψ_r:

$$\psi = \sum_{r=0}^{\infty} a_r \psi_r.$$

Calculate the coefficients a_r using the general method of section 11. The evaluation can easily be done using the form (46) for ψ_r. We can see, by use of successive integration by parts:

$$\int_{-\infty}^{+\infty} e^{\xi\xi_0} \frac{d^r}{d\xi^r}\left(e^{-\xi^2}\right) d\xi = (-1)^r \xi_0^r \int_{-\infty}^{+\infty} e^{\xi\xi_0} e^{-\xi^2} d\xi = (-1)^r \xi_0^r \sqrt{\pi}\, e^{\xi_0^2/4}.$$

Starting with this expansion, show the development of the wave function ψ with time by simply introducing the factor $\exp(-iE_r t/\hbar)$ for each of the eigenfunctions ψ_r. The sum of the series appears, as in the preceding exercise, as a Maclaurin series:

$$\psi(t) = \pi^{-1/4} \exp\left[-\frac{1}{2}(\xi - \xi_0 \cos 2\pi\nu_0 t)^2\right].$$

The state represents a wave packet localized about $\xi = \xi_0 \cos 2\pi\nu_0 t$ and sinusoidally displaced with time. The size of the packet can be examined by using the variable x.

The coefficients of the expansion representing ψ vary with the index r. Show that a pronounced maximum is obtained for a certain value of r which corresponds to the fact that the energy of the packet is defined almost precisely. Express this energy as a function of the amplitude ξ_0 of the packet and show that it corresponds to the classical expression.

Exercise 5. A diatomic molecule may be considered to a first approximation as a system of two point masses M and m at a separation r. Assume that in the neighborhood of the equilibrium separation r_0 (see fig. 13) the potential energy is of the form:

$$V = \frac{1}{2}k(r - r_0)^2,$$

Write the classical equation of motion. Show that by taking $\xi = r - r_0$ as the variable, the kinetic and potential energies take the same form as for a harmonic oscillator characterized by mass μ ($1/\mu = 1/M + 1/m$) and force constant k.

Exercise 6. Using the generating function $\exp(2\xi z - z^2)$, whose expansion in powers of z gives the Hermite polynomials:

$$\exp(2\xi z - z^2) = \sum_{n=0}^{\infty} \frac{z^n}{n!} \mathcal{H}_n(\xi)$$

derive:

$$\mathcal{H}'_n(\xi) = 2n\mathcal{H}_{n-1}(\xi); \qquad \mathcal{H}''_n(\xi) = 4n(n-1)\mathcal{H}_{n-2}(\xi).$$

Using (36), find the recursion relationship:

$$\mathcal{H}_{n+1}(\xi) = 2\xi\mathcal{H}_n(\xi) - 2n\mathcal{H}_{n-1}(\xi).$$

Using this find the first few Hermite polynomials:

$$\begin{aligned}
\mathcal{H}_0 &= 1 \\
\mathcal{H}_1 &= 2\xi \\
\mathcal{H}_2 &= 4\xi^2 - 2 \\
\mathcal{H}_3 &= 8\xi^3 - 12\xi \\
\mathcal{H}_4 &= 16\xi^4 - 48\xi^2 + 12.
\end{aligned}$$

4 DEGENERATE SYSTEMS

The Plane Rotor 15

We now turn to an idealized model but one which is useful in the discussion of real systems, namely, a point mass m constrained to move in a circular trajectory of constant radius r (fig. 16). The Schrödinger equation governing this system is gotten from (1-8) using a zero potential at all points. The laplacian is taken as a

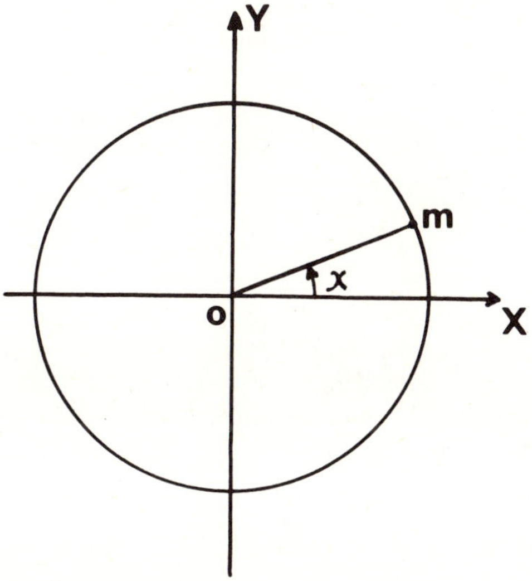

Figure 16. Diagram of the plane rotor; the position of the mass m is defined by the angle χ.

particular case of the general expression in spherical coordinates (Appendix IV) or, more directly, by a change in variables:

$$x = r \cos \chi \, ; \qquad y = r \sin \chi,$$

where:

$$\tan \chi = \frac{y}{x}.$$

This allows us to write for the function of the single variable, χ, which defines the position of the mass on the circle:

$$\frac{\partial}{\partial x} = \frac{d}{d\chi} \cdot \frac{\partial \chi}{\partial x} = -\frac{\sin \chi}{r} \frac{d}{d\chi}$$

where:

$$\frac{\partial^2}{\partial x^2} = -\frac{\sin\chi}{r}\frac{d}{d\chi}\left(-\frac{\sin\chi}{r}\frac{d}{d\chi}\right) = \frac{\sin\chi\cos\chi}{r^2}\frac{d}{d\chi} + \frac{\sin^2\chi}{r^2}\frac{d^2}{d\chi^2}$$

and in the same way:

$$\frac{\partial^2}{\partial y^2} = -\frac{\sin\chi\cos\chi}{r^2}\frac{d}{d\chi} + \frac{\cos^2\chi}{r^2}\frac{d^2}{d\chi^2}$$

which gives:

$$\Delta = \frac{1}{r^2}\frac{d^2}{d\chi^2}.$$

Schrödinger's equation thus takes the form:

$$-\frac{\hbar^2}{2mr^2}\frac{d^2\psi}{d\chi^2} = E\psi$$

which contains the moment of inertia of the system $I = mr^2$.

We now have a second order differential equation with constant coefficients, which has as a solution:

$$\psi = ce^{\alpha\chi} \qquad \text{with} \qquad E = -\frac{\alpha^2\hbar^2}{2mr^2}.$$

The energy E of the system, which is all kinetic ($V = 0$), is necessarily positive, so that $\alpha^2 \ll 0$. Therefore we find $\alpha = ki$. The values of k can only be those which are well defined at every point on the circular trajectory, which implies:

$$\psi(\chi) = \psi(\chi + 2\pi) \qquad \text{where} \qquad e^{2k\pi i} = 1.$$

The number k must therefore be an integer. The solution of the problem is then written:

$$(1) \qquad \psi_k = ce^{ik\chi} \qquad E_k = \frac{k^2\hbar^2}{2I}$$

each state being characterized by a value of the quantum number k. The normalization coefficient c is calculated by integration around the circle:

$$\int_0^{2\pi} |\psi_k^2|\, d\chi = 2\pi c^2 = 1.$$

The expressions (1) show that we get two distinct states k and $-k$ corresponding to the same value of energy (if $k \neq 0$). These values of energy are said to be degenerate. We stress that the states defined by the two opposite values of k are distinct and the corresponding wave functions are orthogonal:

$$\int_0^{2\pi} \psi_k^* \psi_{-k} d\chi = c^2 \int_0^{2\pi} e^{-2ik\chi} d\chi = 0.$$

The existence of an energy degeneracy suggests that knowledge of the energy does not suffice to completely define the state. It is possible to make the development more complete by considering the value L_z of the angular momentum, classically written:

$$L_z = mvr$$

and this is related to the energy by:

$$\text{(2)} \qquad E = \frac{1}{2} mv^2 = \frac{L_z^2}{2mr^2} = \frac{L_z^2}{2I}.$$

Passage to the wave mechanical treatment requires the use of the operator $\mathbf{L}_z$ associated with L_z. This operator is easily constructed using its classical expression in cartesian coordinates:

$$L_z = m(xv_y - yv_x) = xp_y - yp_x.$$

The variables x and p_y are compatible and the definition of the operator associated with their product can be gotten without difficulty (Chapter 3). The same is true for the second term yp_x and we get:

$$\mathbf{L}_z = \frac{\hbar}{i}\left(x\frac{\partial}{\partial y} - y\frac{\partial}{\partial x}\right),$$

which can then be applied to functions of the single variable χ to give:

$$\text{(3)} \qquad \mathbf{L}_z = \frac{\hbar}{i}\left(r \cos \chi \cdot \frac{\cos \chi}{r} + r \sin \chi \cdot \frac{\sin \chi}{r}\right)\frac{d}{d\chi} = \frac{\hbar}{i}\frac{d}{d\chi}.$$

Applying this operator to the function ψ_k defined by (1) we see that:

$$\mathbf{L}_z\psi_k = \frac{\hbar}{i}\frac{d}{d\chi} ce^{ik\chi} = k\hbar\psi_k.$$

ψ_k is thus an eigenfunction of $\mathbf{L}_z$ which shows (Chapter 3) that in the state defined by ψ_k the variable L_z has well-defined values equal to $k\hbar$. This value is changed to $-k\hbar$ in the state ψ_{-k} which corresponds to the rotation in the negative (clockwise) sense; $k = 0$ is an exception corresponding to zero velocity and is thus non-degenerate. We note also that equation (2) yields:

$$E = \frac{k^2\hbar^2}{2I},$$

in accordance with (1), and we find the usual situation that a classical relation remains valid in quantum theory when it involves only compatible variables.

16 The Rigid Rotor

This model constitutes a generalization of the plane rotor, the point mass m moves freely on a sphere of radius r. This system corresponds to a diatomic molecule when the distance between the nuclei is fixed.

The Schrödinger equation for the rigid rotor can be written:

$$-\frac{\hbar^2}{2m}\Delta\psi = E\psi$$

taking $V = 0$ at all points.

The constant r can be treated most easily using spherical co-ordinates in which the laplacian can be written:

$$\Delta\psi = \frac{1}{r^2}\frac{\partial}{\partial r}\left(r^2\frac{\partial\psi}{\partial r}\right) + \frac{1}{r^2}\Lambda\psi$$

where Λ is defined by equation 16 (Appendix IV). The function ψ is independent of r, with the result that the Schrödinger equation takes the form:

$$\Lambda\psi + \frac{2EI}{\hbar^2}\psi = 0.$$

This equation has the same form as the defining equation of the spherical functions (Appendix V-10), thus:

$$\Lambda Y_l + l(l+1)Y_l = 0.$$

The comparison of these two equations shows that the possible energy values are of the form:

$$E_l = l(l+1)\frac{\hbar^2}{2I}, \qquad \text{with} \qquad l = 0, 1, 2, \ldots$$

For the most general spherical function of order l, Y_l is a linear combination with arbitrary coefficients of $2l + 1$ functions Y_l^m (see Appendix V). Thus for a possible value of the energy in the preceding form, we have $2l + 1$ states defined respectively by the functions:

$$\psi_{l,m} = Y_l^m(\theta, \chi) = \frac{1}{\sqrt{2\pi}} e^{im\chi} N_l^m (\sin\theta)^{-m} f_l^m(t)$$

(5)

with $t = \cos\theta$.

Recalling the definition of the factors N_l^m and f_l^m (Appendix V-18) we see:

$$(6) \quad N_l^m = \left[\frac{2l+1}{2} \cdot \frac{(l+m)!}{(l-m)!}\right]^{\frac{1}{2}} \qquad f_l^m(t) = \frac{1}{2^l \cdot l!} \frac{d^{l-m}}{dt^{l-m}} (t^2 - 1)^l.$$

We have shown (Appendix VI-6*a*, 6*b*) that the functions Y_l^m are simultaneously eigenfunctions of the operators $\mathbf{L}^2$ and $\mathbf{L}_z$, associated respectively with the square of the length of the angular momentum and its vector component on one of the reference axes. Thus we have:

$$(7) \qquad \mathbf{L}^2\psi_{l,m} = l(l+1)\hbar^2\psi_{l,m} \qquad \mathbf{L}_z\psi_{l,m} = m\hbar\psi_{l,m}.$$

The definition of the state of the rotor requires knowledge of two variables each fixed by a quantum number. The first is the magnitude of the angular momentum, related to the energy in the same way as in the case of the plane rotor, and having quantum number l. The second, the projection of this vector on some reference axis, taken here as the z-axis, is fixed by the quantum number m, and the inequality $|m| \leqslant l$ expresses the fact that the projection cannot be larger than the original vector. The presence of degeneracy requires that the second variable be introduced in order to define the state.

We note finally (Chapter 3), that the equations (7) lead to the result that the wave functions $\psi_{l,m}$ and $\psi_{l',m'}$—differing in at least one of the indices l or m—are necessarily orthogonal. Since we know that the functions Y_l^m are normalized on a sphere of unit radius, we can write, using the Kronecker deltas $\delta_{ll'}$, $\delta_{mm'}$:

$$\int \psi_{l,m}^* \psi_{l',m'} d\sigma = \delta_{l,l'} \delta_{m,m'}.$$

As a final step, we can determine the values of the matrix elements of the coordinates (ξ, η, z) of the particle, which are required for the calculation of transition probabilities. We get the following (Appendix V):

$$\langle l+1, m \,|\, z \,|\, l, m \rangle = r\left[\frac{(l+1+m)(l+1-m)}{(2l+3)(2l+1)}\right]^{\frac{1}{2}}$$

$$\langle l-1, m \,|\, z \,|\, l, m \rangle = r\left[\frac{(l+m)(l-m)}{(2l+1)(2l-1)}\right]^{\frac{1}{2}}$$

$$\langle l+1, m+1 \,|\, \xi \,|\, l, m \rangle = -r\left[\frac{(l+2+m)(l+1+m)}{(2l+3)(2l+1)}\right]^{\frac{1}{2}}$$

$$\langle l-1, m+1 \,|\, \xi \,|\, l, m \rangle = r\left[\frac{(l-m)(l-1-m)}{(2l+1)(2l-1)}\right]^{\frac{1}{2}}$$

$$\langle l+1, m-1 \,|\, \eta \,|\, l, m \rangle = r\left[\frac{(l+2-m)(l+1-m)}{(2l+3)(2l+1)}\right]^{\frac{1}{2}}$$

$$\langle l-1, m-1 \,|\, \eta \,|\, l, m \rangle = -r\left[\frac{(l+m)(l-1+m)}{(2l+1)(2l-1)}\right]^{\frac{1}{2}}$$

where we have used Dirac's notation $\langle l'm'|z|lm\rangle$ rather than $z_{l'm';lm}$ to designate the matrix elements of z. All the other matrix elements are zero, and thus, transitions other than $l \to l \pm 1$ for the pure rotational motion of a polar diatomic molecule subjected to an electromagnetic field, are forbidden.

17 The One-Electron Atom without Spin

We will now study the system consisting of a fixed point nucleus with charge Ze around which moves an electron of charge $-e$ assumed to be a point mass, m_0. The potential energy of the system is a function of the electron-nuclear distance r and is given by:

$$V(r) = -\frac{Ze^2}{4\pi\varepsilon_0 r} \quad \text{in MKSA units.}$$

The Schrödinger equation of the system is written:

$$-\frac{\hbar^2}{2m_0}\Delta\psi - \frac{Ze^2}{4\pi\varepsilon_0 r}\psi = E\psi.$$

The form of the potential energy suggests to the use of spherical coordinates (r, θ, χ) and the equation has the form (Appendix IV-16):

$$-\frac{\hbar^2}{2m_0}\left[\frac{1}{r^2}\frac{\partial}{\partial r}\left(r^2\frac{\partial\psi}{\partial r}\right) + \frac{1}{r^2}\Lambda\psi\right] - \frac{Ze^2}{4\pi\varepsilon_0 r}\psi = E\psi.$$

It is useful to replace r by a dimensionless variable. This is done by substituting $\rho = \alpha r$.

As a result $\partial/\partial r = \alpha\,\partial/\partial\rho$, so that we can rewrite the equation in terms of the variables (ρ, θ, χ) as:

$$\frac{1}{\rho^2}\frac{\partial}{\partial\rho}\left(\rho^2\frac{\partial\psi}{\partial\rho}\right) + \frac{1}{\rho^2}\Lambda\psi + \frac{2m_0 Ze^2}{4\pi\varepsilon_0\hbar^2\alpha\rho}\psi + \frac{2m_0 E}{\alpha^2\hbar^2}\psi = 0.$$

The parameter α is generally taken so that:

$$(8) \qquad \frac{2m_0 E}{\alpha^2\hbar^2} = -\frac{1}{4}.$$

The equation depends only on the dimensionless parameter:

$$(9) \qquad \lambda = \frac{2m_0 Ze^2}{4\pi\varepsilon_0\hbar^2\alpha}$$

and the Schrödinger equation takes the form:

$$(10) \qquad \frac{1}{\rho^2}\frac{\partial}{\partial\rho}\left(\rho^2\frac{\partial\psi}{\partial\rho}\right) + \frac{1}{\rho^2}\Lambda\psi + \left(\frac{\lambda}{\rho} - \frac{1}{4}\right)\psi = 0.$$

The function $\psi(\rho, \theta, \chi)$ can be expanded on the complete orthogonal system of the spherical functions $Y_l^m(\theta, \chi)$. Thus we can write:

$$\psi = \sum_{l,m} R_{l,m}(\rho) \cdot Y_l^m(\theta, \chi) \tag{11}$$

where m varies from $-l$ to $+l$ and l from 0 to $+\infty$. The functions Y_l^m satisfy (Appendix V-10):

$$\Lambda Y_l^m + l(l+1)Y_l^m = 0.$$

Substituting the expansion (11) in (10) we get:

$$\sum_{l,m} \left\{ \frac{1}{\rho^2}\frac{d}{d\rho}\left(\rho^2 \frac{dR_{l,m}}{d\rho}\right) + \left[\frac{\lambda}{\rho} - \frac{l(l+1)}{\rho^2} - \frac{1}{4}\right] R_{l,m} \right\} Y_l^m(\theta, \chi) = 0.$$

The orthonormality of the Y_l^m pertains to the different values of l or m (Appendix II), consequently the coefficients of the expansion should all be zero and we can write the equation in the form:†

$$\frac{1}{\rho^2}\frac{d}{d\rho}\left(\rho^2 \frac{dR_l}{d\rho}\right) + \left[\frac{\lambda}{\rho} - \frac{l(l+1)}{\rho^2} - \frac{1}{4}\right] R_l = 0. \tag{12}$$

Asymptotic Behavior of the Solutions

When ρ passes to the limit, the coefficient of R_l goes to $-\frac{1}{4}$. The asymptotic form of $R_l(\rho)$ which we call $R_l^{(\infty)}(\rho)$ satisfies the equation:

$$\frac{d^2R_l^{(\infty)}}{d\rho^2} + \frac{2}{\rho}\frac{dR_l^{(\infty)}}{d\rho} - \frac{1}{4}R_l^{(\infty)} = 0.$$

The homogeneous form of this equation allows us to make $R_l^{(\infty)} = e^v$ and the equation then takes the form:

$$\rho(v'' + v'^2) + 2v' - \frac{\rho}{4} = 0.$$

We will look for a solution of the form $v = C\rho^n$. On substitution this gives:

$$n(n+1)C\rho^{n-1} + n^2C^2\rho^{2n-1} - \frac{\rho}{4} = 0.$$

To obtain an asymptotic solution, it is necessary that n be such that the exponents of the two terms of the equation are equal, and the third is lower. Of the three possible combinations, $n-1 = 2n-1$; $n-1 = 1$ and $2n-1 = 1$, the last is the only convenient one, since it leads to $n = 1$ and $C^2 = \frac{1}{4}$, so that $C = \pm\frac{1}{2}$. The solution must be normalizable and must vanish at infinity, so that we discard the positive solution and take:

$$R_l^{(\infty)} \sim e^{-\frac{\rho}{2}}$$

† The value of m does not appear in (12) so that it is not necessary to make use of it in the index and we can simply write $R_l(\rho)$.

which leads to:

(13) $$R_l = w_l(\rho)e^{-\frac{\rho}{2}}.$$

Substitution into equation (12) gives the differential equation which $w_l(\rho)$ must satisfy:

(14) $$w_l'' + w_l'\left(\frac{2}{\rho} - 1\right) + \left(\frac{\lambda - 1}{\rho} - \frac{l(l+1)}{\rho^2}\right)w_l = 0.$$

Laguerre Polynomials. The form of equation (14) shows the possibility of a singularity in the solution for $\rho = 0$ so that we will seek a solution near the origin with the form $w_l \sim \rho^s$. Substitution of this in (14) gives:

$$\rho^{s-2}[s(s+1) - l(l+1)] + (\lambda - 1 - s)\rho^{s-1} = 0.$$

This equation should be an identity for ρ near $\rho = 0$. The principal part of this expression is formed by the term of lowest degree, here ρ^{s-2}, whose coefficient should be zero. There are two possibilities:

$$s_1 = l; \qquad s_2 = -(l+1).$$

The solution s_2 is incompatible with the normalization condition if $l \geqslant 1$; when $l = 0$ it leads to a function discontinuous at the origin, and therefore we discard it. Thus we concern ourselves only with the first solution, and take:

(15) $$w_l(\rho) = \rho^l \mathscr{L}_l(\rho)$$

where $\mathscr{L}_l(\rho)$ represents a function finite at the origin. Substitution into (14) leads to the differential equation defining $\mathscr{L}_l(\rho)$:

(16) $$\rho\mathscr{L}_l'' + [2(l+1) - \rho]\mathscr{L}_l' + (\lambda - l - 1)\mathscr{L}_l = 0.$$

We next seek a solution in the form of an infinite series:

(17) $$\mathscr{L}_l(\rho) = \sum_{k=0}^{\infty} \alpha_k \rho_k.$$

We take the coefficient of ρ_r identically zero in the form obtained in expanding equation (16):

(18) $$(r+1)(r+2l+2)a_{r+1} = (r+l+1-\lambda)a_r.$$

The recurrence relationship thus obtained allows the determination of all the coefficients a_r beginning with a_0. The asymptotic behavior for the terms of high index is given by:

$$a_r \sim \frac{1}{r}a_{r-1} \qquad \text{so that} \qquad a_r \sim \frac{1}{r!}$$

so that:

$$\mathscr{L}_l(\rho) \sim e^{\rho}, \qquad \text{and} \qquad R^l(\rho) = \rho^l \mathscr{L}_l(\rho)\, e^{-\rho/2} \sim \rho^l\, e^{\rho/2}.$$

The solution we get is not normalizable and cannot conform to a real physical situation. We can only have an acceptable solution $\mathscr{L}_l(\rho)$ if one of the coefficients of the expansion (17) is zero, and then the recursion relation (18) insures that all the succeeding coefficients will vanish.

The condition that the series truncate in a term of order r and yield a simple polynomial, called a Laguerre polynomial, is written:

$$r + l + 1 - \lambda = 0.$$

The parameter λ introduced in (9) must then take an integral value n from which the Schrödinger equation yields physically acceptable solutions. The numbers r and l are positive or zero, and we see that:

$$n = r + l + 1 \geqslant 1.$$

This number n which governs the possible values of α through (9) and the energy through (8) is called the principle quantum number. For a given value of n we must have:

$$r = n - l - 1 \geqslant 0 \qquad \text{so that} \qquad l \leqslant n - 1.$$

Finally, the state of an atom is defined by three integers n, l, and m, known as the quantum numbers:

(19)

$n \geqslant 1$	principal quantum number
l from 0 to $(n - 1)$	azimuthal quantum number
m from $(-l)$ to $(+l)$	magnetic quantum number.

The value of α defined by (9) is:

$$\alpha = \frac{2m_0Ze^2}{4\pi\varepsilon_0\hbar^2} \cdot \frac{1}{n} \cdot$$

α is inversely proportional to the dimensions of the atom and we can introduce a quantity:

$$a_0 = 4\pi\varepsilon_0 \frac{\hbar^2}{m_0e^2} \tag{20}$$

which is called the principal or first Bohr radius. Thus:

$$\alpha = \frac{2Z}{na_0}$$

and the energy is given by:

$$E_n = -\frac{1}{(4\pi\varepsilon_0)^2} \cdot \frac{m_0Z^2e^4}{2\hbar^2} \cdot \frac{1}{n^2} \cdot \tag{21}$$

This energy depends only on the principal quantum number n. For a given value of n, l can take the values 0, 1, ..., $(n-1)$, and for each value of l, there are $(2l+1)$ possible values of m from $-l$ to $+l$. The number of states corresponding to a value of n is:†

(22) $$1 + 3 + \ldots + (2n - 1) = n^2.$$

In each of these states, the electron distribution is given by the wave function:

(23) $$\psi_{n,l,m} = R_{n,l}(\rho) \cdot Y_l^m(\theta, \chi) = N_r \rho^l \mathfrak{L}_{n,l}(\rho) \cdot e^{-\frac{\rho}{2}} \cdot Y_l^m(\theta, \chi).$$

The coefficient N_r is determined by the normalization condition of the wave function:

$$\int |\psi_{n,l,m}|^2 d\tau = 1$$

which, by our writing in spherical coordinates and introducing the element of solid angle $d\sigma$, gives:

$$\int_0^\infty R_{n,l}^2 r^2 dr \int |Y_l^m|^2 d\sigma = 1.$$

By convention, the spherical functions are normalized and the radial function, $R_{l,n}(\rho)$, which is real, must satisfy the normalization condition:

$$\int_0^\infty R_{n,l}^2 r^2 dr = 1$$

which, with the dimensionless variable ρ, is:

(24) $$\alpha^3 \int_0^\infty R_{n,l}^2 \rho^2 d\rho = \left(\frac{2Z}{na_0}\right)^3 \int_0^\infty R_{n,l}^2 \rho^2 d\rho = 1$$

Now for several additional comments on the Laguerre polynomials introduced in the study of $R_{n,l}$.

We recall that the polynomials satisfy (16):

(16) $$\rho \mathscr{L}''_{n,l} + [2(l+1) - \rho]\, \mathscr{L}'_{n,l} + (n - l - 1)\, \mathscr{L}_{n,l} = 0.$$

We can generate these polynomials starting with the differential equation:

$$xy' + (x - k)y = 0$$

where the solution is, to within a constant factor:

$$y = x^k e^{-x}.$$

† The sum of the odd numbers is an arithmetic progression of ratio 2 which has sum n^2.

We differentiate this equation $k + 1$ times. Taking account of the formula giving the derivative of a product, we immediately obtain from the recurrence relationship:

$$\frac{d^r}{dx^r}(xy) = x\frac{d^r y}{dx^r} + r\frac{d^{r-1}y}{dx^{r-1}}.$$

We get:

$$x\frac{d^{k+2}y}{dx^{k+2}} + (x+1)\frac{d^{k+1}y}{dx^{k+1}} + (k+1)\frac{d^k y}{dx^k} = 0$$

and by using:

$$u = \frac{d^k y}{dx^k}$$

$$xu'' + (x+1)u' + (k+1)u = 0.$$

We can associate a polynomial with the function u thus defined and show:

$$\mathscr{L} = u e^x \qquad \text{or} \qquad u = \mathscr{L} e^{-x}.$$

This polynomial, which we call $\mathscr{L}_k$, satisfies the following equation obtained by substitution:

$$x\mathscr{L}_k'' + (1-x)\mathscr{L}_k' + k\mathscr{L}_k = 0$$

and we get:

$$\mathscr{L}_k = e^x u = e^x \frac{d^k}{dx^k}(x^k e^{-x}).$$

Differentiating this final equation p times:

$$x\frac{d^{p+2}}{dx^{p+2}}\mathscr{L}_k + (p+1-x)\frac{d^{p+1}}{dx^{p+1}}\mathscr{L}_k + (k-p)\frac{d^p}{dx^p}\mathscr{L}_k = 0$$

which we compare to equation (16) for the Laguerre polynomials:

$$\rho\mathscr{L}_{n,l}'' + [2(l+1) - \rho]\mathscr{L}_{n,l}' + (n-l-1)\mathscr{L}_{n,l} = 0.$$

These equations are identical if we take:

$$2(l+1) = p+1\,; \qquad n-l-1 = k-p$$

so that:

$$p = 2l+1\,; \qquad k = n+l.$$

We thus get the general expression for the Laguerre polynomials:

$$(25) \qquad \mathscr{L}_{n,l} = \frac{d^{2l+1}}{d\rho^{2l+1}}\mathscr{L}_{n+l} = \frac{d^{2l+1}}{d\rho^{2l+1}}\left[e^\rho \frac{d^{n+l}}{d\rho^{n+l}}(\rho^{n+l}e^{-\rho})\right].$$

Nomenclature of the States and the Wave Functions

The states are characterized by three quantum numbers n, l, m and traditionally these are written by first giving the value of n, followed by a letter characterizing the value of l, using $s(l=0)$, $p(l=1)$, $d(l=2)$ and $f(l=3)$. The value of m is indicated by a subscript.

The wave function in state n, l, m is given by (23):

$$\psi_{n,l,m} = \mathrm{R}_{n,l}(\rho)\,\mathrm{Y}_l^m(\theta,\chi) = \mathrm{N}_r\,\rho^l\,\mathscr{L}_{n,l}(\rho)\,e^{-\rho/2}\,\mathrm{Y}_l^m(\theta,\chi)$$

where the Laguerre polynomial is taken from (25). The normalization coefficient (24) is computed in each case using the eulerian integral:

$$\int_0^\infty \rho^r e^{-\lambda\rho} d\rho = (-1)^r \frac{d^r}{d\lambda^r}\int_0^\infty e^{-\lambda\rho} d\rho = \frac{r!}{\lambda^{r+1}}.$$

We will write down the functions $\psi_{n,l,m}$ for the first few values of n by giving the spherical functions $\mathrm{Y}_l^m(\theta,\chi)$ and the radial functions $\mathrm{R}_{n,l}(\rho)$.

Spherical Functions:†

(s), $l=0$ $\quad \mathrm{Y}_0^0 = \dfrac{1}{\sqrt{4\pi}}$ $\qquad s_0$

(p), $l=1$:

$\mathrm{Y}_1^{\pm1} = \dfrac{1}{\sqrt{2\pi}}\dfrac{\sqrt{3}}{2}\,e^{\pm i\chi}\sin\theta$ $\qquad p_{\pm1}$

$\mathrm{Y}_1^0 = \dfrac{1}{\sqrt{2\pi}}\dfrac{\sqrt{6}}{2}\cos\theta$ $\qquad p_0$

(d), $l=2$:

$\mathrm{Y}_2^{\pm2} = \dfrac{1}{\sqrt{2\pi}}\dfrac{\sqrt{15}}{4}\,e^{\pm 2i\chi}\sin^2\theta$ $\qquad d_{\pm2}$

$\mathrm{Y}_2^{\pm1} = \dfrac{1}{\sqrt{2\pi}}\dfrac{\sqrt{15}}{2}\,e^{\pm i\chi}\sin\theta\cos\theta$ $\qquad d_{\pm1}$

$\mathrm{Y}_2^0 = \dfrac{1}{\sqrt{2\pi}}\dfrac{\sqrt{10}}{4}(3\cos^2\theta-1)$ $\qquad d_0$

† These spherical functions are calculated by application of the equations (Appendix V-18) but are written with positive sign so that the same expression can give Y_l^m and Y_l^{-m}. In fact, application of the equations (Appendix V-18) gives a negative sign when m is odd and positive, which otherwise can be expressed by introducing a factor $(-1)^{(m+|m|)/2}$ valid for all m. Introduction of the sign is only necessary when we wish to use the recurrence relations between spherical functions (Appendix V-21, 22, 23).

Radial Functions:

n = 1. We have $\rho = \frac{2Zr}{a_0}$:

The only case $l = 0$ (1s state)

$$R_{1,0} = \left(\frac{Z}{a_0}\right)^{3/2} \cdot 2e^{-\frac{\rho}{2}}.$$

n = 2. We have $\rho = \frac{Zr}{a_0}$:

for $l = 0$ (2s state)

$$R_{2,0} = \left(\frac{Z}{a_0}\right)^{3/2} \cdot \frac{1}{2\sqrt{2}}(2 - \rho)e^{-\frac{\rho}{2}};$$

for $l = 1$ (2p state)

$$R_{2,1} = \left(\frac{Z}{a_0}\right)^{3/2} \cdot \frac{1}{2\sqrt{6}}\rho e^{-\frac{\rho}{2}}.$$

n = 3. We have $\rho = \frac{2Zr}{3a_0}$:

for $l = 0$ (3s state)

$$R_{3,0} = \left(\frac{Z}{a_0}\right)^{3/2} \cdot \frac{1}{9\sqrt{3}}(6 - 6\rho + \rho^2)e^{-\frac{\rho}{2}};$$

for $l = 1$ (3p state)

$$R_{3,1} = \left(\frac{Z}{a_0}\right)^{3/2} \cdot \frac{1}{9\sqrt{6}}\rho(4 - \rho)e^{-\frac{\rho}{2}};$$

for $l = 4$ (3d state)

$$R_{3,2} = \left(\frac{Z}{a_0}\right)^{3/2} \cdot \frac{1}{9\sqrt{30}}\rho^2 e^{-\frac{\rho}{2}}.$$

n = 4. We have $\rho = \frac{Zr}{2a_0}$:

for $l = 0$ (4s state)

$$R_{4,0} = \left(\frac{Z}{a_0}\right)^{3/2} \cdot \frac{1}{96}(24 - 36\rho + 12\rho^2 - \rho^3)e^{-\frac{\rho}{2}};$$

for $l = 1$ (4p state)

$$R_{4,1} = \left(\frac{Z}{a_0}\right)^{3/2} \cdot \frac{1}{32\sqrt{15}}\rho(20 - 10\rho + \rho^2)e^{-\frac{\rho}{2}};$$

for $l = 2$ (4d state)

$$R_{4,2} = \left(\frac{Z}{a_0}\right)^{3/2} \cdot \frac{1}{96\sqrt{5}}\rho^2(6 - \rho^2)e^{-\frac{\rho}{2}};$$

for $l = 3$ (4*f* state)

$$R_{4,3} = \left(\frac{Z}{a_0}\right)^{3/2} \cdot \frac{1}{96\sqrt{35}} \rho^3 e^{-\frac{\rho}{2}}.$$

Graphical Representation of the Wave Functions

The function $\psi_{n,l,m}$ appears as a product of factors which it is convenient to treat separately.

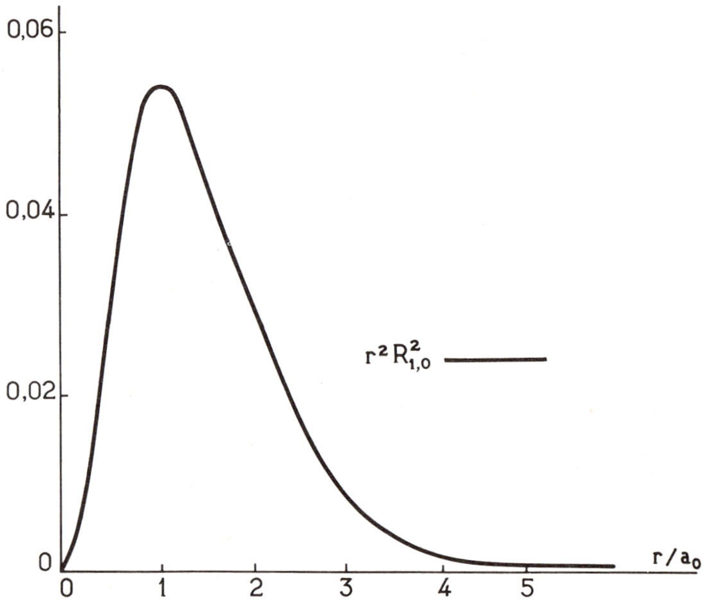

Figure 17a. Radial probability density for the 1*s* state of hydrogen (after Pluvinage).

The radial term $R_{n,l}$ determines the probability of finding the electron between spheres of radius r and $r + dr$ so that:

$$dp = R^2_{n,l} r^2 dr \int |Y_l^m|^2 d\sigma = \alpha^3 R^2_{n,l} \rho^2 d\rho.$$

Diagram 17 represents dp/dr for the first four levels of the hydrogen atom ($Z = 1$), taking as the unit of length a_0 the first Bohr radius, chosen so that it corresponds to the radius of the sphere of maximum probability density in the 1*s* state. We note that for $n > 1$ the curves relative to increasing values of l defining the electron distributions extend to the nucleus and this fact is important in understanding the distributions of energy levels in many-electron atoms.

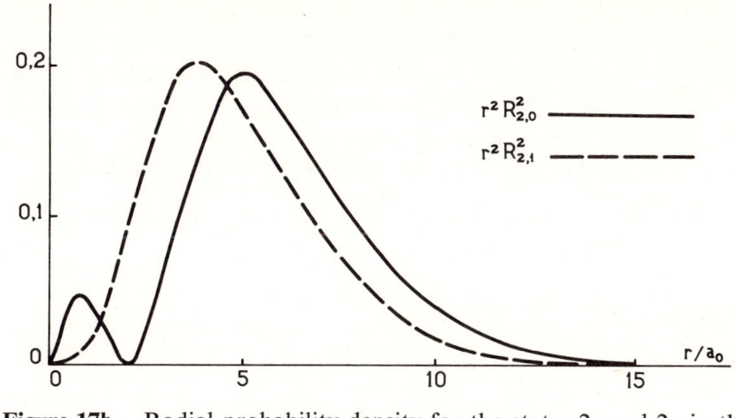

Figure 17b. Radial probability density for the states 2*s* and 2*p* in the hydrogen atom (after Pluvinage).

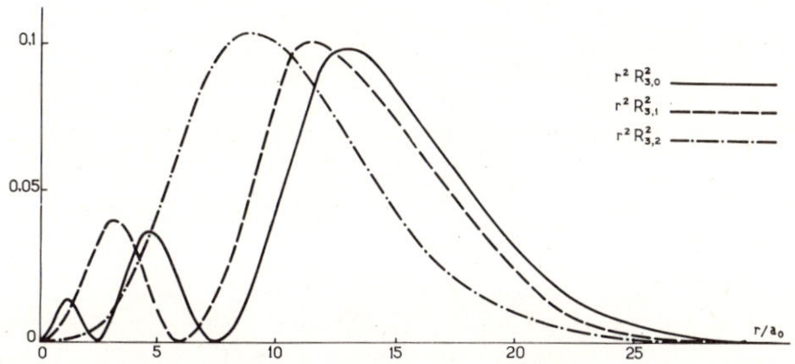

Figure 17c. Radial probability density for the state 3*s*, 3*p*, and 3*d*, in the hydrogen atom (after Pluvinage).

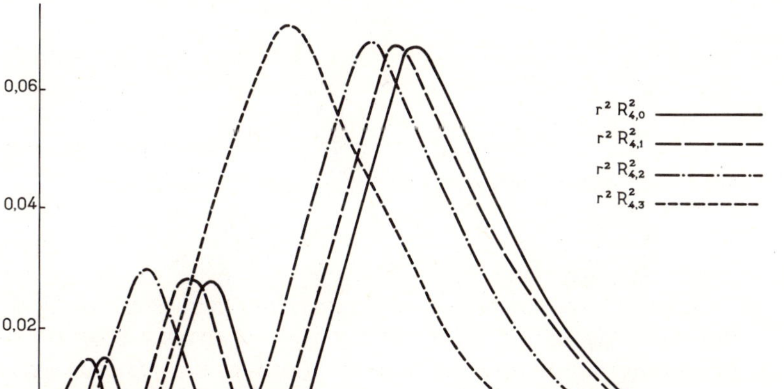

Figure 17d. Radial probability density for the states 4*s*, 4*p*, 4*d* and 4*f*, in the hydrogen atom (after Pluvinage).

We proceed now to the study of the variation of $\psi_{n,l,m}$ on a sphere of radius r, this variation being determined by the spherical functions. The results, needed for the case of the directed valence, lead to the use of a graphical representation in polar coordinates. This is obtained by taking in direction OA a radius vector equal to the modulus of the wave function at point A. When we have to compare many diagrams of this kind we take the normalized functions

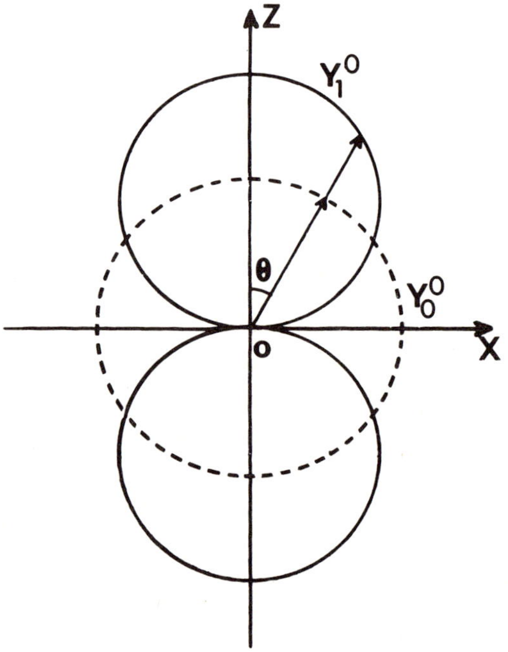

Figure 18. Diagram in polar coordinates of the amplitude variations of the wave function in the state $2p_0$ as a function of the angle for constant r, accordingly, for $|Y_1^0|$. The same diagram contains $|Y_0^0|$ (dotted) for purposes of comparison.

on a sphere of unit radius in such a way that the radius r plays no part. Figure 18 relates to the state $s(Y_0^0)$ and $p_0(Y_1^1)$. Since these are surfaces of revolution about the z-axis, it is sufficient to indicate their cross-section on the xz-plane. If we limit ourselves to real p functions, as we always do, the z-axis serves as a reference axis.

We conclude by showing (fig. 19) the d orbitals,† indicating in each case the circle corresponding to the spherical s function with the

† See note page 96.

same normalization on a unit sphere. We note in particular that only the orbital $d_0(Y_2^0)$ has a non-zero extension in the z-direction and is the only orbital that can help to form a chemical bond in this direction.

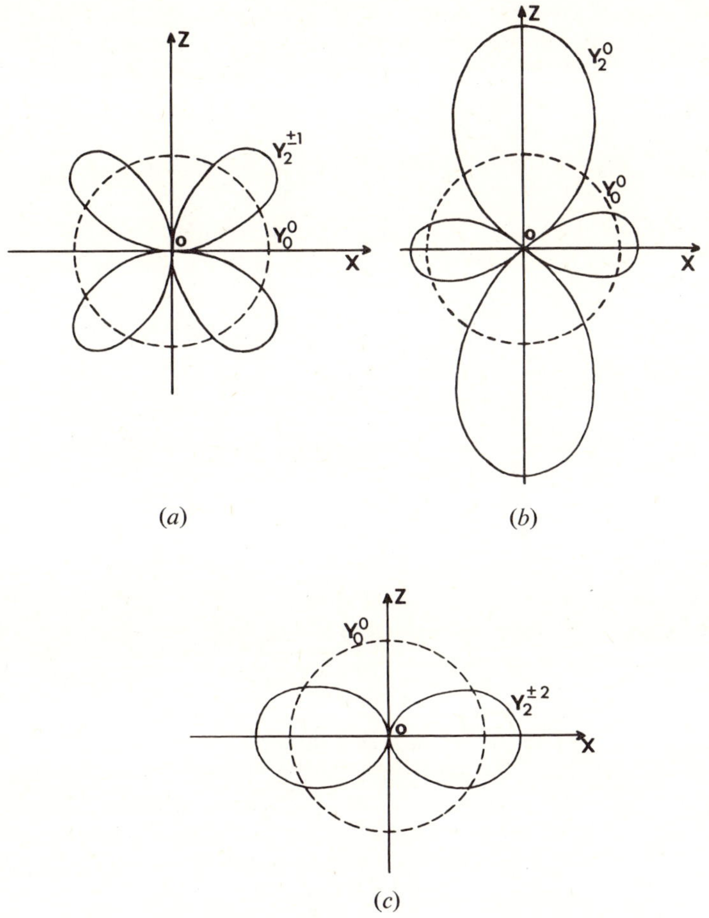

Figure 19. Diagrams in polar coordinates of the variations of: (*a*) $|Y_2^{\pm2}|$, (*b*) $|Y_2^{\pm1}|$, and (*c*) $|Y_2^0|$, compared in each case to $|Y_0^0|$.

The Nature of the Degeneracy of the Energy Levels

We once again stress the existence of a degeneracy in the energy and its significance. This degeneracy is expressed by the fact that three quantum numbers are necessary to define a state while the energy depends on only one of them.

As a result, two other quantities besides the energy are necessary for the definition of a state. In studying the rigid rotor, we introduced the angular momentum whose square has the value $l(l+1)\hbar^2$, and the projection of the angular momentum on the z-axis with value $m\hbar$. The difference between this and the case of the rotor is that the energy is no longer a function of the angular momentum and requires the introduction of a quantum number n independent of l.

18 Spin

The preceding study of the one-electron atom leads to a collection of states each defined by a set of values of n, l, and m. Comparison of this theory with physical systems to which it can be applied (spectra of the alkali metals, for example) shows that the predicted number of states is too small. In particular, the model does not predict the existence of doublets in the spectra of the alkali metals.

As a result, the model of an atom consisting of an electron acting as a point charge is insufficient. Using the classical concept, the system is most simply thought of with the point particle (electron) as an undeformable sphere of radius ρ. The motion of the center-of-mass is exactly that which would be anticipated for a point mass. This extended electron admits to the possibility of a proper rotation about an axis passing through the center of the electron. This rotation is called the spin. We will first discuss this system from the classical viewpoint.

Classical mechanics characterizes the proper rotation of the electron by the vector angular momentum or, in short, the *spin moment*. This is defined relative to a reference system passing through the center-of-mass of the electron, and must be distinguished from the vector of the same nature which we treated in the previous paragraph taken relative to the center-of-mass of the atom. The magnitude of the spin moment is a function of the moment of inertia I and the angular speed ω through the classical expression $I\omega$ with corresponding kinetic energy $I\omega^2/2$. It is illustrative to thoroughly examine this model† by giving a distribution of the mass in the sphere, with the result that we are able to assume $I \sim m_0 \rho_e^2$ for the order of magnitude of the moment of inertia. We can get an idea of the value of the radius ρ_e of the electron by assuming

† All the more so since it leads to the fact that the points on the surface of the electron are displaced with a velocity $v = \hbar(m_0 \rho) > 10^{13}$ cm/sec which seems incompatible with the laws of special relativity.

that its energy is calculated by the same formula as that of a spherical capacitor of the same radius and carrying the charge e, namely $e^2/2\rho_e$. If we write this energy equal to that which results from application of Einstein's relation, we see:

$$\frac{e^2}{2\rho_e} \sim m_0c^2 \qquad \text{so that} \qquad \rho_e \sim \frac{e^2}{2m_0c^2} \sim 10^{-13}\ \text{cm}.$$

From less simple considerations, starting from the fact that the kinetic energy of the moving electron can be interpreted as the energy of the magnetic field established by the displacement of the charge, we are lead to a result of the same order of magnitude.

To extend the argument we use the results relative to the quantization of the rotor and write the spin moment of the order of Planck's constant so that:

$$\frac{1}{2}\,\text{I}\omega^2 = \frac{(\text{I}\omega)^2}{2\text{I}} = \frac{\hbar^2}{2m_0\rho_e^2} = \frac{2\hbar^2 m_0 c^4}{e^4} \sim 1\ \text{erg}, \quad \text{so that} \quad \sim 10^{12}\ \text{eV}.$$

These considerations lead to a very important result: knowledge of the huge order of magnitude which we are lead to attribute to the kinetic energy of the proper rotation of the electron. This allows us to state that this energy must be quantized and is not susceptible to modification by any effect and should therefore be regarded as invariant.

Quantum Mechanical Picture. The transition from the classical to the quantum model is not immediate and the group of ideas previously treated relative to a point particle are not easily carried over. The only analogy which can be exploited relates to the rigid rotor which is the most simple rotating system and whose study leads to state functions which are none other than the spherical function (Appendix V). In the case of spin, however, we do not immediately see which function can be used. A safer guide can be found by setting aside the state functions and by giving attention solely to the quantity which leads to a comparison of the spin angular momentum **S** with **L**. In order to do this we must introduce the operators associated with these variables (as developed in Appendix VI). We postulate that the operators representing the spin components $\mathbf{S}_x$, $\mathbf{S}_y$, and $\mathbf{S}_z$ obey the same anticommutation laws (Appendix VI-2) as the operators associated with the angular momentum, so that:

$$\begin{aligned} \mathbf{S}_x\mathbf{S}_y - \mathbf{S}_y\mathbf{S}_x &= i\hbar\mathbf{S}_z \\ \mathbf{S}_y\mathbf{S}_z - \mathbf{S}_z\mathbf{S}_y &= i\hbar\mathbf{S}_x \\ \mathbf{S}_z\mathbf{S}_x - \mathbf{S}_x\mathbf{S}_z &= i\hbar\mathbf{S}_y\,. \end{aligned} \tag{26}$$

This generalization is possible from the fact that these relations involve no hypothesis related to the functions on which they act. The study made in Appendix VI shows that we can have a system of common eigenvalues of $\mathbf{S}^2 = \mathbf{S}_x^2 + \mathbf{S}_y^2 + \mathbf{S}_z^2$ and of $\mathbf{S}_z$ so that:

for $\mathbf{S}^2$: $l(l+1)\hbar^2$ with either $l = k$ or $l = k + \frac{1}{2}$, k a positive or zero integer.

for $\mathbf{S}_z$: $m\hbar$ with $m = l, l-1, \ldots, -l$.

The quantization of $\mathbf{S}^2$ involves the energy and we have given a reason why the energy has only one observable value. Experiment shows that $l = 1/2$ so that:

$$S^2 = \frac{3}{4}\hbar^2; \qquad S_z = \sigma\hbar \quad \text{with} \quad \sigma = +\frac{1}{2} \quad \text{or} \quad -\frac{1}{2}.$$

Representation of the Spin States

The introduction of spin makes it possible to observe two definite states of the spin component S_z with values $\hbar/2$ and $-\hbar/2$. In a given state we can be certain that measurement of $\mathbf{S}_z$ leads to $\hbar/2$ or $-\hbar/2$ and no other possibility exists. For the representation of these states it is particularly convenient to use a formalism† parallel to those state functions which serve to define the state of a point particle. This is the reason for introducing spin functions.

In the case of a point particle, the probability dp of finding it in element $d\tau$ is calculated from the function $\psi(x,y,z)$ by $dp = |\psi^2|\, d\tau$. In a similar fashion, we suppose that any rotational state of the electron can be described by a spin function $\eta(S_z)$ defined for only two values $S_z' = \hbar/2$ and $S_z'' = -\hbar/2$ and permitting calculation of the probability of observing these values so that:

$$p' = |\eta(S_z')|^2 \quad \text{and} \quad p'' = |\eta(S_z'')|^2$$

which implies:

$$p' + p'' = |\eta(S_z')|^2 + |\eta(S_z'')|^2 = 1$$

corresponding to the normalization condition. There is a parallel between the cases of a state function $\psi(x,y,z)$ and a spin function $\eta(S_z)$ (where we now write $\eta(\sigma)$ with $S_z = \sigma\hbar$ where $\sigma = \pm\frac{1}{2}$), the only difference being the fact that x, y, z can take all possible values while σ can only have two values. All of the ideas introduced in

† It is, in fact, a postulate that the probabilities of getting $S_z = \pm\hbar/2$ are calculable from a state function, as is the case with the position coordinates.

the case of continuous variables are going to be carried over, the integrals being changed to a sum over the two values of the variable σ. Thus the orthogonality of the two spin functions gives:

$$(27)\quad \sum_{\sigma} \eta_1^*(\sigma)\eta_2(\sigma) = \eta_1^*\left(\frac{1}{2}\right)\eta_2\left(\frac{1}{2}\right) + \eta_1^*\left(-\frac{1}{2}\right)\eta_2\left(-\frac{1}{2}\right) = 0.$$

Given this, knowledge of the spin function involves understanding how these spin functions transform under the action of the spin operators.

Beginning with $\mathbf{S}_z$ which has $\hbar/2$ and $-\hbar/2$ for eigenvalues, and calling $\alpha(\sigma)$ and $\beta(\sigma)$ the corresponding eigenfunctions, we have:

$$\mathbf{S}_z\alpha(\sigma) = \frac{\hbar}{2}\,\alpha(\sigma)\,; \qquad \mathbf{S}_z\beta(\sigma) = -\frac{\hbar}{2}\,\beta(\sigma).$$

The function $\alpha(\sigma)$ represents a state in which we have exclusively $S_z = \hbar/2$ and which has zero probability of having the other value:

$$\alpha\left(\frac{1}{2}\right) = 1\,; \qquad \alpha\left(-\frac{1}{2}\right) = 0$$

and inversely:

$$\beta\left(\frac{1}{2}\right) = 0\,; \qquad \beta\left(-\frac{1}{2}\right) = 1.$$

The action of the operators has been deduced (Appendix V) in a treatment which intentionally does not specify the state equation. The equations (Appendix VI-10, 11) apply to the functions $\alpha(\sigma)$ and $\beta(\sigma)$ which correspond respectively to $m = \frac{1}{2}$ and $m = -\frac{1}{2}$. The only non-zero coefficients $a_{m,m+1}$, and $b_{m+1,m}$ calculated from (Appendix VI-15) have absolute value:

$$|a_{-\frac{1}{2},\frac{1}{2}}| = |b_{\frac{1}{2},-\frac{1}{2}}| = 1.$$

We can show these coefficients to be equal to unity by setting aside the complex factor of modulus one, which corresponds to the arbitrary factor which always appears in the state equation. Thus we see explicitly:

$$(28)\quad \begin{array}{ll} (\mathbf{S}_x - i\mathbf{S}_y)\alpha = \hbar\beta\,; & (\mathbf{S}_x + i\mathbf{S}_y)\alpha = 0 \\ (\mathbf{S}_x - i\mathbf{S}_y)\beta = 0\,; & (\mathbf{S}_x + i\mathbf{S}_y)\beta = \hbar\alpha. \end{array}$$

The action of the operators $\mathbf{S}_x$ and $\mathbf{S}_y$ can be found by addition and subtraction to be:

$$\mathbf{S}_x\alpha = \frac{\hbar}{2}\,\beta\,; \qquad \mathbf{S}_y\alpha = \frac{i\hbar}{2}\,\beta$$

$$\mathbf{S}_x\beta = \frac{\hbar}{2}\,\alpha\,; \qquad \mathbf{S}_y\beta = -\frac{i\hbar}{2}\,\alpha.$$

In conclusion, the action of the operators $\mathbf{S}_x$, $\mathbf{S}_y$, and $\mathbf{S}_z$ on the spin functions $\alpha(\sigma)$ and $\beta(\sigma)$ is expressed by means of the following three matrices known as the Pauli matrices $[J_r]$; we have written the first of these to illustrate the order in which the elements are taken:

$$\begin{bmatrix} \left\langle \frac{1}{2} \middle| S_x \middle| \frac{1}{2} \right\rangle & \left\langle \frac{1}{2} \middle| S_x \middle| -\frac{1}{2} \right\rangle \\ \left\langle -\frac{1}{2} \middle| S_x \middle| \frac{1}{2} \right\rangle & \left\langle -\frac{1}{2} \middle| S_x \middle| -\frac{1}{2} \right\rangle \end{bmatrix} = \frac{\hbar}{2}\begin{pmatrix} 0 & 1 \\ 1 & 0 \end{pmatrix}.$$

We have then:

$$[S_r] = \frac{\hbar}{2}[J_r].$$

(29)

$$[J_1] = \begin{pmatrix} 0 & 1 \\ 1 & 0 \end{pmatrix}; \quad [J_2] = \begin{pmatrix} 0 & -i \\ i & 0 \end{pmatrix}; \quad [J_3] = \begin{pmatrix} 1 & 0 \\ 0 & -1 \end{pmatrix}.$$

We note the hermitian character of these matrices since this is necessary for the representation of real quantities.

State Functions of the Atom with Spin

We are now going to treat the problem of the one-electron atom with spin. To begin with, the Schrödinger equation is not modified in the first approximation. On the one hand it is not necessary to introduce a term corresponding to the kinetic energy of proper rotation, since this must be regarded as a constant. On the other hand, the assumption of a finite dimension for the electron must be unimportant if we limit ourselves to considering the attractive electrostatic forces due to the nucleus.†

As a consequence of the natural assumption of a symmetric distribution of charge on the electron, it follows that the resultant of the applied forces acts at the center-of-mass of the electron and cannot modify the spin state. In this first approximation there is a complete separation of the description of the orbital‡ and of the spin. The state of the electron can be defined by knowledge of four compatible variables x, y, z, and S_z (or σ), with the result that the state function can be written $\psi(x, y, z, \sigma)$. We specify the meaning of such a function by stating that it determines the probability density when dealing with continuous variables x, y, and

† A secondary approximation (4, §24) introduces a coupling as a result of a magnetic field tied to the electron displacement.

‡ The term orbital, synonymous with the wave function describing a stationary state, is introduced to preserve the analogy with the non-quantum mechanical description of states.

z, and the probability when the variable can only take discrete values (σ). Under these conditions $|\psi(x,y,z,\frac{1}{2})|^2 d\tau$ represents the probability that the electron is in $d\tau$ in the neighborhood of point (x,y,z), with a spin component of $\sigma = \frac{1}{2}$ and likewise for $|\psi(x,y,z,-\frac{1}{2})|^2 d\tau$. Thus the normalization condition can be written:

$$(30) \quad \sum_{\sigma} \int | \psi(x, y, z, \sigma |^2 d\tau$$

$$= \int \left| \psi\left(x, y, z, \frac{1}{2}\right) \right|^2 d\tau + \int \left| \psi\left(x, y, z, -\frac{1}{2}\right) \right|^2 d\tau = 1$$

combining integration over space and summation on the spin σ.

Like all functions of σ, any such function $\psi(x,y,z,\sigma)$ can be broken down on the base functions α and β, but the coefficients must only be functions of x, y, and z:

$$(31) \qquad \psi(x, y, z, \sigma) = \psi_1(x, y, z)\alpha(\sigma) + \psi_2(x, y, z)\beta(\sigma).$$

As we have indicated, this function $\psi(x,y,z,\sigma)$ must satisfy the same Schrödinger equation without spin, which implies that the functions ψ_1 and ψ_2 must be solutions of the Schrödinger equation having the same form as those we obtained in the preceding section. In conclusion, with this approximation the states of the system are represented by the functions $\psi_{n,l,m}\alpha$ and $\psi_{n,l,m}\beta$. The first of these represents a state of known energy E_n given by (21), orbital angular momentum $\sqrt{l(l+1)}\hbar$, whose projection on the z-axis is $m\hbar$ and which has a spin component $\frac{1}{2}\hbar$. The state is thus defined by the four quantum numbers n, l, m, and $\frac{1}{2}$. The second function corresponds to the state defined by n, l, m, and $-\frac{1}{2}$.

5 APPROXIMATION METHODS IN QUANTUM MECHANICS

The Variational Method 19

As we have already seen (Chapter 3), the mean energy in a state defined by the function $\psi(x, y, z)$ is given by the integral:

(1) $$\mathrm{E} = \int \psi^* \mathbf{H} \psi d\tau,$$

where the function ψ is normalized, that is to say ψ satisfies:

(2) $$\int \psi^* \psi d\tau = 1.$$

We now consider a family of normalized functions $\psi(x, y, z, \xi)$, dependent on a parameter ξ, such that for some value $\xi = \xi_0$ the function is a rigorous solution of the Schrödinger equation. Under this assumption, the integral (1) is a function of ξ, and we have shown (Chapter 3) that we must have:

(3) $$\left(\frac{d\mathrm{E}}{d\xi}\right)_{\xi=\xi_0} = 0.$$

In other words, the energy levels correspond to the extremum of the mean energy defined by (1).

This treatment of the variational method obviously does not correspond to the conditions under which it is utilized. As a matter of fact, the application is to families of functions which do not contain the rigorous solution to the problem, and condition (3) allows us to select only those functions which represent the best possible approximation. The essential point is to start with a family of functions which we have reason to believe will contain approximately the exact solution. The importance of this method rests on the facility with which we can apply it and the possibility which it offers for discussion of numerous problems in a qualitative manner (partial ionic character of bonds, mesomerism, etc.).

We will apply precisely this method to the common situation where we know two approximate solutions ψ_a and ψ_b of the Schrödinger equation of the system which correspond to mean energies close to the true energy. If these two functions are rigorous solutions, having the same energy eigenvalue E, all linear combinations of these functions in the form:

(4) $$\psi = c_a \psi_a + c_b \psi_b$$

would also be rigorous solutions of this equation belonging to the same value of energy. We are thus justified in thinking that if ψ_a and ψ_b are the *approximate solutions* corresponding to neighboring values of the *mean energy*, the linear combination (4) also represents an approximate solution. We thus make use of a family of approximate solutions dependent on the coefficients c_a and c_b which, however, are not independent, since the function ψ must be normalized, hence the condition:

$$1 = c_a^2 + c_b^2 + 2c_ac_b\int \psi_a\psi_b d\tau$$

where we assume that the functions ψ_a and ψ_b are real and normalized, which allows us to limit ourselves to consideration of real coefficients c_a and c_b. We often write this condition in the form:

$$(5) \qquad \mathrm{N} = c_a^2 + c_b^2 + 2c_ac_b\mathrm{S} = 1 \quad \text{with:} \quad \mathrm{S} = \int \psi_a\psi_b d\tau.$$

The integral S is called the *overlap integral* of ψ_a and ψ_b. It has an absolute value between zero and unity which is an immediate consequence of the relationship:

$$\int(\psi_a - \psi_b)^2 d\tau = 2 - 2\mathrm{S} \geqslant 0$$

likewise:

$$\int(\psi_a + \psi_b)^2 d\tau = 2 + 2\mathrm{S} \geqslant 0$$

hence,

$$-1 \leqslant \mathrm{S} \leqslant 1.$$

Subsequently we will only consider the case where the overlap integral is positive.

The mean energy in the state given by (4) is found from (1) which we write always assuming that the coefficients c_a and c_b are real.

$$\mathrm{E} = \int(c_a\psi_a + c_b\psi_b)\mathbf{H}(c_a\psi_a + c_c\psi_b)d\tau,$$

which we write:

$$(6) \qquad \mathrm{E} = c_a^2\mathrm{H}_{aa} + c_b^2\mathrm{H}_{bb} + 2c_ac_b\mathrm{H}_{ab}$$

with the usual notation:

$$H_{aa} = \int \psi_a \mathbf{H} \psi_a d\tau ;$$

$$H_{bb} = \int \psi_b \mathbf{H} \psi_b d\tau ;$$

$$H_{ab} = \int \psi_a \mathbf{H} \psi_b d\tau = \int \psi_b \mathbf{H} \psi_a d\tau$$

the final equality being a consequence of the hermitian character of the operator **H**.

The extremum condition of E, taking into account the normalization condition (5), is written by introducing Lagrange's multiplier λ (Appendix III):

$$\frac{\partial E}{\partial c_a} = \lambda \frac{\partial N}{\partial c_a}; \qquad \frac{\partial E}{\partial c_b} = \lambda \frac{\partial N}{\partial c_b}$$

so that explicitly:

$$H_{aa}c_a + H_{ab}c_b = \lambda(c_a + Sc_b)$$
$$H_{ab}c_a + H_{bb}c_b = \lambda(c_b + Sc_a).$$

By multiplying the first equation by c_a, the second by c_b, and adding, we get:

$$H_{aa}c_a^2 + H_{bb}c_b^2 + 2H_{ab}c_a c_b = \lambda(c_a^2 + c_b^2 + 2Sc_a c_b),$$

so that:

$$\lambda = E.$$

The advantage of introducing the Lagrange multiplier rests essentially in this relation which gives λ a simple meaning.

The energy cannot be arbitrary since the two linear homogeneous equations which determine the coefficients c_a and c_b cannot have solutions which are not trivial (all zero) unless they satisfy a condition of compatibility which we get by setting the determinant of the coefficients of the c's equal to zero, so that:

$$\begin{vmatrix} H_{aa} - E & H_{ab} - ES \\ H_{ab} - ES & H_{bb} - E \end{vmatrix} = 0$$

and

(7) $$f(E) = (H_{aa} - E)(H_{bb} - E) - (H_{ab} - ES)^2 = 0.$$

We have $f(H_{aa}) < 0$ and $f(H_{bb}) < 0$, with the result that the two roots E′ and E″ of the second order equation which fix the energy E, lie across the mean energy values H_{aa} and H_{bb}. This result agrees with the principle that the energy should represent an extremum in a real state so that the improvement of the solution when we go from ψ_a to ψ is naturally expressed by lowering of the energy level.

This method has given rise to a vocabulary frequently used in chemical theory. The state defined by ψ is called a hybrid of the states represented by ψ_a and ψ_b. However, this expression unfortunately introduces a confusion which we must keep in mind: it leads us to the incorrect assumption that ψ_a and ψ_b define real

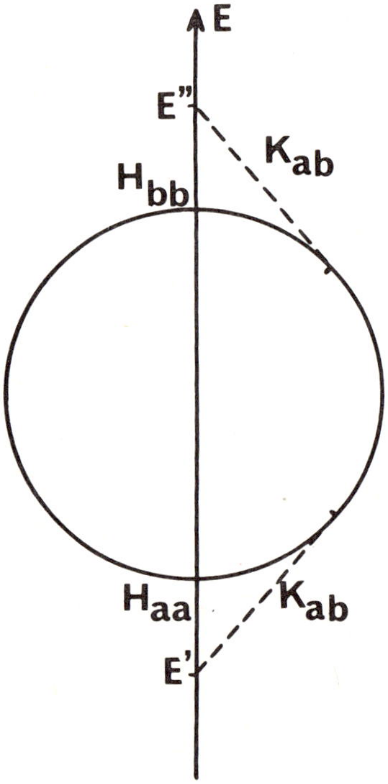

Figure 20. Graphical construction of the two energy levels E′ and E″, defined by the hybridization of the states of mean energy H_{aa} and H_{bb}.

states when they are actually only the elements of description of the system. The risk of confusion is also increased by use of the term *resonance energy* frequently employed to designate the quantity $\Delta = H_{aa} - E' > 0$. Δ represents the change in energy corresponding to the improvement in solution when we go from ψ_a to ψ, for which the term hybridization energy is certainly more appropriate.

A frequently used approximation can be made by replacing the second term of (7) by its principal part:

$$H_{ab} - ES \simeq H_{ab} - H_{aa}S = K_{ab},$$

with the result that (7) is written:

(8) $$(H_{aa} - E)(H_{bb} - E) = K_{ab}^2.$$

The determination of the roots E′ and E″ of this equation is reduced to finding points on the energy axis from which we can draw the tangents of length K_{ab} to the circle whose diameter is defined by H_{aa} and H_{bb}. This construction (fig. 20) allows us to note that Δ only takes positive values for given K_{ab} when the mean energies H_{aa} and H_{bb} are contiguous, which is, moreover, the *a priori* condition for applying this method.

Exercise 1. Apply (7) to the case where one of the basis functions, ψ_a for example, is an exact solution of the Schrödinger equation, and show that the corresponding eigenvalue is obtained as one of the two roots of the corresponding equation.

Exercise 2. See page 55.

Perturbation Calculations: Non-Degenerate Case 20

The problem treated here is more specialized than the previous case. We assume that we know the solution to the Schrödinger equation for an initial system and we want to find the solution for a second system, differing from the first by a relatively small variation in the potential energy, which we call the perturbation.

The equation for the first system is written:

(9) $$\mathbf{H}_0 \psi_r^{(0)} = E_r^{(0)} \psi_r^{(0)}.$$

We assume that the set of eigenvalues $E_r^{(0)}$ is known and is non-degenerate, as are the eigenfunctions which we take as normalized.

The hamiltonian of the second system differs from the first by the potential energy and we take:

(10) $$\mathbf{H}_1 = \mathbf{H}_0 + V_1,$$

where $V_1(x, y, z)$ represents the perturbation term.† To establish the fact that this term is small we write it in the form:

(11) $$V_1 = \varepsilon W,$$

where W represents a finite function and ϵ a coefficient sufficiently small so that the series expansions introduced in the calculations converge.

† Following an earlier remark concerning the x-coordinate operator, it is useful to use this notation for the operators V, whose action on ψ is an ordinary algebraic multiplication.

We now consider ϵ as a variable. The eigenfunctions and eigenvalues of $\mathbf{H}$ are going to vary with ϵ, and for $\epsilon = 0$ will revert to those of the first system. We assume that for sufficiently small ϵ the variations are continuous, with the result that it is possible to establish an unambiguous correspondence between the levels of the second system and those of the first. We can then *a priori* take an expansion of the form:

$$\begin{aligned} \psi_r &= \psi_r^{(0)} + \varepsilon\psi_r' + \varepsilon^2\psi_r'' + \dots \\ E_r &= E_r^{(0)} + \varepsilon E_r' + \varepsilon^2 E_r'' + \dots \end{aligned} \tag{12}$$

We assume that these expansions satisfy the Schrödinger equation of the second system:

$$(\mathbf{H}_0 + \varepsilon W)(\psi_r^{(0)} + \varepsilon\psi_r' + \varepsilon^2\psi_r'')$$
$$= (E_r^{(0)} + \varepsilon E_r' + \varepsilon^2 E_r'')(\psi_r^{(0)} + \varepsilon\psi_r' + \varepsilon^2\psi_r''),$$

where we have limited ourselves to terms of the second order in ϵ. We will assume that the equation is satisfied, whatever ϵ may be, and will then equate the coefficients of the successive powers of ϵ. For the constant term ($\epsilon = 0$) we find the equation of the initial system which is satisfied by hypothesis. For the following terms we get:

Term in ϵ:

$$\mathbf{H}_0\psi_r' + W\psi_r^{(0)} = E_r^{(0)}\psi_r' + E_r'\psi_r^{(0)}. \tag{13}$$

Term in ϵ^2:

$$\mathbf{H}_0\psi_r'' + W\psi_r' = E_r^{(0)}\psi_r'' + E_r'\psi_r' + E_r''\psi_r^{(0)}. \tag{14}$$

First order perturbations. This is defined by the first term in the series (12) and calculated by means of equation (13). We begin by the calculation of E_r'. To do this we multiply (13) by $\psi_r^{(0)\star}$ and integrate over all space.

$$\int \psi_r^{(0)*}\mathbf{H}_0\psi_r' d\tau + \int \psi_r^{(0)*} W \psi_r^{(0)} d\tau = E_r^{(0)} \int \psi_r^{(0)*}\psi_r' d\tau + E_r'.$$

The hermitian character of $\mathbf{H}_0$ allows us to transform the first integral on the left, which we write:

$$\int \psi_r^{(0)*}\mathbf{H}_0\psi_r' d\tau = \int \psi_r'\mathbf{H}_0\psi_r^{(0)*} d\tau = E_r^{(0)} \int \psi_r^{(0)*}\psi_r' d\tau$$

and which becomes equal to the integral term on the right, with the result that we finally get:

$$E_r' = \int \psi_r^{(0)*} W \psi_r^{(0)} d\tau = W_{rr}, \tag{15}$$

that is to say, the matrix element of the function W. To find the variation in energy we reintroduce ϵ, and get:

$$(16) \qquad \varepsilon E_r' = \int \psi_r^{(0)*} \varepsilon W \psi_r^{(0)} d\tau = \int \psi_r^{(0)*} V_1 \psi_r^{(0)} d\tau.$$

The integral represents (3-22), the mean value of the perturbation potential energy in a state defined by $\psi_r^{(0)}$.

We now want the expression for ψ_r'. This function can be represented by a series on the orthonormal system of $\psi_k^{(0)}$ (assumed complete), so that:

$$(17) \qquad \psi_r' = \sum_k c_{kr} \psi_k^{(0)}.$$

Substituting this into equation (13):

$$\sum_k c_{kr} \mathbf{H}_0 \psi_k^{(0)} + W \psi_r^{(0)} = E_r^{(0)} \sum_k c_{kr} \psi_k^{(0)} + W_{rr} \psi_r^{(0)}.$$

The function $W\psi_r^{(0)}$ can also be expanded on the $\psi_k^{(0)}$, the coefficients of expansion forming the matrix elements of this function so that:

$$W\psi_r^{(0)} = \sum_k W_{kr} \psi_k^{(0)}$$

by substituting in the preceding equation, we get:

$$\sum_k \left(c_{kr} E_k^{(0)} + W_{kr} - c_{kr} E_r^{(0)}\right) \psi_k^{(0)} = W_{rr} \psi_r^{(0)}.$$

The expansion brings in nothing more than the numerical coefficients involving the orthonormal system. As a result we can form an identity where $k \neq r$:

$$c_{kr}(E_k^{(0)} - E_r^{(0)}) + W_{kr} = 0$$

where:

$$c_{kr} = \frac{W_{kr}}{E_r^{(0)} - E_k^{(0)}}, \qquad \text{with } k \neq r.$$

This equation is indeterminate when $k = r$. We can thus choose the coefficient c_{rr} so that the wave function is normalized. This is written to within the second order in ϵ:

$$\int |\psi_r|^2 d\tau = 1 + \varepsilon\left(\int \psi_r^{(0)*} \psi_r' d\tau + \int \psi_r^{(0)} \psi_r'^* d\tau\right) = 1 + \varepsilon(c_{rr} + c_{rr}^*) = 1$$

which allows us to set $c_{rr} = 0$.

To reiterate, the first order perturbation calculation leads to the following expressions for the eigenvalues and eigenfunctions, on the assumption that these eigenvalues are non-degenerate:

$$(18)\quad \begin{cases} E_r = E_r^{(0)} + (V_1)_{rr} \qquad \text{with} \qquad (V_1)_{rr} = \int \psi_r^{(0)*} V_1 \psi_r^{(0)} d\tau \\ \psi_r = \psi_r^{(0)} + \sum_{k \neq r} \dfrac{(V_1)_{kr}}{E_r^{(0)} - E_k^{(0)}} \psi_k^{(0)}. \end{cases}$$

The form obtained for ψ_r shows that the approximation is only useful to the extent that the correspondence between ψ_r and $\psi_r^{(0)}$ is unambiguous, which implies:

$$|(V_1)_{kr}| \ll |E_r^{(0)} - E_k^{(0)}|.$$

Second order perturbation. We limit ourselves here to the calculation of the energy whose second expansion term E_r'' is gotten by the same method as E_r' but starting from (14). The integration following multiplication by $\psi_r^{(0)\star}$ gives, again using the hermitian character of **H**:

$$E_r'' = \int \psi_r^{(0)*} W \psi_r' d\tau.$$

The coefficient of E_r' vanishes in this calculation since ψ_r' given by the expansion (17) is orthogonal to $\psi_r^{(0)}$ since $c_{rr} = 0$. We get finally:

$$E_r'' = \sum_{k \neq r} c_{kr} \int \psi_r^{(0)*} W \psi_k^{(0)} d\tau = \sum_{k \neq r} \frac{W_{kr} W_{rk}}{E_r^{(0)} - E_k^{(0)}}$$

The definition of the matrix elements W_{kr} employs the hermiticity condition which can be shown directly:

$$W_{rk} = W_{kr}^{\star}, \qquad \text{hence} \qquad W_{kr} W_{rk} = |W_{kr}|^2.$$

The second order contribution to the energy thus takes the form:

$$(19)\qquad \varepsilon^2 E_r'' = \sum_{k \neq r} \frac{|(V_1)_{kr}|^2}{E_r^{(0)} - E_k^{(0)}}.$$

Note. The potential energy of the perturbation can depend not only on the position coordinate of the electron but also on its momentum components, and introduction of this into the Schrödinger equation necessitates the use of the associated operator. This is the case in LS coupling or in the Zeeman effect. The equations established here remain valid in these cases.

Application to the Anharmonic Oscillator 21

Besides its relevance to our treatment of the vibrational spectra of diatomic molecules, the anharmonic oscillator also provides a particularly apt example of the general points discussed above.

Description of the system. A first approximation of the variation of potential energy of a diatomic molecule about the nuclear equilibrium distance r_e is only good for small deformations. The experimental curve (fig. 13) is distinctly different from a parabola: it exhibits a much more rapid rise for small values of the nuclear distance r (which represents the incompressibility of the electronic clouds of the bound atoms), while for large r it tends toward a limit defined by the dissociation energy, D_e, of the molecule.

Various analytic expressions have been proposed to represent this behavior, including the Morse potential, which has the advantage of simplifying the calculations and which has the form:

$$(20) \qquad V(x) = D_e(1 - e^{-ax})^2 \qquad \text{with} \qquad x = r - r_e.$$

For $x = 0$ we find $V = 0$ and for $x = \infty$, $V = D_e$.

The behavior near equilibrium ($x = 0$) of the potential can be found from the limit of the power series expansion which we write:

$$(21) \qquad V = D_e a^2 x^2\left(1 - ax + \frac{7}{12}a^2x^2\right) = \frac{1}{2}kx^2 - fx^3 + gx^4$$

with:

$$k = 2D_e a^2; \qquad f = D_e a^3; \qquad g = \frac{7}{12}D_e a^4.$$

We shall study the energy levels of this system, treating it as a perturbed harmonic oscillator.

First order perturbation terms. We shall use the equations from the preceding section taking $V_1 = -fx^3$, which leads to the following expression for the first order perturbation energy:

$$\int \psi_r^{(0)*} V_1 \psi_r^{(0)} dx = -f \int \psi_r^{(0)*} x^3 \psi_r^{(0)} dx$$

where the $\psi_r^{(0)}$ are the eigenfunctions of the harmonic oscillator (§14). It now becomes a question of finding the matrix elements of x^3 which is done using the usual rules for the product of matrices. We start from:

$$x\psi_r^{(0)} = \sum_k x_{kr}\psi_k^{(0)}$$

from which we get:

$$x^2\psi_r^{(0)} = \sum_k x_{kr}\,.\,x\psi_k^{(0)} = \sum_k x_{kr}\sum_i x_{ik}\psi_i^{(0)} = \sum_i\Big(\sum_k x_{ik}x_{kr}\Big)\psi_i^{(0)}$$

which defines the matrix elements of x^2:

$$(x^2)_{ir} = \sum_k x_{ik}x_{kr}$$

and another application of this gives:

$$(x^3)_{rs} = \sum_{i,j} x_{ri}x_{ij}x_{js}. \tag{22}$$

The calculation here is particularly easy because the only elements of the harmonic oscillator matrix which are not zero are:

$$\begin{aligned} &x_{r,r+1} = x_0\sqrt{(r+1)} \\ &x_{r,r-1} = x_0\sqrt{r} \qquad \text{with} \qquad x_0 = \frac{x_1}{\sqrt{2}} = \frac{1}{2\pi}\sqrt{\frac{h}{2m\nu_0}}. \end{aligned} \tag{23}$$

Equation (22) implies that the only elements of the x^3 matrix which can be non-zero are those where $i - r = \pm 1$; $j - i = \pm 1$; $s - j = \pm 1$, which requires $s - r$ to be equal to an odd integer which can only be ± 1 or ± 3. Significantly we must have $(x^3)_{rr} = 0$, with the result that the corresponding contribution to the energy is zero.

Had the preceding resulted in an energy correction other than zero, we would have had reason to stop right then. Since this did not happen, we find it useful to consider the second order correction due to V_1 on the one hand, and the first order correction due to the higher order term in the potential expansion, gx^4, on the other hand. If we take $V = V_0 + V_1 + V_2$ with $V_0 = \frac{1}{2}kx^2$, $V_1 = -fx^3$, and $V_2 = gx^4$, we see that we have:

$$\left|\frac{V_2}{V_1}\right| \sim \left|\frac{V_1}{V_0}\right| \sim ax \qquad \text{so that} \qquad \left|\frac{V_2}{V_0}\right| \sim \left|\frac{V_1}{V_0}\right|^2.$$

We are thus lead to believe that the two energy contributions considered above—which vary together approximately as V_1^2 and V_2—should be of the same order of magnitude.

Second order perturbation due to V_1**.** The use of (19) requires knowledge of the matrix elements $(V_1)_{kr}$ which are non-zero. We assume that it is sufficient to consider the following values: $k = r - 3$, $r - 1$, $r + 1$, and $r + 3$. The fact that for $r = 0$, 1, and 2, certain of these elements do not exist, is unimportant, the formula which we will obtain gives a value of zero for the corresponding matrix elements. Equation (22) yields the following values:

$$(x^3)_{r-3,r} = x_{r-3,r-2} \cdot x_{r-2,r-1} \cdot x_{r-1,r} = x_0^3\sqrt{r(r-1)(r-2)}$$

$$(x^3)_{r-1,r} = x_{r-1,r-2} \cdot x_{r-2,r-1} \cdot x_{r-1,r} + x_{r-1,r} \cdot x_{r,r+1} \cdot x_{r+1,r}$$
$$+ x_{r-1,r} \cdot x_{r,r-1} \cdot x_{r-1,r} = 3x_0^3\sqrt{r^3}.$$

By replacing r by $r + 3$ in the first equation, and r by $r + 1$ in the second, we get—since the matrices are hermitian and therefore symmetrical as well as real:

$$(x^3)_{r+3,r} = (x^3)_{r,r+3} = x_0^3\sqrt{(r+3)(r+2)(r+1)}$$

$$(x^3)_{r+1,r} = (x^3)_{r,r+1} = 3x_0^3\sqrt{(r+1)^3}.$$

The increase in energy due to the second order perturbation is gotten from (19):

$$f^2\sum_k \frac{(x^3)^2_{k,r}}{E_r^{(0)} - E_k^{(0)}} \qquad \text{where:} \quad k = r-3,\ r-1,\ r+1,\ r+3.$$

The energy values in the non-perturbed state are given by (3-39) $E_r^{(0)} = (r + \frac{1}{2})h\nu_0$ so that the denominators in the previous equation are respectively, $E_r^{(0)} - E_k^{(0)} = 3h\nu_0$, $h\nu_0$, $-h\nu_0$, and $-3h\nu_0$. We get by arithemtic manipulation:

$$-\frac{30f^2x_0^6}{h\nu_0}\left(r^2 + r + \frac{11}{30}\right).$$

First order perturbation from V_2. The calculation is similar to that for V_1, but requires the matrix element $(x^4)_{rr}$ given by extension of (22), so that:

$$(x^4)_{rr} = \sum_{i,j,k} x_{ri}x_{ij}x_{jk}x_{kr}.$$

The only combinations which appear are shown in the table below:

i	j	k
$r+1$	$r+2$	$r+1$
$r-1$	$r-2$	$r-1$
$r+1$	r	$r+1$
$r-1$	r	$r+1$
$r+1$	r	$r-1$
$r-1$	r	$r-1$

hence, following reduction:

$$(x^4)_{rr} = 6x_0^4\left(r^2 + r + \frac{1}{2}\right).$$

To conclude these calculations we write the following expression for the energy level with index r:

$$E_r = \left(r + \frac{1}{2}\right)h\nu_0 - \frac{30f^2x_0^6}{h\nu_0}\left(r^2 + r + \frac{11}{30}\right) + 6gx_0^4\left(r^2 + r + \frac{1}{2}\right).$$

We note that the dependence of E_r on r involves two second order trinomials which are both of the form $(r + \frac{1}{2})^2$ to within a constant

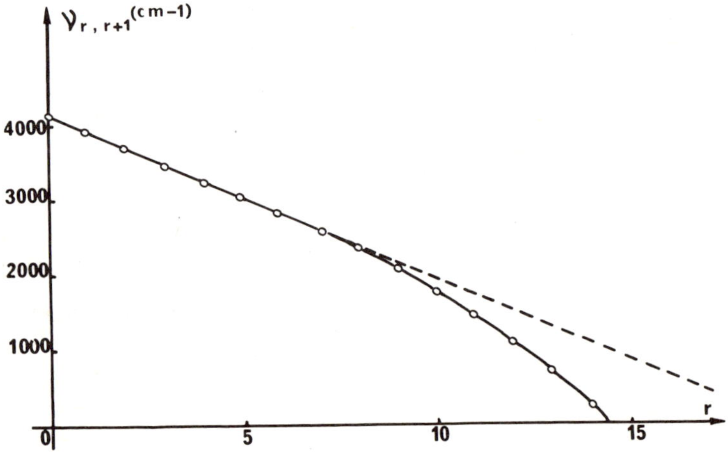

Figure 21. Frequencies of the successively observed transitions in the vibrational spectrum of H_2 (after Herzberg, *Spectra of Diatomic Molecules*).

term. Thus, by setting aside the somewhat small correction ε_0 on the ground state ($r = 0$), we can use the following equivalent equation to get the values of the energy levels of the anharmonic oscillator:

$$E_r = \varepsilon_0 + \left(r + \frac{1}{2}\right)h\nu_0 - \left(r + \frac{1}{2}\right)^2 . 6x_0^4\left(\frac{5f^2x_0^2}{h\nu_0} - g\right).$$

By using equations (21) for k, f, and g, and (23) for x_0, as well as the expression $\nu_0 = (1/2\pi)\sqrt{k/m}$ for the frequency of the classical harmonic oscillator, we get, after reduction:

$$E_r = \varepsilon_0 + \left(r + \frac{1}{2}\right)h\nu_0 - \left(r + \frac{1}{2}\right)^2 \frac{h^2\nu_0^2}{4D_e}$$

or:

$$E_r = \varepsilon_0 + \left(r + \frac{1}{2}\right)h\nu_0 - \varkappa\left(r + \frac{1}{2}\right)^2 h\nu_0$$

by introducing the coefficient of anharmonicity,

$$\varkappa = \frac{h\nu_0}{4D_e}.$$

As a result, the transition frequencies $r \to r + 1$ arising from the absorption of infra-red radiation are in arithmetic progression, which effectively represents the real situation, at least for the first few terms, the decrease becoming more rapid with the higher terms (fig. 21). This demonstrates that the Morse potential is only an approximation to the potential energy curve of the diatomic molecule. In particular, its behavior is certainly defective for small values of r, since it tends toward a large but finite limit as r tends to zero.

Perturbation Calculations: Degenerate Case 22

We return to the problem treated in section 20, but now we will assume that certain levels of the initial system are degenerate. If E_r designates one such eigenvalue, Schrödinger's equation will be satisfied by ν independent functions which we call $\psi_{r1}^{(0)}, \psi_{r2}^{(0)}, \ldots, \psi_{r\nu}^{(0)}$, all corresponding to the same value of $E_r^{(0)}$. It follows that if we consider a linear combination of these functions with arbitrary coefficients c_i:

$$\psi_r^{(0)} = \sum_{i=1}^{\nu} c_i \psi_{ri}^{(0)} \tag{24}$$

the function $\psi_r^{(0)}$ is also a solution of the Schrödinger equation for this same value $E_r^{(0)}$:

$$\mathbf{H}_0 \psi_r^{(0)} = E_r^{(0)} \psi_r^{(0)}. \tag{25}$$

This function $\psi_r^{(0)}$ depends on ν parameters, since the $\psi_{ri}^{(0)}$ are independent functions, which means that it is impossible to choose coefficients to satisfy the following identity:

$$\sum_{i=1}^{\nu} c_i \psi_{ri}^{(0)} \equiv 0, \tag{26}$$

unless these coefficients are all zero.

The function $\psi_r^{(0)}$ defined by (24) can also be expressed by replacing the basis functions $\psi_{ri}^{(0)}$ by linearly independent combinations, exactly as one changes reference systems in vector spaces. The basis functions $\psi_{ri}^{(0)}$ constitute the most general elements describing the state (24) and are replaceable by any other system of ν independent functions of this set. It is convenient to choose these basis functions so that they are orthogonal to each other, thus forming an orthogonal system of ν functions. At the end of this section we will introduce a procedure which allows us to make this choice. We assume that the functions $\psi_{ri}^{(0)}$ will eventually fulfill this condition provided that condition (26) is automatically satisfied. All that is needed, in effect, is to integrate over all space following multiplication by $\psi_{ri}^{(0)\star}$, which leads to $c_i = 0$, whatever i may be. Any other system of values of the coefficients will not be convenient.

We now turn to the perturbation of the system, using the notation of the previous section, namely, taking the perturbation energy in the form $V_1 = \epsilon W$. We will consider an energy eigenvalue of the system E_r which coincides with $E_r^{(0)}$ for $\epsilon = 0$. The eigenfunction should also tend toward a solution of the initial problem, but the existence of the degeneracy prevents us from knowing which one, and we can only say that this solution should be of the form (24), with coefficients c_i which have yet to be determined.

We begin with the calculation of the first order perturbations, starting from equation (13). This is written:

$$\text{(13)} \qquad \mathbf{H}_0\psi_r' + W\psi_r^{(0)} = E_r^{(0)}\psi_r' + E'\psi_r^{(0)}.$$

We once again integrate over all space, following multiplication by $\psi_{ri}^{(0)\star}$, which will give us ν equations. For each of these, the calculation is the same as for a non-degenerate level, with the same simplifications, the difference lying in the evaluation of the two integrals:

$$\int \psi_{ri}^{(0)\star} W \psi_r^{(0)} d\tau = \sum_{j=1}^{\nu} c_j \int \psi_{ri}^{(0)\star} W \psi_{rj}^{(0)} d\tau = \sum_{j=1}^{\nu} W_{ri,rj} c_j$$

$$\int \psi_{ri}^{(0)\star} \psi_r^{(0)} d\tau = \sum_{j=1}^{\nu} c_j \int \psi_{ri}^{(0)\star} \psi_{rj}^{(0)} d\tau = c_i,$$

the $\psi_{ri}^{(0)}$ being orthonormal.

To simplify writing, and taking into account the fact that we consider some fixed level E_r, we drop the index r in the designation of the matrix elements, writing it simply as W_{ij}.

Equation (15) is replaced by:

$$\sum_{j=1}^{\nu} W_{ij} c_j = E_r' c_i.$$

The ν coefficients c_i are roots of a system of ν linear equations. In order to obtain non-zero solutions, the compatibility condition must be satisfied and the determinant of the coefficients of c_i must vanish, so that:

$$(27) \qquad \begin{vmatrix} W_{11} - E_r' & W_{12} & \dots & W_{1\nu} \\ W_{21} & W_{22} - E_r' & \dots & W_{2\nu} \\ \dots & \dots & \dots & \dots \\ W_{\nu 1} & W_{\nu 2} & \dots & W_{\nu\nu} - E_r' \end{vmatrix} = 0.$$

The expansion of this determinant leads to an equation of degree ν in E_r' ,which determines the energy corrections corresponding to the first perturbation. We note that equation (13) relating to the non-degenerate case is reintroduced as a special case of the above equation. An equivalent form of (27) is obtained by introducing the factor ε throughout. This displays the matrix elements of V_1 instead of those calculated from W, the roots of the equation becoming the values $\varepsilon E_r'$ of the energy corrections.

According to the form of V_1, which affects the matrix elements W_{ij}, the roots of (27) can be simple or multiple and two extreme cases can be examined. In the first of these, the roots of (27) are distinct, which makes the ν eigenvalues of the perturbed system non-degenerate. In this case, the application of the perturbation completely removes the degeneracy. In the second case, (27) has a single root, E_r' of the order ν. Under those conditions the perturbation does not modify the degeneracy. In the intermediate case, where there are many multiple roots, the perturbation partially lifts the degeneracy.

We note that, in general, a degeneracy in eigenvalues is coupled to a symmetry of the system, and the various possibilities we have just proposed correspond to the fact that the perturbation can to a greater or lesser degree alter the symmetry of the initial system.

Once the ν roots E_r' of equation (27) are determined, it becomes possible to solve the system of linear equations, giving the c_i more or less arbitrarily depending on whether the roots E_r' are degenerate or non-degenerate. In following our initial procedure, it is possible to change the system of basis functions and to take precisely those functions needed in the calculation which give the first approximation to the perturbation problem. Equation (27) then takes the simple form:

$$(28) \qquad \begin{vmatrix} W_{11} - E'_r & 0 & \dots & 0 \\ 0 & W_{22} - E'_r & \dots & 0 \\ 0 & 0 & \dots & 0 \\ 0 & 0 & \dots & W_{\nu\nu} - E'_r \end{vmatrix} = 0.$$

Only the matrix elements W_{ii} can differ from zero and a certain number of these can be equal, in which case the degeneracy is not completely removed by the perturbation.

This calculation treats the first order perturbation. It can be easily extended to the second order perturbation, and equation (19) remains valid for the calculation of matrix elements in the system of basis functions adapted to the perturbation, such that the equation in E'_r takes the form (28) given previously.

Schmidt Orthogonalization Procedure. The problem is to replace ν functions of x, y, z by an equal number of linear combinations forming an orthogonal system. The method is similar to that used to construct a cartesian system of basis vectors starting from three independent basis vectors forming a trihedral. The first vector of the new system can be taken in an arbitrary direction, in particular in the same direction as the first vector of the original system. The second vector can be chosen so that it lies in the plane defined by the first two vectors of the original system, with only the final vector being completely fixed.

The situation is identical here, except for the number of dimensions, which is not limited to three. We designate the first system of functions by $\psi_1, \psi_2, \dots, \psi_\nu$, and the second system, which is assumed to be orthonormal, by $\theta_1, \theta_2, \dots, \theta_\nu$. We choose arbitrarily:

$$\theta_1 = c_1\psi_1,$$

c_1 being fixed by the normalization condition. We take θ_2 as a linear combination of ψ_2 and θ_1, which turns out to be:

$$\theta_2 = c_2(\psi_2 - \lambda\theta_1).$$

The coefficient λ should be such that θ_2 is orthogonal to θ_1, which implies:

$$\int \theta_1^*\theta_2 d\tau = c_2 \int \theta_1^*(\psi_2 - \lambda\theta_1)d\tau = 0,$$

so that:

$$\lambda = \int \theta_1^*\psi_2 d\tau.$$

The coefficient c_2 is likewise fixed, so that θ_2 is normalized. Additionally, the third vector, θ_3, can be taken in the form:

$$\theta_3 = c_3(\psi_3 - \lambda_1\theta_1 - \lambda_2\theta_2).$$

The coefficients λ_1 and λ_2 should be such that θ_3 is orthogonal to θ_1 and θ_2, which can be written:

$$\lambda_1 = \int \theta_1^* \psi_3 d\tau ; \qquad \lambda_2 = \int \theta_2^* \psi_3 d\tau.$$

The method can be extended step by step, thus allowing the formation of a system of ν orthonormal functions θ_r, with the only condition being that the initial ν functions be independent, that is to say, that the identity (26) can only be satisfied if all of the coefficients c_i are zero.

Application to the Action of an Electric Field on the Plane Rotor 23

We take as an example the plane rotor (§15) governed by the equation:

$$\mathbf{H}\psi = -\frac{\hbar^2}{2I}\frac{d^2\psi}{d\chi^2} = E\psi.$$

We assume that the particle carries an electric charge q, or, in other words, that the rotor has a dipole moment $\mu = qr$, by assuming a charge $-q$ localized at the center so that we have an electrically

Figure 22. Diagram of the plane rotor subject to an electrical field f. The system is defined by the angle χ between the permanent moment $\vec{\mu}$ and the field.

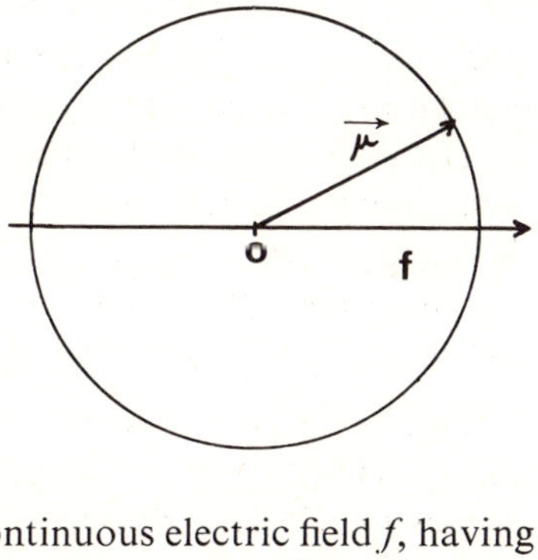

neutral system. Application of a continuous electric field f, having the direction taken along the origin of the angles (fig. 22), leads to a potential energy:

$$V = -\vec{\mu}\cdot\vec{f} = -\mu f \cos \chi.$$

We shall treat the problem of the action of the field as a perturbation using the equations deduced in the previous paragraph. We rewrite the expression for the eigenfunctions and eigenvalues of energy in the absence of a perturbation:

$$\psi_k^{(0)} = \frac{1}{\sqrt{2\pi}} e^{ik\chi}$$

$$E_k^{(0)} = \frac{k^2\hbar^2}{2I} \qquad \text{with} \qquad E_k^{(0)} = E_{-k}^{(0)}.$$

The perturbation calculation requires the matrix elements:

$$V_{rs} = -\frac{\mu f}{2\pi}\int_0^{2\pi} e^{i(s-r)\chi} \cos \chi d\chi.$$

Expressing $\cos\chi$ by means of the complex exponentials allows us to write:

$$V_{rs} = -\frac{\mu f}{4\pi}\int_0^{2\pi} (e^{i(s-r+1)\chi} + e^{i(s-r-1)\chi})d\chi.$$

The only non-zero matrix elements are those for which we have $s = r \pm 1$, and their common value is:

$$V_{r,r\pm 1} = -\frac{\mu f}{2}.$$

We then use the formulas deduced in our study of perturbation calculations.

First order perturbation. The energy $E_k^{(0)}$ is degenerate, the two states defined by k and $-k$ having the same energy. The matrix elements involved in the calculation are $V_{k,k}$, $V_{-k,-k}$, and $V_{k,-k}$, and are necessarily zero from the previous calculation. We thus have no first order perturbation, with the result that application of the field does not lift the degeneracy of the plane rotor, this being related to the fact that this perturbation does not alter the symmetry with regard to the meridian plane which is responsible for this degeneracy. Thus we can continue to use the functions $\psi_k^{(0)}$ and $\psi_{-k}^{(0)}$ for the second order perturbation calculation.

Second order perturbation. As we saw earlier, the only non-zero matrix elements V_{kr}, correspond to $k = r \pm 1$. Thus, by using expression (19) for the energy of the perturbed rotor:

$$E_r = E_r^{(0)} + \frac{|V_{r,r-1}|^2}{E_r^{(0)} - E_{r-1}^{(0)}} + \frac{|V_{r,r+1}|^2}{E_r^{(0)} - E_{r+1}^{(0)}}$$

which becomes, using the value obtained for $V_{r,r\pm 1}$:

$$E_r = \frac{r^2\hbar^2}{2I} + \frac{I\mu^2 f^2}{\hbar^2(4r^2-1)}, \qquad \begin{matrix}\text{avec}\\ \text{with}\end{matrix} \qquad r > 0.$$

This calculation ceases to be valid for $r = 0$, since only one element of the matrix $V_{0,1}$ can be defined. By treating separately the calculation for this case, we get:

$$E_0 = -\frac{I\mu^2 f^2}{2\hbar^2}.$$

Figure 23. Energy levels of the plane rotor and their displacement under the action of an electrical field.

The energy levels (fig. 23) depend quadratically on the field strength f and we can write for each level:

$$E_r = E_r^{(0)} - \frac{1}{2}\alpha f^2$$

where α defines the *electrical polarizability* in state r. The polarizability is only positive in the rest state ($r = 0$). This calculation can be completed by a statistical treatment, taking account of the fraction of the rotors of energy E_e in thermal equilibrium at temperature T. As the temperature is elevated, there are gradually fewer and fewer rotors in the ground state, with the result that the mean polarizability decreases with increasing temperature. This corresponds to the Debye equation, with the reservation that Debye's equation corresponds to a spatial rotor. We will not go through the calculation for the case of the plane rotor since it is uninteresting for the representation of real phenomena. We conclude this discussion by comparing the result of this perturbation calculation with that which can be obtained classically. It should be mentioned that perturbation calculations were first introduced in classical mechanics, and more particularly in celestial mechanics.

Calculation of classical mechanical perturbations. The classical equation of motion is written:

$$I\ddot{\chi} = -\mu f \sin \chi.$$

In the absence of a field, the motion is uniform, and by using a convenient origin for time we have:

$$\chi = \omega t.$$

We will show that the solution of the problem, following the application of the field, can be taken in the form:

$$\chi = \omega t + \chi_1(t).$$

We put this in the equation of motion, assuming that with the supposedly small perturbation term we can limit ourselves to its principal part $\chi = \omega t$:

$$I\ddot{\chi}_1 = -\mu f \sin \omega t.$$

The integration can be done immediately; the solution suitable for the perturbation problem, such that χ_1 does not increase indefinitely with time, is written:

$$\chi_1 = \frac{\mu f}{I\omega^2} \sin \omega t.$$

Comparison with the quantum treatment should in principle be based on the energy value, but the actual approximation is perhaps inadequate since we are dealing with a second order perturbation. It is much easier to base the comparison on the value of the polarizability which we can evaluate classically by calculating the mean value of the component of the dipole moment μ_f in the field direction. We have:

$$\mu_f = \mu \cos \chi = \mu \cos (\omega t + \chi_1) \simeq \mu \left(\cos \omega t - \frac{\mu f}{I\omega^2} \sin^2 \omega t\right)$$

and, taking the mean value:

$$\bar{\mu}_f = -\frac{\mu^2 f}{2I\omega^2}.$$

This value is proportional to the field *f*, which permits calculation of the corresponding polarizability:

$$\alpha = -\frac{\mu^2}{2I\omega^2} = -\frac{\mu^2}{4E}$$

while in the state defined by the quantum number *r*, the preceding theory yielded

$$\alpha_r = -\frac{2I\mu^2}{\hbar^2(4r^2 - 1)}\cdot$$

By using the energy $E_r = r^2\hbar^2/2I$ we can see that the two methods tend toward the same result for high values of the quantum number *r*.

LS Coupling 24

Section 18, devoted to the concept of spin, dealt with the first approximation of the Bohr atom, which does not require the introduction of the spin variable into the potential energy expression. The interpretation of the coupling phenomena requires a second approximation in which it is no longer possible to assume that the spin and the orbital motion of the electron are independent. The mechanism of the interaction, known as LS coupling, is introduced by the following two properties which we will justify qualitatively:

(*a*) The phenomenon of spin that is, the rotation of a charged particle, implies the existence of a magnetic moment proportional to the electron angular momentum:

$$\vec{\mu} = \gamma\vec{S} \tag{29}$$

where γ is the *gyromagnetic ratio*.

(*b*) The electron is moving in the electrical field of the nucleus and is thus in the same situation as if it were placed in a time-varying electrical field with which we can associate a magnetic field.

The rigorous calculation of this effect is beyond the scope of this book and we will limit ourselves to giving an approximate treatment which allows determination of the form of the interaction and an estimate of its magnitude.

(*a*) The derivation of equation (29) can be presented in a simple manner by considering a point electric charge moving on an invariant trajectory with a periodic motion of frequency ν. During this motion (fig. 24) the total charge per second passing any point M on the trajectory is νq, with the result that the system is similar to an electric circuit placed on this trajectory and carrying a current of intensity:

$$i = \nu q = \frac{q}{\mathrm{T}}$$

where $\mathrm{T} = 1/\nu$ designates the period of the motion of the charge. This equivalence allows us to attribute to this system a magnetic moment, which in the case of a plane trajectory of area A is given by:

$$\vec{\mu} = \vec{n} i \mathrm{A} = \frac{q\vec{n}\mathrm{A}}{\mathrm{T}}$$

$\vec{n}$ being the unit normal to the trajectory oriented by reference to the direction of the current i.

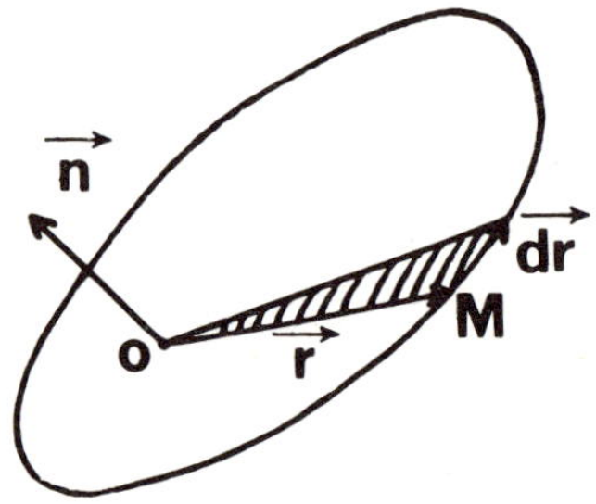

Figure 24. Planar trajectory with the unit normal $\vec{n}$ and with the cross-hatches indicating the element of area $d\mathrm{A}$ defined by $\vec{r} \wedge d\vec{r} = 2\vec{n} d\mathrm{A}$.

Another vector can be introduced by assuming that the angular momentum $\vec{\mathrm{L}}$ is constant in time, which is always the case for a particle moving in a field derived from a central force. Under these conditions we can write:

$$\vec{\mathrm{L}} dt = \vec{r} \wedge m\vec{v} dt = m\vec{r} \wedge d\vec{r}.$$

The vector product represents twice the area indicated in the figure, with the result that by integrating over one period:

$$\vec{\mathrm{L}}\mathrm{T} = 2m\vec{n}\mathrm{A}.$$

Comparison with the expression for $\vec{\mu}$ gives:

$$\vec{\mu} = \frac{q}{2m} \vec{\mathrm{L}}.$$

Equation (29) is thus justified with $\gamma = q/2m$ as the gyromagnetic ratio.

One would think that the proper rotation of the electron should, in an analogous manner, involve the existence of a magnetic moment, and that, in effect, is what happens; but the simplistic character of the model, which causes the spin to resemble a classical rotation, does not admit to a quantitative expression. The complete Dirac theory, confirmed by experimental evidence, shows that the gyromagnetic ratio is twice that obtained for the orbital motion, i.e., q/m. In conclusion, a one-electron atom (with charge $-e$ and mass m_0) has a magnetic moment with the following components:

(30)
$$\text{orbital moment:} \quad \vec{\mu}_L = -\frac{e}{2m_0}\vec{L}$$
$$\text{spin moment:} \quad \vec{\mu}_S = -\frac{e}{m_0}\vec{S}.$$

(*b*) An elementary but inexact approach starts with the calculation of the magnetic field acting on the electron moving with velocity $\vec{v}$, and consists in showing that the effect should be the same as if we consider the electron fixed, and the nucleus—the source of the electric field—as having velocity $-\vec{v}$. We know that such electromagnetic phenomena require use of the laws of special relativity and it will be necessary to make use of the Lorentz relations. Given this reservation, we can assume that from the point of view of the magnetic effect the charge Ze behaves like the current element:

$$i\vec{dl} = -Ze\vec{v}$$

and the magnetic induction acting on the electron can be calculated by Laplace's law:

$$\vec{B} = \frac{\mu_0}{4\pi} i\vec{dl} \wedge \frac{\vec{r}}{r^3} = -\frac{\mu_0}{4\pi} Ze\vec{v} \wedge \frac{\vec{r}}{r^3}.$$

The system, consisting of an electron with spin magnetic moment $\vec{\mu}_s$ placed in a magnetic field, has potential energy:

$$V = -\vec{\mu}_S . \vec{B} = -\frac{\mu_0}{4\pi} . \frac{Ze^2}{m_0}\left(\vec{v} \wedge \frac{\vec{r}}{r^3}\right).\vec{S}.$$

This formula gives the angular momentum:

$$\vec{L} = \vec{r} \wedge m_0\vec{v},$$

or in the equivalent form:

$$V = \frac{\mu_0 Ze^2}{4\pi m_0^2 r^3}\vec{L}.\vec{S}.$$

The rigorous theory leads to half this value:

$$V = \frac{\mu_0 Ze^2}{8\pi m_0^2 r^3}\vec{L}.\vec{S}$$

which in electrostatic units is:

$$V = \frac{Ze^2}{2m_0^2 c^2 r^3}\vec{L}.\vec{S}.$$

LS coupling can be treated by means of perturbation theory. To write the Schrödinger equation, taking the interaction into account, it is necessary to introduce the operator associated with the coupling energy which has just been calculated. This is not difficult since the components of $\vec{L}$ and $\vec{S}$ are compatible and are also compatible with the radius vector r. We can then take:

$$\mathbf{V} = \frac{\mu_0 Z e^2}{8\pi m_0^2 r^3} (\mathbf{L}_x \mathbf{S}_x + \mathbf{L}_y \mathbf{S}_y + \mathbf{L}_z \mathbf{S}_z). \tag{31}$$

We now have a typical perturbation calculation for a degenerate level. We can examine this by taking the example of a p level, which in no way alters the generality of the reasoning. In Chapter 4, in the treatment of an atom with spin but with no coupling, we showed that the basis functions can be formed from the spinless atomic functions which we call ψ_1, ψ_0 and ψ_{-1} simply by indicating the value of the quantum number m which distinguishes them and the spin functions α and β. We can list these functions:

$$\begin{aligned} &\theta_1 = \psi_1\alpha\,; \quad &\theta_3 = \psi_0\alpha\,; \quad &\theta_5 = \psi_{-1}\alpha \\ &\theta_2 = \psi_1\beta\,; \quad &\theta_4 = \psi_0\beta\,; \quad &\theta_6 = \psi_{-1}\beta. \end{aligned} \tag{32}$$

The calculation of the matrix elements V_{ij} is most easily accomplished by introducing the operators $\mathbf{L}_x \pm i\mathbf{L}_y$ and $\mathbf{S}_x \pm i\mathbf{S}_y$. We get:

$$\mathbf{L}_x\mathbf{S}_x + \mathbf{L}_y\mathbf{S}_y + \mathbf{L}_z\mathbf{S}_z$$
$$= \frac{1}{2}[(\mathbf{L}_x + i\mathbf{L}_y)(\mathbf{S}_x - i\mathbf{S}_y) + (\mathbf{L}_x - i\mathbf{L}_y)(\mathbf{S}_x + i\mathbf{S}_y)] + \mathbf{L}_z\mathbf{S}_z\,.$$

As an example we will give the details of the calculation for the matrix elements V_{r2}, the others being obtained in the same way. We have to treat the function $\mathbf{V}\theta_2$. Starting with equations (10), (11), and (15) Appendix VI, we can write:

$$(\mathbf{L}_x + i\mathbf{L}_y)\psi_1 = 0$$
$$(\mathbf{L}_x - i\mathbf{L}_y)\psi_1 = \hbar a_{0,1}\psi_0 = \sqrt{2}\hbar\psi_0.$$

The equations (4-28) give the action of the spin operator on the function β, namely:

$$(\mathbf{S}_x + i\mathbf{S}_y)\beta = \hbar\alpha$$
$$(\mathbf{S}_x - i\mathbf{S}_y)\beta = 0.$$

This set of equations permits us to write:

$$\frac{1}{2}[(\mathbf{L}_x + i\mathbf{L}_y)(\mathbf{S}_x - i\mathbf{S}_y) + (\mathbf{L}_x - i\mathbf{L}_y)(\mathbf{S}_x + i\mathbf{S}_y)]\theta_2 = \frac{\sqrt{2}}{2}\hbar^2\theta_3.$$

In addition, ψ_2 and β are respectively eigenfunctions of $\mathbf{L}_z$ and $\mathbf{S}_z$ so that $\mathbf{L}_z\psi_1 = \hbar\psi_1$ and $\mathbf{S}_z\beta = -\hbar\beta$. We have then:

$$\mathbf{V}\theta_2 = \frac{K}{r^3}(\sqrt{2}\theta_3 - \theta_2) \qquad \text{with} \qquad K = \frac{\mu_0 Z e^2 \hbar^2}{16\pi m_0^2}.$$

To form the matrix element, we must multiply by $\theta_r^\star$ and integrate over the space variables and sum over the spin variables (4-30). We get:

$$V_{r,2} = \sum_\sigma \int \theta_r^\star V\theta_2 \, d\tau.$$

To effect the integration note that the functions ψ_m all contain the same factor $R_{n,1}(\rho)$ dependent on the radius vector, with the result that all elements of the desired matrix contain the factor:

$$\left\langle \frac{1}{r^3} \right\rangle = \int_0^\infty \frac{1}{r^3} R_{n,1}^2 r^2 dr.$$

In addition, the second factor of ψ_m (4-11) is formed by one of the spherical functions Y_1^m, a set of which forms an orthonormal system on the sphere of unit radius. We then see that all of the matrix elements $V_{r,2}$ are zero with the exception of:

$$V_{2,2} = -K\left\langle \frac{1}{r^3} \right\rangle; \qquad V_{3,2} = K\sqrt{2}\left\langle \frac{1}{r^3} \right\rangle.$$

By operating in the same way for the other matrix elements we can write equation (27) giving the energy values of the first order perturbation. We take:

$$E_r' = \lambda K \left\langle \frac{1}{r^3} \right\rangle$$

and the equation is written:

$$\begin{vmatrix} 1-\lambda & 0 & 0 & 0 & 0 & 0 \\ 0 & -1-\lambda & \sqrt{2} & 0 & 0 & 0 \\ 0 & \sqrt{2} & -\lambda & 0 & 0 & 0 \\ 0 & 0 & 0 & -\lambda & \sqrt{2} & 0 \\ 0 & 0 & 0 & \sqrt{2} & -1-\lambda & 0 \\ 0 & 0 & 0 & 0 & 0 & 1-\lambda \end{vmatrix} = 0.$$

This determinant is easily broken down into factors with the result that the equation can be written:

$$(1-\lambda)^2(\lambda^2+\lambda-2)^2 = (\lambda-1)^4(\lambda+2)^2 = 0.$$

We get two multiple roots so that the degeneracy is partially lifted, and this results in two levels, one of fourth order, and the other of second order, whose energies are written, using the notation of equation (12) as:

$$\varepsilon E'_{(1)} = \frac{\mu_0 Ze^2\hbar^2}{16\pi m_0^2}\left\langle\frac{1}{r^3}\right\rangle \qquad \text{(order 4)}$$

$$\varepsilon E'_{(2)} = -2\frac{\mu_0 Ze^2\hbar^2}{16\pi m_0^2}\left\langle\frac{1}{r^3}\right\rangle \qquad \text{(order 2).}$$

These two corrections act in opposite directions and the mean energy is not modified by the introduction of this interaction.

Having outlined the application of the classical perturbation calculation, we will now develop a method based on the fact that the initial degeneracy can be removed in the same way by using the potential energy V, as above, or a more simple form of the energy, which we will derive.

Total angular momentum. We first present the classical point of view. In the absence of coupling between the spin and the displacement of the electron, the orbital angular momentum $\vec{L}$ and the spin momentum $\vec{S}$ remain constant over time. The existence of a coupling no longer allows the separation of the momenta and it is the total angular momentum defined by:

$$\vec{J} = \vec{L} + \vec{S},$$

which remains constant.

These considerations can be justified from the quantum viewpoint by using the operators associated with the components of $\vec{J}$, these being easily formed, since the addition of two operators is always permitted. Thus we get:

$$\mathbf{J}_x = \mathbf{L}_x + \mathbf{S}_x\,; \qquad \mathbf{J}_y = \mathbf{L}_y + \mathbf{S}_y\,; \qquad \mathbf{J}_z = \mathbf{L}_z + \mathbf{S}_z.$$

The operators $\mathbf{L}_x$, $\mathbf{L}_y$, and $\mathbf{L}_z$ on the one hand, and $\mathbf{S}_x$, $\mathbf{S}_y$, and $\mathbf{S}_z$ on the other, satisfy the same relations (Appendix IV-2) to within a factor $\hbar$. As a result, since $\mathbf{L}_i$ and $\mathbf{L}_j$ are compatible with each other, these equations are also satisfied by $\mathbf{J}_x$, $\mathbf{J}_y$, and $\mathbf{J}_z$. Consequently, the indicated properties of the operators $\mathbf{L}_i$ (Appendix VI) remain true for the $\mathbf{J}_i$'s. In particular, it is possible to define $\mathbf{J}^2$ and to characterize the states of the system by the simultaneous eigenfunctions of $\mathbf{H}$, $\mathbf{J}^2$ and $\mathbf{J}_z$.

The following problem then arises: knowing the solution of the one-electron atomic problem in the absence of LS coupling, find the solution of the perturbed problem, that is, for the set of states defined by J^2 and J_z, the wave functions, and the corresponding energies. The first part can be resolved by the simple argument which we present next.

The properties of the operators $\mathbf{J}_i$ imply, considering the foregoing, that the eigenvalues of $\mathbf{J}^2$ are of the form $j(j+1)$. For each possible value of j such as j' we have $2j'+1$ values of J_z from $-j'\hbar$ to $+j'\hbar$. The values of J^2 and J_z, being quantized, are not modified when the magnitude of the LS coupling is varied and thus remain the same when this interaction is negligible. But in this limiting case, L_z and S_z are known, with the result that we can immediately arrange the list of possible values of J_z as shown in the table below for a typical p level.

	θ_1	θ_2	θ_3	θ_4	θ_5	θ_6
L_z	1	1	0	0	− 1	− 1
S_z	1/2	− 1/2	1/2	− 1/2	1/2	− 1/2
J_z	3/2	1/2	1/2	− 1/2	− 1/2	− 3/2

For each of the basis states of the non-perturbed system, the table indicates the wave function defined by (32) with the eigenvalues of the three operators $\mathbf{L}_z$, $\mathbf{S}_z$, and $\mathbf{J}_z$, given to within a factor $\hbar$.

Once we have arranged this table we examine the highest value of J_z, here being $\frac{3}{2}\hbar$ which can only be related to a value of J^2 corresponding to $j' = \frac{3}{2}$. We thus get the highest value of J^2 to which four values of J_z $(\frac{3}{2}, \frac{1}{2}, -\frac{1}{2}, -\frac{3}{2})$ correspond to within a factor $\hbar$. This decomposition does not exhaust the table since the values of $\pm\frac{1}{2}\hbar$ are duplicated. We can thus extend the preceding reasoning by saying that these values can only be associated with a value of J^2 for which $j'' = \frac{1}{2}$.

Calculation of the wave functions. This second part allows us to refine the preceding intuitive reasoning. To simplify the calculations we will first indicate several fundamental properties.†

(*a*) The linear space constructed from the basis functions $\theta_1, \ldots, \theta_i$, is unchanged by the action of the operator $\mathbf{J}^2$.

This is due to the fact that the three operators $\mathbf{J}_r$ transform each of the basis functions θ_r into a function belonging to the defined space.

(*b*) The linear sub-space defined by the basis functions belonging to the same eigenvalues of $\mathbf{J}_z$ is not altered by $\mathbf{J}^2$.

---

† The linear space defined by the functions $\theta_1, \ldots, \theta_n$ is spanned by the set of linear combinations $\sum_1^n c_r\theta_r$ with constant coefficients c_r. A sub-space is defined in the same way starting from a limited number of functions θ_r.

This is a consequence of the commutability of $\mathbf{J}^2$ and $\mathbf{J}_z$. As a result, if we take one of the basis functions θ_r of this sub-space, we can write on the one hand $\mathbf{J}_z\theta_r = \lambda\theta_r$, and on the other hand:

$$\mathbf{J}_z(\mathbf{J}^2\theta_r) = \mathbf{J}^2(\mathbf{J}_z\theta_r) = \lambda\mathbf{J}^2\theta_r.$$

$\mathbf{J}^2\theta_r$ corresponds to the same eigenvalue λ of $\mathbf{J}_z$ as θ_r and belongs to the subspace defined by this eigenvalue.

(*c*) If the function χ_a is an eigenfunction of $\mathbf{J}^2$, and if χ_b is orthogonal to χ_a, then the function $\mathbf{J}^2\chi_b$ is also orthogonal to χ_a.

This property is related to the hermitian character of $\mathbf{J}^2$, which allows us to write:

$$\sum_{\sigma}\int \chi_b^*(\mathbf{J}^2\chi_a)d\tau = \sum_{\sigma}\int (\mathbf{J}^2\chi_b)^*\chi_a d\tau.$$

The integral on the left is zero, since $\mathbf{J}^2\chi_a = \lambda\chi_a$, and since the functions χ_a and χ_b are orthogonal by hypothesis and thus the property is established.

Given these facts, examination of the preceding table leads to the following conclusions.

(1) θ_1 is the unique eigenfunction of $\mathbf{J}_z$ belonging to the eigenvalue $3\hbar/2$. From (*b*) the function $\mathbf{J}^2\theta_1$ must give θ_1 to within a factor, so that θ_1 can be simultaneously an eigenfunction of $\mathbf{J}_z$ and $\mathbf{J}^2$ and there can be no higher eigenvalue. In addition, the reasoning developed in Appendix VI for $\mathbf{L}_z$ and $\mathbf{L}^2$ shows that the eigenvalue of $\mathbf{J}^2$ is in the form $j'(j'+1)\hbar^2$ with the maximum eigenvalue of $\mathbf{J}_z$ equal to $j'\hbar$ which means in this case $j' = \frac{3}{2}$.

(2) The different eigenfunctions of $\mathbf{J}_z$ belonging to the same eigenvalue of $\mathbf{J}^2$ and to eigenvalues of $\mathbf{J}_z$ equal respectively to $\hbar/2$, $-\hbar/2$, and $-3\hbar/2$ are gotten from θ_1 by the action of the operator $J_x - iJ_y$ by using equation (Appendix IV-10):

$$(\mathbf{J}_x - i\mathbf{J}_y)\chi_m = \hbar a_{m-1,m}\chi_{m-1}$$

with $$a_{m-1,m} = \sqrt{(j'+m)(j'-m+1)}.$$

Thus we get the desired four eigenfunctions ($j' = \frac{3}{2}$):

$$J_z = 3\hbar/2: \qquad \chi'_{3/2} = \theta_1$$

$$J_z = \hbar/2: \qquad \chi'_{1/2} = \frac{1}{\sqrt{3}}(\mathbf{J}_x - i\mathbf{J}_y)\chi'_{3/2} = \frac{1}{\sqrt{3}}(\theta_2 + \sqrt{2}\theta_3)$$

$$J_z = -\hbar/2: \qquad \chi'_{-1/2} = \frac{1}{2}(\mathbf{J}_x - i\mathbf{J}_y)\chi'_{1/2} = \frac{1}{\sqrt{3}}(\theta_5 + \sqrt{2}\theta_4)$$

$$J_z = -3\hbar/2: \qquad \chi'_{-3/2} = \frac{1}{\sqrt{3}}(\mathbf{J}_x - i\mathbf{J}_y)\chi'_{-1/2} = \theta_6.$$

(3) We now consider the sub-space of eigenfunctions belonging to the eigenvalue $\hbar/2$ of $\mathbf{J}_z$. We have found an eigenfunction of $\mathbf{J}^2$ belonging to this sub-space, namely, $\chi'_{1/2} = (1/\sqrt{3})(\theta_2 + \sqrt{2}\theta_3)$. The function in this sub-space, which is orthogonal, is of the form $c_1\theta_2 + c_2\theta_3$. The orthogonality and normalization conditions are satisfied by taking $\chi''_{1/2} = (1/\sqrt{3})(\sqrt{2}\theta_2 - \theta_3)$. Property (*c*) shows that $\chi''_{1/2}$ should be an eigenfunction of $\mathbf{J}^2$. The corresponding eigenvalue is of the form $j''(j''+1)\hbar^2$ with $j'' = \frac{1}{2}$ since the eigenvalue of $\mathbf{J}_z$ is $\hbar/2$ and since it cannot be higher. Then again, the second function belonging to the same eigenvalue of $\mathbf{J}^2$ is gotten from the first by the action of the operator $\mathbf{J}_x - i\mathbf{J}_y$, the coefficient $a_{m-1,m}$ this time being equal to unity. It becomes:

$$\chi''_{-1/2} = (\mathbf{J}_x - i\mathbf{J}_y)\chi''_{1/2} = \frac{1}{\sqrt{3}}(\theta_4 - \sqrt{2}\theta_5)$$

and we can verify as above that $\chi'_{-1/2}$ and $\chi''_{-1/2}$ are two orthogonal functions.

In conclusion, we have succeeded in resolving the initial six-dimensional space, which is generated by the problem, and which is necessarily the same as that obtained by the classical perturbation calculation. This point can also be verified directly by noting that the product of the operators appearing in the perturbation expression can be put in the equivalent form:

$$\mathbf{L}_x\mathbf{S}_x + \mathbf{L}_y\mathbf{S}_y + \mathbf{L}_z\mathbf{S}_z = \frac{1}{2}(\mathbf{J}^2 - \mathbf{L}^2 - \mathbf{S}^2).$$

All the functions of this space are eigenfunctions of $\mathbf{L}^2$ with eigenvalue $2\hbar^2$, and of $\mathbf{S}^2$ with eigenvalue $3\hbar^2/4$. As a result, for all the functions of the first space which are eigenfunctions of $\mathbf{J}^2$ belonging to the eigenvalue $15\hbar^2/4$, we have:

$$\frac{1}{2}(\mathbf{J}^2 - \mathbf{L}^2 - \mathbf{S}^2)\chi' = \frac{\hbar^2}{2}\chi'$$

and likewise for a function in the second space:

$$\frac{1}{2}(\mathbf{J}^2 - \mathbf{L}^2 - \mathbf{S}^2)\chi'' = -\hbar^2\chi''.$$

Thus it appears that the two methods should lead to the same result. The expression for the correction corresponding to the first order perturbation can be found immediately and the values derived have already been given.

Exercises. The preceding calculations supply the material for some simple exercises which will allow the reader to become familiar with the use of operators.

(1) Write the anticommutation relations for $\mathbf{L}_r$ and $\mathbf{S}_r$. From this find the relations for $\mathbf{J}_r$. Show that the operators $\mathbf{J}_r$ commute with the operator $\mathbf{V}$ from the LS coupling.
(2) Show that in the previous paragraph there is no eigenvalue of $\mathbf{J}_z$ having a value greater than $3\hbar/2$ (apply $\mathbf{J}_x + i\mathbf{J}_y$ to θ_1).
(3) Verify directly that θ_1 is an eigenfunction of $\mathbf{J}^2$ by forming this operator from $\mathbf{J}_x \pm i\mathbf{J}_y$ and $\mathbf{J}_z$ (see Appendix VI for the analogous treatment of $\mathbf{L}^2$).
(4) Show by use of the operators $\mathbf{J}_x \pm i\mathbf{J}_y$ that there is no other eigenfunction of $\mathbf{J}^2$ besides $\chi''_{1/2}$ and $\chi'_{-1/2}$ which belongs to the same eigenvalue of this operator.

The vector model. We analyzed LS coupling by taking the example of a p level, but this is easily transposed to the case of any value of the quantum number l. The method remains the same and we find that the initial level of multiplicity $2(2l + 1)$ separates into two, corresponding to values j' and j'' of the quantum number j which

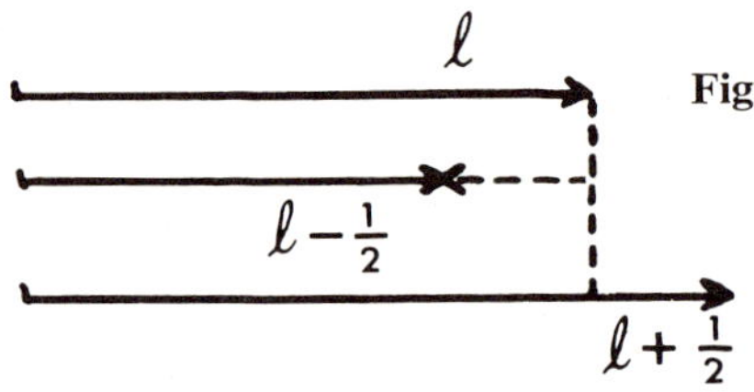

Figure 25. Application of the vector model of LS coupling for a value equal to l for the azimuthal quantum number.

defines $\mathbf{J}^2$. The highest of these, namely, j' is also gotten immediately by taking the maximum eigenvalue of $\mathbf{J}_z$, which corresponds to the unique eigenfunction $\psi_{n,l,m}\,\alpha$ with $m = l$, and which makes $j' = l + \frac{1}{2}$.

Beginning with this first function, we can—as in the example for a p level—derive a set of other eigenfunctions belonging to the same eigenvalue $j'(j' + 1)\hbar^2$ of $\mathbf{J}^2$, and to the eigenvalues of $\mathbf{J}_z$ between $j'\hbar$ and $-j'\hbar$, giving in all, $2j' + 1 = 2l + 3$ functions. We do not exhaust the initial space, and the reasoning developed in the example of a p level shows that there is a second value of $j, j'' = 1 - \frac{1}{2}$, which leads to the introduction of $2j'' + 1 = 2l - 1$ eigenfunctions of J_z. With the first $2l + 3$ this gives the multiplicity $2(2l + 1)$ of the initial space.

The possibility of one such resolution can be examined by considering the *vector model* of the atom, which is in fact stated in an intuitive language from a general theorem of the theory of three-dimensional rotation groups. The law states: the possible values of j are obtained by combining two vectors of lengths l and s (equal to $\frac{1}{2}$) respectively, yielding the list $l + s, l + s - 1, \ldots, |l - s|$.

In the case under discussion ($s = \frac{1}{2}$), we get the two values $l + \frac{1}{2}$ and $l - \frac{1}{2}$ if l is non-zero (fig. 25) and only $j = \frac{1}{2}$, when dealing with an s level ($l = 0$).

The BWK Method 25

(L. Brillouin, G. Wentzel, and H. A. Kramers)

This method allows us to get an asymptotic solution of the Schrödinger equation in the one-dimensional case. We write that equation in the form:

(33)

$$\psi'' + \frac{2m}{\hbar^2}(E - V)\psi = \psi'' + \frac{1}{\hbar^2}u(x)\,\psi = 0 \quad \text{with} \quad u = 2m(E - V).$$

The form of this equation suggests that we take ψ in the form:†

$$\psi = \exp\left(\int_{x_0}^{x} W dx\right), \tag{34}$$

from which we get the equation satisfied by W:

$$W' + W^2 + \frac{1}{\hbar^2}u(x) = 0. \tag{35}$$

We know that wave mechanics must correspond to classical mechanics if Planck's constant is equal to zero, and this suggests that we can seek solutions of (35) in a power series on the successive powers of $\hbar$, the approximation consisting of taking the first terms of this series. The form of (35) requires us to take the first term in $1/\hbar$, and thus we write:

$$W = \frac{1}{\hbar}W_0 + W_1 \ldots \tag{36}$$

By limiting ourselves to the first two terms, equation (35) can be written:

$$\frac{1}{\hbar^2}(W_0^2 + u(x)) + \frac{1}{\hbar}(W_0' + 2W_0W_1) \simeq 0$$

which implies:

$$\begin{aligned} W_0^2 + u(x) &= 0 \\ W_0' + 2W_0W_1 &= 0. \end{aligned} \tag{37}$$

† In each case we establish the origin x_0 at some arbitrary position.

We will treat the calculation in detail in the case where the potential curve $V(x)$ takes the form of a potential valley (fig. 26). We assume first that E is known, and we designate the points where $V(x) = E$ as x_a and x_b. The axis is thus divided into three regions in a manner analogous to the treatment of the potential well which we treated in section 9. The form of the solutions in each of the three regions, called respectively (1), (2), and (3), is as follows:
$x < x_a$ (region 1): The equations (37) yield:

$$W_0^{(1)} = \pm \sqrt{(-u_1)} \qquad \text{or} \qquad -u_1 > 0$$

and

$$W_1^{(1)} = -\frac{1}{2}\frac{W_0^{(1)\prime}}{W_0^{(1)}} = -\frac{1}{4}\frac{(-u_1')}{(-u_1)}.$$

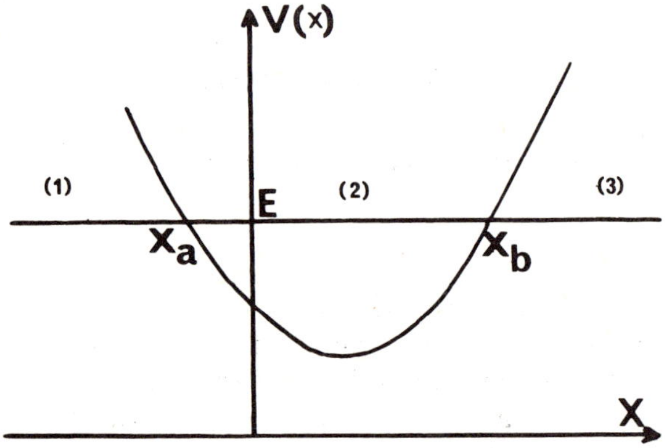

Figure 26. Energy diagram of a potential well indicating the three regions considered in the BWK method for energy E.

The wave function (34) is obtained by integration with arbitrary origin which we take at point x_a and an integration constant γ_1. We then have:

$$\text{Log}\ \psi_1 = \int_{x_0}^{x} W^{(1)} dx$$

(38)

$$= \pm \frac{1}{\hbar}\int_{x_a}^{x} \sqrt{2m(V-E)}dx - \frac{1}{4}\text{Log}\ [2m(V-E)] + \gamma_1.$$

We generally set aside the question of the sign of the first term, which is very simple here since it must be taken so that $\psi_1 \to 0$ as $x \to -\infty$.

$x_a < x < x_b$ (region 2): The preceding calculation leads to complex W_0 just as in section 9, since $u_2 > 0$. We are thus lead to take:

$$W_0^{(2)} = \pm i\sqrt{u_2}$$

$$W_1^{(2)} = -\frac{1}{2}\frac{W_0^{(2)'}}{W_0^{(2)}} = -\frac{1}{4}\frac{u_2'}{u_2}$$

so that after integration and with the same origin as in the first case:

$$\begin{aligned} \text{Log}\, \psi_2 &= \int_{x_0}^{x} W^{(2)} dx \\ &= \pm \frac{i}{\hbar}\int_{x_a}^{x} \sqrt{2m(E - V)}dx - \frac{1}{4}\,\text{Log}\,[2m(E - V)] + \gamma_2. \end{aligned} \tag{39}$$

$x_b < x$ (region 3): The treatment is the same as for region 1 and we have:

$$W_0^{(3)} = \pm \sqrt{(-u_3)}$$

$$W_1^{(3)} = -\frac{1}{2}\frac{W_0^{(3)'}}{W_0^{(3)}} = -\frac{1}{4}\frac{(-u_3')}{(-u_3)}$$

$$\begin{aligned} \text{Log}\, \psi_3 &= \int_{x_0}^{x} W^{(3)} dx \\ &= \pm \frac{1}{\hbar}\int_{x_b}^{x} \sqrt{2m(V - E)}dx - \frac{1}{4}\,\text{Log}\,[2m(V - E)] + \gamma_3, \end{aligned} \tag{40}$$

this time taking x_b as the limit of integration.

If we want to continue following the method for the potential well, as in section 9, we have to write the boundary conditions ($\psi_1 = \psi_2$; $d\psi_1/dx = d\psi_2/dx$) at point x_a, separating the first two regions, and likewise at x_b. However, there is a difference between the case in section 9 and this one, because in this instance we not only have a discontinuity of the potential energy but we have a change in the analytic expression of the solution, due to the method of approximation. The problem remains the same in spite of this, use of the function W defined by (34) permitting only the replacement of the conditions described previously by the continuity of W and of ψ at points x_a and x_b.

The method is thus completely natural, but its application is tricky because the approximation used ceases to be valid precisely at the boundaries where the wave amplitude becomes infinite (equations 38, 39, and 40) at the point where we ought not to have any irregularity in the exact solution. The original part of the BWK method consists in the formulation of the boundary conditions of the solution which, as in the case of section 9, led to the energy quantization.

We present the principle of the technique in Appendix VII, here limiting ourselves to the result:

$$2\int_{x_a}^{x_b}\sqrt{2m(E-V)}dx=\left(r+\frac{1}{2}\right)h \tag{41}$$

r being a positive integer or zero.

Relation to classical mechanics. Our preceding study has lead to two solutions for the wave function ψ_2 in region 2, which will be the region of classical motion of the particle of energy E. For the sake of convenience, we will confine ourselves to this region in examining the analogies between the wave picture and that of classical mechanics. We rewrite equation (39) in the form:

$$\psi_2'=C[2m(E-V)]^{-1/4}\cdot\exp\left[\frac{i}{\hbar}\int_{x_a}^{x}\sqrt{2m(E-V)}dx+\gamma_2'\right]$$

$$\psi_2''=C[2m(E-V)^{-1/4}]\cdot\exp\left[-\frac{i}{\hbar}\int_{x_a}^{x}\sqrt{2m(E-V)}dx+\gamma_2''\right].$$

We recognize in these two expressions the form of the Maupertuis action (Appendix I-17):

$$S=\int_{x_a}^{x}mv dx=\pm\int_{x_a}^{x}\sqrt{2m(E-V)}dx.$$

The other part, the amplitude of these wave functions, is expressed as a function of the classical speed of the particle by using the principle of conservation of energy:

$$E-V=\frac{1}{2}mv^2 \quad \text{where} \quad \sqrt{2m(E-V)}=mv.$$

The functions ψ_2' and ψ_2'' can thus be given in the equivalent form:†

$$\psi_2'=\frac{C}{\sqrt{v}}\exp\left(\frac{iS'}{\hbar}\right);\qquad \psi_2''=\frac{C}{\sqrt{v}}\exp\left(\frac{iS''}{\hbar}\right),$$

where:

$$S'=\int_{x_a}^{x}\sqrt{2m(E-V)}dx,\qquad S''=-\int_{x_a}^{x}\sqrt{2m(E-V)}dx.$$

The exponential factor is none other than that used to introduce the analogy between mechanics and wave propagation, and

† The normalization condition in region (2) imposes the same value of the modulus C for ψ_2' and ψ_2''. The phases are arbitrary and we make the simplest choice, $\gamma_2'=\gamma_2''=0$.

the expression for the wave amplitude can equally well be related to those classical considerations. We can show that in such motion, the probability of finding a particle on the element dx of the trajectory, is proportional to the time dt of passage of the particle over dx, so that:

$$dp \sim dt = \frac{dt}{dx}dx = \frac{dx}{v} \sim |\psi_2^2|\, dx$$

and we again find the expression calculated by beginning from ψ_2' or ψ_2''.

In conclusion, the two solutions obtained by the BWK method in region (2) are exactly the wave functions which classical mechanics (Appendix I) permits us to write for the two inverse motions on $x_a x_b$, where evolution over time is gotten by replacing S by $S - Et$. From this point of view, we see the attraction of this method as a means for demonstrating the relationship between classical and wave mechanics.

When classical mechanics is applied to the motion of the body, that is, when the associated wave is either ψ_2' or ψ_2'', wave mechanics shows that the general solution of the wave equation is given by the linear combination:

$$\psi_2 = c_2'\psi_2' + c_2''\psi_2'' .$$

In the state represented by ψ_2, the probability of finding the particle moving in one or the other sense ought to be the same, which shows $|c_2'| = |c_2''|$. In addition, since the wave equation is real, the solutions are real, having the form:

$$\psi_2 = \frac{C}{\sqrt{v}} \cos\left(\frac{1}{\hbar}\int_{x_a}^{x} \sqrt{2m(E - V)}dx + \gamma\right).$$

This amplitude can vanish between the limits x_a and x_b, and we thus find the classical description of a standing wave. From this point of view, the quantum condition (41) can be written:

$$(42) \qquad 2\int_{x_a}^{x_b} \frac{mv}{h} dx = r + \frac{1}{2} \qquad \text{or} \qquad 2\int_{x_a}^{x_b} \frac{dx}{\lambda} = r + \frac{1}{2}$$

by introducing the de Broglie wavelength. This condition is, to within a minor factor, identical to the elementary relation which allows us to write for a system of stationary waves on a string of length L, with a constant wavelength λ: $L = k\lambda/2$, and which is generalized to:

$$2\int_{x_a}^{x_b} \frac{dx}{\lambda} = k \qquad \text{or} \qquad 2\int_{x_a}^{x_b} mv dx = kh.$$

This condition can again be put in the following form by carrying out the integration on a complete period of the motion:

$$\oint p dx = kh.$$

This is precisely the quantum condition from the old quantum theory.

We note that the final integral cannot be written by replacing the momentum p by $\sqrt{2m(V - E)}$, whose sign remains constant, which leads to a zero value for the cyclic integral.

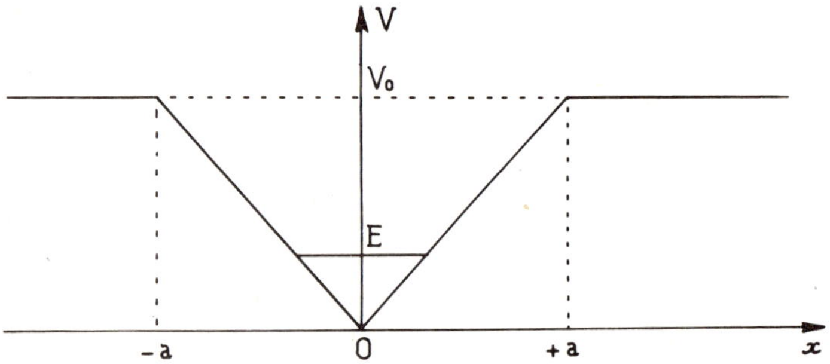

Figure 27. Triangular potential well. The origin of the energy is taken at the bottom of the well.

Exercise 5. Apply the BWK method to find the energy levels of a system made up of a particle of mass m moving on the axis in the potential well represented by figure 27. Do this by taking $ma^2 V_0/h^2 = 2$. To what value of V_0 does this condition correspond, assuming we are dealing with an electron and taking $a = 1$ Å?

PART TWO

THE TREATMENT OF MANY-ELECTRON SYSTEMS IN THE ONE-ELECTRON APPROXIMATION

6 THE MANY-ELECTRON ATOM

Generalities: The Pauli Principle 26

We turn now to real systems of greater complexity than the hydrogen atom, in general, systems containing several electrons. Their rigorous treatment requires a generalization of the Schrödinger equation which we have so far written for a single electron only, reserving full treatment for Part III. Here we limit ourselves to the first approximation methods—ignoring the interaction between the electrons, which can be treated separately—so that the problem reduces to the one-electron problem. In the second approximation we will take into account the electron interactions, using the methods we have studied.

We begin by examining the problem of the many-electron atom, and will confine ourselves to this example since our study of the one-electron atom has already prepared us for this. By following the general procedure outlined above, we can immediately, in the first approximation, put together the results obtained for the various electrons of the atom under discussion, each of which occupies a state defined by the four quantum numbers n, l, m, and σ. The problem, then, is to understand the way in which the electrons are distributed among the different possible states, and especially, to determine the distribution corresponding to the minimum energy for the system and thus to its state of greatest stability. In the study of the one-electron atom we expressed the state energy as a function of the quantum number n, the lowest state corresponding to $n = 1$. If the electrons were independent we know that they would all occupy this $1s$ state. However, in reality, the occupation of the possible levels by the electrons is governed by a new postulate, the *Pauli principle*, which we will state generally in Part III. Here we limit ourselves to the particular form of the Pauli principle best suited to the problem at hand. *Two electrons cannot occupy the same state defined by a system of four quantum numbers.*

We present our first example of this rule by writing the most stable electronic structure of the lightest atoms in the order of the number of their electrons, indicating how they are distributed in the different states by means of index numbers. The possibility of having, for example, two electrons in the $1s$ state is tied to the existence of the spin quantum number which allows us to have distinct states defined respectively by the two sets of n, l, m, and σ $(1,0,0,\frac{1}{2})$ and $(1,0,0,-\frac{1}{2})$. We get:

$$\begin{aligned} &\mathrm{H} : (1s)^1 \\ &\mathrm{He} : (1s)^2 \\ &\mathrm{Li} : (1s)^2\,(2s,\ 2p)^1 \\ &\mathrm{Be} : (1s)^2\,(2s,\ 2p)^2. \end{aligned}$$

We see that the Pauli principle alone is insufficient to allow us to determine, beginning with the third element if the electron is situated in a $2s$ or $2p$ state, this being determined only when we take the electronic interactions into account. Meanwhile, it is significant that, even ignoring the effect of these interactions, the Pauli principle—by indicating a distribution of the electrons on different levels—leads to their spatial separation. This diminishes the importance of their interactions and facilitates the following approach. To extend the treatment we need to give a basic approach for including the interactions, and that can be presented in sufficiently rigorous fashion only in the most simple case—helium —which we now treat in some detail.

27 The Helium Atom (One-Electron Approximations)

By taking the two possible spin values into account, we have shown that the Pauli principle permits the $1s$ level to be occupied by two electrons, and this corresponds to the ground state which will be our concern here. We propose to study the behavior of one of these electrons, designated by the index l, for which we write the Schrödinger equation:

$$\text{(1)} \qquad \mathbf{H}\psi = -\frac{\hbar^2}{2m_0}\Delta\psi - \frac{Ze^2}{4\pi\varepsilon_0 r}\psi + \mathrm{V}(r)\psi = \mathrm{E}\psi.$$

We reserve the letter Z to designate the atomic number, which we take as equal to two for the numerical solutions. The function V(r), which we will determine, represents the interaction with the second electron.

We begin with the fact that the electronic distribution of electron number 2 has a spherical symmetry and we can imagine that it forms a kind of *electrostatic screen* on the nucleus, which then behaves as if it has a charge $Z^{\star}e$ which is less than the real charge Ze. We are then lead, *a priori*, to take the same form for the orbital as in the one-electron case, in this instance a $1s$ state:

$$\text{(2)} \qquad \psi = \mathrm{N}e^{-\frac{\alpha r}{2}} \qquad \text{with} \qquad \frac{8\pi\mathrm{N}^2}{\alpha^3} = 1.$$

The parameter α which we are going to find is tied to the *apparent atomic number* $Z^\star$ and is therefore different from that obtained when we ignore the electron-electron interaction.

Our hypothesis about the form of ψ allows us to begin calculating the function $V(r)$ from the electronic distribution of the second electron, which must have the same wave function (2) as the first electron. The rather artificial method we follow imposes different roles on the two electrons in the system, so that it is useful to distinguish them by using α' as the parameter of (2) relative to the second electron. The equality $\alpha' = \alpha$ arises only in the final result of the calculations.

The function $V(r)$ can be calculated by assuming that the action of the second electron is equivalent to that of a distribution of electron charge with the density:

$$\frac{dq}{d\tau} = -e \mid \psi^2 \mid = -eN^2.\exp(-\alpha' r).$$

The electrical potential at any point can be derived by integrating Poisson's equation. We can proceed to a direct calculation by first evaluating the interior charge for a sphere of radius r using the relation:

$$q(r) = -4\pi e N^2 \int_0^r \exp(-\alpha' \rho).\rho^2 d\rho.$$

This is a eulerian integral which is easily evaluated and leads to:

$$\text{(3)} \qquad q(r) = -e\left[1 - \left(1 + \alpha' r + \frac{1}{2}\alpha'^2 r^2\right).\exp(-\alpha' r)\right].$$

We can satisfy ourselves that we get the electron charge for large values of r and that q goes to zero as r does.

The electrical field f generated at a distance r from the origin by such a charge distribution is calculated as if the charge $q(r)$ is concentrated at the origin, and leads to:

$$f = -\frac{e}{4\pi\varepsilon_0 r^2}\left[1 - \left(1 + \alpha' r + \frac{1}{2}\alpha'^2 r^2\right).\exp(-\alpha' r)\right].$$

The electrical potential u at this point is calculated from the radial field f ($f = -du/dr$) and we can see that we get:

$$u = -\frac{e}{4\pi\varepsilon_0 r}\left[1 - \left(1 + \frac{\alpha' r}{2}\right).\exp(-\alpha' r)\right]$$

and from this the potential energy of the electron interaction:

$$\mathrm{V} = -eu = \frac{e^2}{4\pi\varepsilon_0 r}\left[1 - \left(1 + \frac{\alpha' r}{2}\right).\exp\left(-\alpha' r\right)\right].$$

It is evident that a function ψ of the form (2) could not be a rigorous solution of equation (1) with such a potential energy function. The alternative then, is an approximate solution, and we saw earlier (§19) that when we have a family of functions dependent on a parameter, the member of this family which approaches the best state function must satisfy the extremum condition on the mean energy:

$$\mathrm{E} = \int \psi^* \mathbf{H} \psi d\tau,$$

the functions ψ being assumed normalized.

The value of E is the sum of three terms corresponding to the three terms appearing in expression (1) for $\mathbf{H}$, and in their evaluation we use the eulerian integrals to facilitate the calculation. The first of these terms represents the mean kinetic energy in the state ψ (4-25):

$$\langle \mathrm{T} \rangle = \int \psi^*\left(-\frac{\hbar^2}{2m_0}\Delta\psi\right)d\tau = -\frac{\hbar^2}{2m_0}\int \psi^*\left(\frac{d^2\psi}{dr^2} + \frac{2}{r}\frac{d\psi}{dr}\right)\cdot 4\pi r^2 dr$$

taking the normalization as in (2) we get:

$$\langle \mathrm{T} \rangle = \frac{\alpha^2\hbar^2}{8m_0}.$$

The second term represents the mean potential energy of the interaction of electron 1 and the nucleus, and we get:

$$\langle \mathrm{V}_1 \rangle = -\frac{Ze^2}{4\pi\varepsilon_0}\int \frac{|\psi^2|}{r}d\tau = -\frac{Ze^2}{4\pi\varepsilon_0}\cdot 4\pi \mathrm{N}^2 \int_0^\infty e^{-\alpha r} r dr = -\frac{\alpha Z e^2}{8\pi\varepsilon_0}.$$

The final term is the mean energy of the interaction of the two electrons, giving:

$$\langle \mathrm{V}_{1,2} \rangle = \int_0^\infty \mathrm{V}|\psi^2|\cdot 4\pi r^2\, dr$$

so that by distinguishing α', which appears in the potential energy $\mathrm{V}(r)$, from α, which appears in the function ψ:

$$(3') \quad \langle \mathrm{V}_{1,2} \rangle = \frac{e^2}{8\pi\varepsilon_0}\left(\alpha - \frac{\alpha^3}{(\alpha+\alpha')^2} - \frac{\alpha^3\alpha'}{(\alpha+\alpha')^3}\right) = \overline{\mathrm{V}}_{1,2}(\alpha, \alpha').$$

Finally, by combining the three contributions to the mean energy, we get:

$$\mathrm{E}(\alpha) = \frac{\hbar^2\alpha^2}{8m_0} - \frac{Ze^2\alpha}{8\pi\varepsilon_0} + \overline{\mathrm{V}}_{1,2}(\alpha, \alpha').$$

The value of α is determined by the extremum condition:

$$\frac{dE}{d\alpha} = \frac{\hbar^2\alpha}{4m_0} - \frac{Ze^2}{8\pi\varepsilon_0} + \frac{\partial \overline{V_{1,2}}}{\partial\alpha} = 0.$$

In the last term of the sum† it is necessary to take $\alpha' = \alpha$, which leads to:

$$\left(\frac{\partial \overline{V_{1,2}}}{\partial\alpha}\right)_{\alpha'=\alpha} = \frac{5}{16}\cdot\frac{e^2}{8\pi\varepsilon_0}.$$

We finally get the value of α:

$$\alpha = \frac{2e^2m_0}{4\pi\varepsilon_0\hbar^2}\left(Z - \frac{5}{16}\right)$$

which we can conveniently compare to that obtained (§17) in the case of a one-electron atom in the 1s state. The expressions are the same as long as we introduce the *apparent atomic number*:

$$Z^\star = Z - s \qquad \text{with} \qquad s = 5/16 \simeq 0.31.$$

The correction s, which is independent of Z, represents the *screening effect* of the second electron on the first. By putting this value of α in the expression for E, we see in each of the three terms of the sum the factor E_0 ($E_0 = m_0 e^4/2\hbar^2[4\pi\varepsilon_0]^2$), the ionization energy of the hydrogen atom in the 1s state. We get:

$$E = E_0(Z^{\star 2} - 2ZZ^\star + 4sZ^\star) = -E_0(Z^{\star 2} - 2sZ^\star).$$

We now evaluate the total energy of the helium atom, which is not twice the energy above, since this would include twice the electronic interaction energy $\overline{V_{1,2}} = 4sZ^\star E_0$. We get:

$$E(He) = 2E - 4sZ^\star E_0 = -2Z^{\star 2}E_0.$$

We thus see that $-Z^{\star 2}E_0$, which is the calculated energy for a one-electron atom but which includes the apparent atomic number, represents the *contribution of an electron to the total energy* of the atom. To make the comparison with the experimental values, we must take:

$$Z = 2 \quad \text{hence} \quad Z^\star = 1.69 \quad \text{and} \quad E(He) = -5.72E_0.$$

From it we deduce the ionization energy I(He) of the helium atom by comparing it to the ion energy $E(He)^+ = -2^2 E_0$ which gives:

$$I(He) = E(He^+) - E(He) = 1.72E_0,$$

† The derivation of (3′) must be made before setting $\alpha' = \alpha$.

instead of the measured value of 1.81 E_0.† It is thus possible to get fairly satisfactory approximation by preserving the form of the one-electron function ψ and conveniently adjusting the screening coefficients.

28 The Slater Orbitals

The description of the helium atom using the one-electron orbitals has been extended to the case of an atom with an arbitrary number of electrons by means of semi-empirical considerations due in the most part to J. C. Slater. This involves a first approximation which is very useful in quantum chemistry. The distribution of each of the atomic electrons is calculated starting from an orbital of the form:

$$\psi_{n,l,m} = R_n(r) \,.\, Y_l^m(\theta, \chi). \tag{4}$$

The radial function $R_n(r)$ is here taken as independent of the quantum number l, in contrast to what happens in the one-electron case, and we take:

$$R_n(r) = N r^{n-1} e^{-\frac{\alpha r}{2}}. \tag{5}$$

The function ψ thus formed is the solution of a Schrödinger equation of the form:

$$-\frac{\hbar^2}{2m_0}\Delta\psi + V(r)\psi = E\psi$$

under the condition that the potential energy takes a suitable form which we must specify. Expression of the laplacian in spherical coordinates (Appendix IV-16) allows us to write:

$$\Delta\psi = \frac{1}{r^2}\frac{d}{dr}\left(r^2\frac{dR_n}{dr}\right)\cdot Y_l^m + R_n \cdot \frac{1}{r^2}\Lambda Y_l^m.$$

The spherical function Y_l^m is an eigenfunction of the operator Λ so that:

$$\Lambda Y_l^m = -\,l(l+1)Y_l^m.$$

The operations when worked out taking $R_n(r)$ in the form (5) lead to the following form for the Schrödinger equation:

$$-\frac{\hbar^2}{2m_0}\left[\frac{n(n-1)-l(l+1)}{r^2} - \frac{\alpha n}{r} + \frac{\alpha^2}{4}\right]\psi + V(r)\,\psi = E\psi.$$

† We will compare this with the self-consistent field calculation in (§46).

We obtain an identity by taking:

$$(6)\qquad E = -\frac{\hbar^2\alpha^2}{8m_0}$$
$$V = -\frac{\alpha\hbar^2 n}{2m_0 r} + \frac{(n+l)(n-l-1)}{r^2}\cdot\frac{\hbar^2}{2m_0}.$$

For large values of r the first term of V becomes dominant, with the result that ψ is the limiting solution for the state function with the electron in the coulomb potential of a central charge which we set equal to $Z^\star e$, and we can find α by the condition:

$$\frac{\alpha\hbar^2 n}{2m_0} = \frac{Z^\star e^2}{4\pi\varepsilon_0}$$

where:

$$(7)\qquad \alpha = \frac{2Z^\star m_0 e^2}{4\pi\varepsilon_0\hbar^2 n} = \frac{2Z^\star}{na_0} \quad \text{and} \quad E = -\frac{Z^{\star 2}}{n^2}E_0$$

by introducing the Bohr radius (§17) and the ionization energy E_0 of the hydrogen atom.

The preceding calculation establishes the advantage of the Slater orbitals which give an acceptable approximation to the atomic orbitals when the electron is far from the nucleus, that is, in the domain occurring in problems related to the chemical bond. On the other hand, the meaning of E is much clearer when it is determined under the condition of satisfying the Schrödinger equation in the limiting region. We can conceive of the possibility of having two different systems of $Z^\star$ according to the screening coefficients: one permitting the direct evaluation of the ionization energy of the electron treated by a hydrogenic formula $-Z^{\star 2}E_0/n^2$; the other describing the behavior of the orbitals.†

We can initially assume that the interaction of two electrons, 1 and 2, in the states defined respectively by the orbitals ψ_1 and ψ_2, with quantum numbers n_1 and n_2 does not affect those orbitals in the same way if the quantum numbers are different. In assuming, for example, $n_1 < n_2$, the explicit discussion of the helium atom shows that the presence of electron 1 must have a noticeable screening effect on the action of the nucleus on electron 2, mainly in the exterior part of the orbital, and conversely, the presence of 2 must slightly modify the behavior of the orbital ψ_1. We introduce the Zener method, which tries to find an acceptable form of the

† A very complete discussion of this point is found in: J. C. Slater, *Quantum Theory of Atomic Structure* (New York: McGraw-Hill, 1960), Vol. I.

orbitals in the outer regions of the atom, bringing in the two-electron interaction only between those electrons which are on the average farthest removed from the nucleus and thus with the largest principal quantum numbers, or, if the quantum number is the same for these two electrons, those which have the highest l value. We now indicate the Slater method for calculating the apparent atomic number $Z^\star$ of an atomic electron.

For this purpose, in accordance with the previous qualitative treatment, the atomic electrons are distributed in groups as follows:

(8) $$| \, 1s \, | \, 2s, 2p \, | \, 3s, 3p \, | \, 3d \, | \, 4s, 4p \, | \, 4d \, | \, 4f \, | \, \ldots$$

The apparent atomic number $Z^\star$ of an electron from the group of index i is derived from the real atomic number of the atom Z by:

(9) $$Z_i^\star = Z - \sum_j s_{ji}.$$

The screening coefficient s_{ji} which expresses the interaction of the electron under consideration with an electron of group j is taken as follows:

$j > i$ $\quad s_{ji} = 0$

$j = i$ $\quad s_{ii} = 0.35$ except for the $1s$ state (§27) where we take $s_{11} = 0.31$.

$j < i$ (*a*) If electron i is in an s or p state we take:

$s_{ji} = 0.85$ for all electrons with quantum number $n_i = 1$,

$s_{ji} = 1$ for all electrons in deeper levels.

(*b*) If electron i is in a d or f state we take all $s_{ji} = 1$.

In using this system Slater replaced the quantum number n which appears in (5) and (6) by an apparent quantum number $n^\star$ with the following correspondence:

(9)

n : 1	2	3	4	5	6
$n^\star$: 1	2	3	3.7	4.0	4.2.

Beyond their use in determining the orbitals, the values obtained for $Z^\star$ permit us to calculate the energy. We have already emphasized the fact that the meaning found for E by means of (6) was not unequivocal. Slater has shown that we can nevertheless interpret the hydrogenic terms (6) as representing the contribution of the electron to the total energy of the atom† which we write:

$$(10) \qquad E = -\frac{\hbar^2}{8m_0}\sum_i \alpha_i^2 = -E_0 \sum_i \frac{Z_i^{\star 2}}{n_i^{\star 2}}.$$

Thus we see that when the electrons belong to two different groups, the interaction inserted for the electron of highest index is only computed once. The situation is more complex in the case of two electrons in the same group, but the calculation we have done for helium shows that equation (10) remains valid and that the term $-E_0 Z^{\star 2}/n^{\star 2}$ should not be considered separately but as an element of the total electronic energy.

As a typical example we apply these relationships to the case of the carbon atom ($Z = 6$) with the following distribution of six electrons $(1s)^2(2s, 2p)^4$. Thus two index groups appear in (7) with apparent atomic number:

$$Z_1^\star = 6 - 0.31 = 5.69$$

$$Z_2^\star = 6 - 3 \times 0.35 - 2 \times 0.85 = 3.25.$$

Hence, a total electronic energy:

$$E(C) = -E_0\left[2(5.69)^2 + 4\left(\frac{3.25}{2}\right)^2\right].$$

The same calculation can be done for the ion C^+: $(1s)^2(2s, 2p)^3$. $Z_1^\star$ remains unchanged, but for the second group we have:

$$Z_2^\star = 6 - 2(0.35) - 2(0.85) = 3.60$$

where:

$$E(C^+) = -E_0\left[2(5.69)^2 + 3\left(\frac{3.60}{2}\right)^2\right].$$

By subtraction we can get the ionization energy of carbon into good agreement with the experimental value ($0.829E_0$):

$$I(C) = E(C^+) - E(C) = 0.84E_0.$$

We complete this discussion by returning to the form (5) of the Slater orbitals. The probability dp of finding the electron in the solid angle $d\omega$ between the distances r and $r + dr$ from the nucleus is such that:

$$(11) \qquad \frac{dp}{dr} = R_n^2(r) . r^2 \,|\, Y_l^m \,|^2 d\omega \sim r^{2n} e^{-\alpha r}$$

† This can be justified by use of the virial theorem (§60).

The representative curve of dp/dr as a function of r has a maximum (fig. 28) for:

$$r_M = \frac{2n^*}{\alpha} = \frac{n^{*2}a_0}{Z^*}. \tag{12}$$

For this value of the radius, the two terms of the function V(r) such that the Slater orbital is a rigorous solution of the Schrödinger equation, are of the same order of magnitude. We know that the

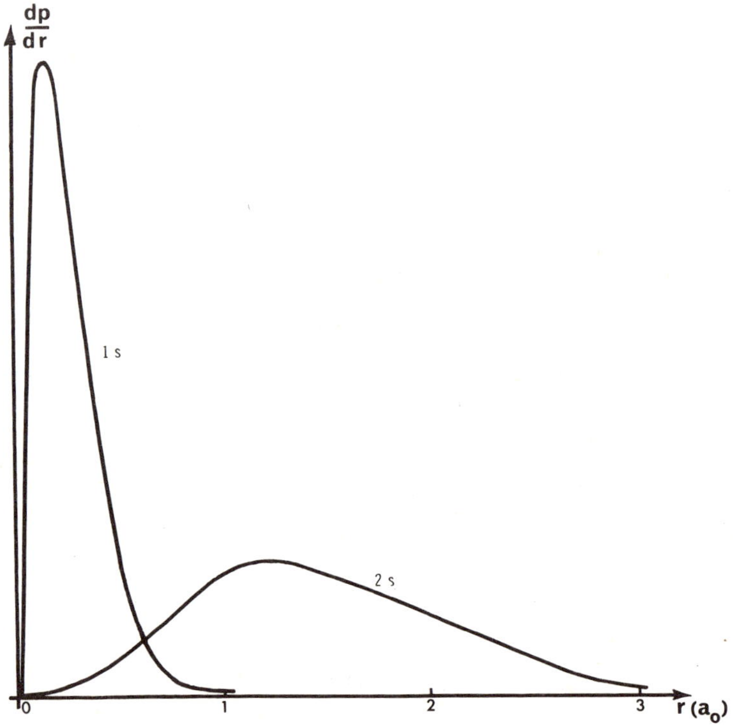

Figure 28. The radial probability densities for the 1s and 2s orbitals in carbon.

approximation is only valid when the first term is predominant so that $r \gg r_n$, which pinpoints the significance of what is meant by the exterior region of the orbital.

Finally we must note that the Slater orbitals corresponding to the same values of l and m but to different values of n are not orthogonal, so that in certain calculations it is preferable to replace them with a system of orthonormal orbitals, derived from the Slater orbitals by the Schmidt orthogonalization procedure (§22).

The Periodic Table of the Elements 29

One can immediately get the electron distribution of an atom in its most stable state by starting from the model without electronic interactions, that is, by progressively filling the one-electron orbitals starting from the $1s$ according to the Pauli principle.

A second approximation, based on the Slater method, accounts for the interactions by means of the screening coefficients. We give an example of this by considering the element potassium ($Z = 19$) for which we compare the two electron distributions:

(*a*) $(1s)^2(2s, 2p)^8(3s, 3p)^8(3d)^1$

(*b*) $(1s)^2(2s, 2p)^8(3s, 3p)^8(4s, 4p)^1$.

The contributions of the electrons in the first three groups to the total electronic energy are the same in both cases, so that we can confine ourselves to considering the last electron. Application of the Slater rules immediately gives:

(*a*) $Z^* = 19 - 18 = 1;\qquad E_a = -\frac{1}{3^2}E_0 = -0.11E_0$

(*b*) $Z^* = 19 - 8 \times 0.85 - 10;\qquad E_b = -\frac{2.2^2}{3.7^2}E_0 = -0.35E_0$.

The electron energy is highest in (*b*) which must therefore represent the structure of potassium.

However, the Slater system does not allow the determination of the order in which the orbitals from the same group are occupied. A first answer is obtained naturally by using what we have found for the one-electron atom (§17). This shows that the electron is on the average closer to the nucleus in the $2s$ state than in the $2p$, for which the screening effect of the electrons of lower or equal index is more important.

In conclusion, for the progressive filling of the different levels by the electrons, which leads to the formulation of the periodic table of the elements,† the order in which the electrons are placed is given by the following scheme (fig. 29). This has only a qualitative

† This description of the electronic structure of the elements by progressive filling of the well-defined one-electron levels presents some difficulty when the atomic numbers are high. In particular, it is known that in the case of the elements $Z = 91$ to $Z = 95$ the electrons represented in the table as $5f$ are in fact distributed on the $5f$, $6d$ hybrid orbitals.

Element	Symbol	Z	1s	2s 2p	3s 3p 3d	4s 4p 4d 4f	5s 5p 5d 5f	6s 6p 6d	7s
Hydrogen	H	1	1						
Helium	He	2	2						
Lithium	Li	3	2	1					
Beryllium	Be	4	2	2					
Boron	B	5	2	2.1					
Carbon	C	6	2	2.2					
Nitrogen	N	7	2	2.3					
Oxygen	O	8	2	2.4					
Fluorine	F	9	2	2.5					
Neon	Ne	10	2	2.6					
Sodium	Na	11	2	2.6	1				
Magnesium	Mg	12	2	2.6	2				
Aluminum	Al	13	2	2.6	2.1				
Silicon	Si	14	2	2.6	2.2				
Phosphorus	P	15	2	2.6	2.3				
Sulfur	S	16	2	2.6	2.4				
Chlorine	Cl	17	2	2.6	2.5				
Argon	A	18	2	2.6	2.6				
Potassium	K	19	2	2.6	2.6	1			
Calcium	Ca	20	2	2.6	2.6	2			
Scandium	Sc	21	2	2.6	2.6.1	2			
Titanium	Ti	22	2	2.6	2.6.2	2			
Vanadium	V	23	2	2.6	2.6.3	2			
Chromium	Cr	24	2	2.6	2.6.5	1			
Manganese	Mn	25	2	2.6	2.6.5	2			
Iron	Fe	26	2	2.6	2.6.6	2			
Cobalt	Co	27	2	2.6	2.6.7	2			
Nickel	Ni	28	2	2.6	2.6.8	2			
Copper	Cu	29	2	2.6	2.6.10	1			
Zinc	Zn	30	2	2.6	2.6.10	2			
Gallium	Ga	31	2	2.6	2.6.10	2.1			
Germanium	Ge	32	2	2.6	2.6.10	2.2			
Arsenic	As	33	2	2.6	2.6.10	2.3			
Selenium	Se	34	2	2.6	2.6.10	2.4			
Bromine	Br	35	2	2.6	2.6.10	2.5			
Krypton	Kr	36	2	2.6	2.6.10	2.6			
Rubidium	Rb	37	2	2.6	2.6.10	2.6	1		
Strontium	Sr	38	2	2.6	2.6.10	2.6	2		
Yttrium	Y	39	2	2.6	2.6.10	2.6.1	2		
Zirconium	Zr	40	2	2.6	2.6.10	2.6.2	2		
Niobium	Nb	41	2	2.6	2.6.10	2.6.4	1		
Molybdenum	Mo	42	2	2.6	2.6.10	2.6.5	1		
Technicium	Tc	43	2	2.6	2.6.10	2.6.5	2		
Ruthenium	Ru	44	2	2.6	2.6.10	2.6.7	1		
Rhodium	Rh	45	2	2.6	2.6.10	2.6.8	1		
Palladium	Pd	46	2	2.6	2.6.10	2.6.10			
Silver	Ag	47	2	2.6	2.6.10	2.6.10	1		
Cadmium	Cd	48	2	2.6	2.6.10	2.6.10	2		
Indium	In	49	2	2.6	2.6.10	2.6.10	2.1		
Tin	Sn	50	2	2.6	2.6.10	2.6.10	2.2		
Antimony	Sb	51	2	2.6	2.6.10	2.6.10	2.3		

Element	Symbol	Z	1s	2s 2p	3s 3p 3d	4s 4p 4d 4f	5s 5p 5d 5f	6s 6p 6d	7s
Tellurium	Te	52	2	2.6	2.6.10	2.6.10	2.4		
Iodine	I	53	2	2.6	2.6.10	2.6.10	2.5		
Xenon	Xe	54	2	2.6	2.6.10	2.6.10	2.6		
Cesium	Cs	55	2	2.6	2.6.10	2.6.10	2.6	1	
Barium	Ba	56	2	2.6	2.6.10	2.6.10	2.6	2	
Lanthanum	La	57	2	2.6	2.6.10	2.6.10	2.6.1	2	
Cerium	Ce	58	2	2.6	2.6.10	2.6.10.1	2.6.1	2	
Praesodymium	Pr	59	2	2.6	2.6.10	2.6.10.2	2.6.1	2	
Neodymium	Nd	60	2	2.6	2.6.10	2.6.10.3	2.6.1	2	
Promethium	Pm	61	2	2.6	2.6.10	2.6.10.4	2.6.1	2	
Samarium	Sm	62	2	2.6	2.6.10	2.6.10.5	2.6.1	2	
Europium	Eu	63	2	2.6	2.6.10	2.6.10.6	2.6.1	2	
Gadolinium	Gd	64	2	2.6	2.6.10	2.6.10.7	2.6.1	2	
Terbium	Tb	65	2	2.6	2.6.10	2.6.10.8	2.6.1	2	
Dyprosium	Ds	66	2	2.6	2.6.10	2.6.10.9	2.6.1	2	
Holmium	Ho	67	2	2.6	2.6.10	2.6.10.10	2.6.1	2	
Erbium	Er	68	2	2.6	2.6.10	2.6.10.11	2.6.1	2	
Thulium	Tm	69	2	2.6	2.6.10	2.6.10.12	2.6.1	2	
Ytterbium	Yb	70	2	2.6	2.6.10	2.6.10.13	2.6.1	2	
Lutetium	Lu	71	2	2.6	2.6.10	2.6.10.14	2.6.1	2	
Hafnium	Hf	72	2	2.6	2.6.10	2.6.10.14	2.6.2	2	
Tantalum	Ta	73	2	2.6	2.6.10	2.6.10.14	2.6.3	2	
Wolfram	W	74	2	2.6	2.6.10	2.6.10.14	2.6.4	2	
Rhenium	Re	75	2	2.6	2.6.10	2.6.10.14	2.6.5	2	
Osmium	Os	76	2	2.6	2.6.10	2.6.10.14	2.6.6	2	
Iridium	Ir	77	2	2.6	2.6.10	2.6.10.14	2.6.7	2	
Platinum	Pt	78	2	2.6	2.6.10	2.6.10.14	2.6.9	1	
Gold	Au	79	2	2.6	2.6.10	2.6.10.14	2.6.10	1	
Mercury	Hg	80	2	2.6	2.6.10	2.6.10.14	2.6.10	2	
Thallium	Tl	81	2	2.6	2.6.10	2.6.10.14	2.6.10	2.1	
Lead	Pb	82	2	2.6	2.6.10	2.6.10.14	2.6.10	2.2	
Bismuth	Bi	83	2	2.6	2.6.10	2.6.10.14	2.6.10	2.3	
Polonium	Po	84	2	2.6	2.6.10	2.6.10.14	2.6.10	2.4	
Astatine	At	85	2	2.6	2.6.10	2.6.10.14	2.6.10	2.5	
Radon	Rn	86	2	2.6	2.6.10	2.6.10.14	2.6.10	2.6	
Francium	Fr	87	2	2.6	2.6.10	2.6.10.14	2.6.10	2.6	1
Radium	Ra	88	2	2.6	2.6.10	2.6.10.14	2.6.10	2.6	2
Actinium	Ac	89	2	2.6	2.6.10	2.6.10.14	2.6.10	2.6.1	2
Thorium	Th	90	2	2.6	2.6.10	2.6.10.14	2.6.10	2.6.2	2
Protactinium	Pa	91	2	2.6	2.6.10	2.6.10.14	2.6.10.2	2.6.1	2
Uranium	U	92	2	2.6	2.6.10	2.6.10.14	2.6.10.3	2.6.1	2
Neptunium	Np	93	2	2.6	2.6.10	2.6.10.14	2.6.10.4	2.6.1	2
Plutonium	Pu	94	2	2.6	2.6.10	2.6.10.14	2.6.10.5	2.6.1	2
Americium	Am	95	2	2.6	2.6.10	2.6.10.14	2.6.10.7	2.6	2
Curium	Cm	96	2	2.6	2.6.10	2.6.10.14	2.6.10.7	2.6.1	2
Berkelium	Bk	97	2	2.6	2.6.10	2.6.10.14	2.6.10.8	2.6.1	2
Californium	Cf	98	2	2.6	2.6.10	2.6.10.14	2.6.10.9	2.6.1	2
Einsteinium	Ei	99	2	2.6	2.6.10	2.6.10.14	2.6.10.10	2.6.1	2
Fermium	Fm	100	2	2.6	2.6.10	2.6.10.14	2.6.10.11	2.6.1	2
Mendelevium	Mv	101	2	2.6	2.6.10	2.6.10.14	2.6.10.12	2.6.1	2
Nobelium	Nm	102	2	2.6	2.6.10	2.6.10.14	2.6.10.13	2.6.1	2

significance since the height of the levels depends on the atomic number of the atom being considered.

We need not stress the advantages of the periodic table which illustrates a periodicity of the properties (volume, electrical and magnetic susceptibility, optical spectra, chemical properties, etc.) in which the outer electrons play a major role. In the last case in particular, it is evident that the extension of the orbital plays an essential role, so that if we take the rare earths, for example, the relation $r_m = (n^{\star 2}/Z^{\star})a_0$ shows that the $4f$ orbital can be considered

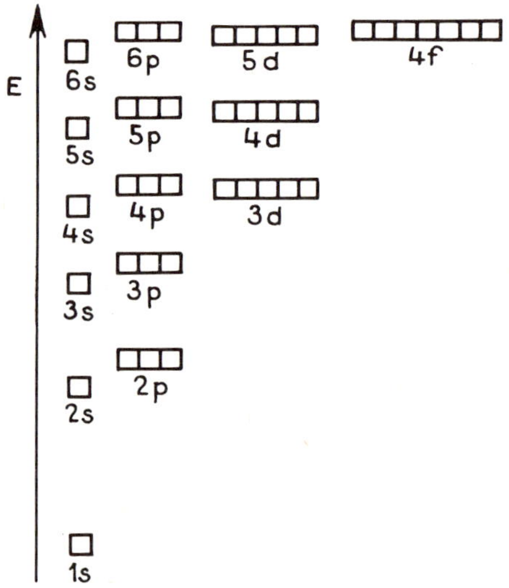

Figure 29. Occupation ordering for the filling of levels in the building up of the periodic table of the elements.

as "interior" in the atom in relation to the $5d$ and $6s$ orbitals. This is responsible for extremely similar chemical properties in all the metals in this family.

We note that for high values of Z, the screening effect of the internal electrons rapidly becomes total ($s = 1$), with the result that the apparent atomic number of the outer electrons does not grow proportionally with Z. As a consequence of this, the distribution of electrons on the more and more elevated levels, due to the Pauli principle, results in the ionization energies rapidly becoming smaller for the heavy elements which are for the most part metallic in character.

However, the approximation which we have just discussed and which allows us to understand the main order of the periodic classification of the elements, is inadequate for analyzing completely the electronic structure of the atoms. The rules we have established do allow us to write $(1s)^2(2s)^2(2p)^2$ for the carbon atom, but we cannot exactly specify how the $2p$ states—numbering six if we take the spin into account—are filled by two electrons. This point can only be discussed through use of the many-electron wave functions which we treat in Part III.

7 THE COVALENT BOND

The Molecular Ion H_2^+ 30

This is the simplest case in which we can introduce the idea of the chemical bond which plays the same fundamental role in the study of diatomic molecules that the hydrogen atom plays in the understanding of atomic structure.

The system is made up of three particles, two protons and an electron, but the large mass difference m_0(electron)/m(proton) = 1/1,840 allows separation of the motion of the different particles (Born-Oppenheimer Approximation §62), and in actual situations allows the system to be treated as composed of a moving electron in the field produced by two protons at a fixed arbitrary distance

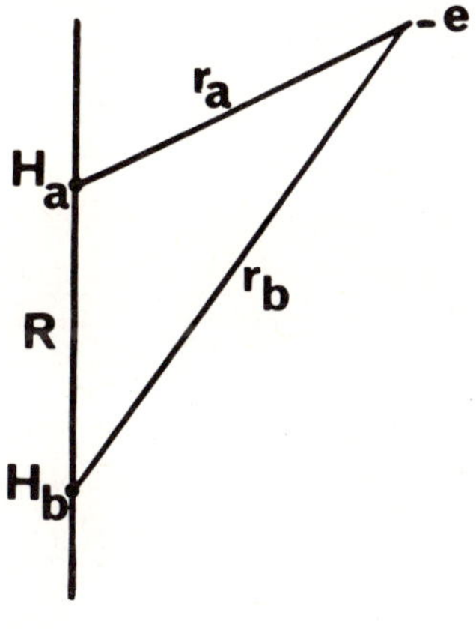

Figure 30. Diagram of the notation used in the study of H_2^+.

R. The Schrödinger equation of the system is written using the notation of figure 30 as:

$$(1) \qquad \mathbf{H}\psi = -\frac{\hbar^2}{2m_0}\Delta\psi - \frac{e^2}{4\pi\varepsilon_0}\left(\frac{1}{r_a} + \frac{1}{r_b} - \frac{1}{R}\right)\psi = E\psi.$$

We introduce the energy of repulsion of the nuclei ($e^2/4\pi\varepsilon_0 R$) so that for the zero energy level we can take the state in which the three particles are infinitely separated.

Many methods of solution are possible, the most direct method uses equation (1) written in elliptical coordinates (Appendix IV). We will limit ourselves to a variational method currently used in quantum chemistry. The basic idea is that for an infinite value of the nuclear separation R, the most stable state of the system is formed by a hydrogen atom and a proton without interaction, with two possible combinations H_a, H_b^+ and H_a^+, H_b. In either case, the behavior of the electron is rigorously that of a hydrogen atom in the $1s$ state, with the wave function (see §17):

$$(2) \qquad \psi_a = Ne^{-\frac{\alpha r_a}{2}}; \qquad \psi_b = Ne^{-\frac{\alpha r_b}{2}}, \quad \text{with} \quad 8\pi N^2/\alpha^3 = 1$$

the energy being that of hydrogen in the ground state, namely $-E_0$. The level thus defined is degenerate, that is, for the same energy value $-E_0$ the Schrödinger equation has as rigorous solutions all the linear combinations:

$$(3) \qquad \psi = c_a\psi_a + c_b\psi_b.$$

The coefficients c_a and c_b can be assumed real without altering the generality of the treatment since the orbitals ψ_a and ψ_b are themselves real. The normalization condition for ψ is written:

$$\int \psi^2 d\tau = c_a^2 + c_b^2 = 1$$

the two orbitals ψ_a and ψ_b being orthogonal because of the nuclear separation. The rules given for interpretation of ψ (§11) allow us to say that in the state represented by (3) we have as the probability of finding the electron on nucleus a or b respectively c_a^2 or c_b^2.

Given the above, when the nuclei approach each other, the functions (2) are no longer rigorous solutions of equation (1), but can always be considered as approximate solutions. The same will be true for the linear combination (3), except that these functions are no longer orthogonal; the normalization conditions will involve the overlap integral, and we write:

$$(4) \qquad \int \psi^2 d\tau = c_a^2 + c_b^2 + 2Sc_ac_b = 1 \qquad \text{with} \qquad S = \int \psi_a\psi_b d\tau.$$

These are the conditions for use of the previously described variational method (§19) and we will work out the solution because of the practical importance of this example.

We will first do the calculation formally, without being explicit as to the form of **H**, and then we will reproduce exactly the general theory of section 19. It is only a question of writing the extremum condition of the mean energy in state ψ, so that setting aside the sign for the conjugate, which is unimportant here:

(5) $$E = \int \psi(\mathbf{H}\psi)d\tau = c_a^2 H_{aa} + c_b^2 H_{bb} + 2c_a c_b H_{ab}$$

with:

$$H_{aa} = \int \psi_a(\mathbf{H}\psi_a)d\tau;$$

$$H_{bb} = \int \psi_b(\mathbf{H}\psi_b)d\tau;$$

$$H_{ab} = \int \psi_a(\mathbf{H}\psi_b)d\tau = \int \psi_b(\mathbf{H}\psi_a)d\tau.$$

The Lagrange method of undetermined multiplier λ (Appendix III) allows us to get for two linear equations which must be satisfied by the coefficients of ψ and which are written, taking $\lambda = E$:

(6) $$\begin{aligned} (H_{aa} - E)c_a + (H_{ab} - ES)c_b &= 0 \\ (H_{ab} - ES)c_a + (H_{bb} - E)c_b &= 0. \end{aligned}$$

The symmetry of $\mathbf{H}$, given in (1) relative to the nuclei a and b, makes $H_{aa} = H_{bb}$, with the result that the compatibility condition (5-7) which determines the energy is written:

$$(H_{aa} - E)^2 = (H_{ab} - ES)^2.$$

Thus we get two solutions to the problem:

(1) The symmetric solution:

$$H_{aa} - E = -(H_{ab} - ES)$$

from which:

(7) $$E_S = \frac{H_{aa} + H_{ab}}{1 + S}; \qquad c_a = c_b; \qquad \psi_S = \frac{\psi_a + \psi_b}{\sqrt{2(1 + S)}}.$$

(2) The antisymmetric solution:

$$H_{aa} - E = H_{ab} - ES$$

which gives:

(8) $$E_A = \frac{H_{aa} - H_{ab}}{1 - S}; \qquad c_a = -c_b; \qquad \psi_A = \frac{\psi_a - \psi_b}{\sqrt{2(1 - S)}}.$$

We note here, and we will show later (§31), that we are able to predict *a priori* that the symmetry of $\mathbf{H}$ implies the symmetric or antisymmetric character of the wave functions from which it is possible to obtain the solutions (7) and (8) directly without the variational calculation.

Discussion of the Solutions

The significance of the preceding two solutions appears in the calculation of the quantities which were introduced in the previous discussion, namely, $H_{aa} = H_{bb}$, H_{ab}, and S. The first step consists in expressing the operator $\mathbf{H}$ is such a way as to use the operator $\mathbf{H}_a$ relative to the hydrogen atom H_a, for which ψ_a is the rigorous solution, thus:

$$\mathbf{H} = \mathbf{H}_a - \frac{e^2}{4\pi\varepsilon_0}\left(\frac{1}{r_b} - \frac{1}{R}\right)$$

$$\text{with } \mathbf{H}_a\psi_a = E_H\psi_a \quad \text{and} \quad E_H = -E_0.$$

As an immediate result of the definitions of H_{aa} and H_{bb}, we get:

$$\text{(9)} \quad H_{aa} = E_H + \frac{e^2}{4\pi\varepsilon_0 R} - I \qquad \text{with} \qquad I = \frac{e^2}{4\pi\varepsilon_0}\int \frac{\psi_a^2}{r_b}\, d\tau$$

$$\text{(10)} \quad H_{ab} = \left(E_H + \frac{e^2}{4\pi\varepsilon_0 R}\right) S - J \quad \text{with} \quad J = \frac{e^2}{4\pi\varepsilon_0}\int \frac{\psi_a\psi_b}{r_b}\, d\tau.$$

The quantities E_S and E_A then take the following form:

$$\text{(11)} \quad \begin{aligned} E_S &= E_H + \frac{e^2}{4\pi\varepsilon_0 R} - \frac{I + J}{1 + S} \\ E_A &= E_H + \frac{e^2}{4\pi\varepsilon_0 R} - \frac{I - J}{1 - S}. \end{aligned}$$

We now compare this result with that obtained from a purely classical discussion by considering the approach of a proton to a hydrogen atom, allowing the atom to remain undeformed in the field of the proton, in other words, to be non-polarizable. Within this assumption, the variation of the potential energy starting from E_H relative to the isolated atom involves, respectively, the contributions of the nucleus and the electron, the latter being calculated from the electron distribution ψ_a by an integral which turns out to be I. We get:

$$\text{(12)} \qquad E' = E_H + \frac{e^2}{4\pi\varepsilon_0 R} - I.$$

The expression we get is very close to the mean energy as calculated from (11):

$$\overline{E} = \frac{1}{2}(E_S + E_A) = E_H + \frac{e^2}{4\pi\varepsilon_0 R} - \frac{I - JS}{1 - S^2}$$

especially when S is small, that is, for large values of the nuclear distance R.

The quantum solution requires replacement of the single classical solution by two others which differ as a result of the integral J, called the *exchange integral.* In contrast to I, J has no classical counterpart and yet is large. The calculation shows that the order of magnitude of J is the same as that of I, and from this we see its predominant role in this solution, since as a result (fig. 31), the level E_S is lower than E_A and must therefore correspond to the normal state of the system.

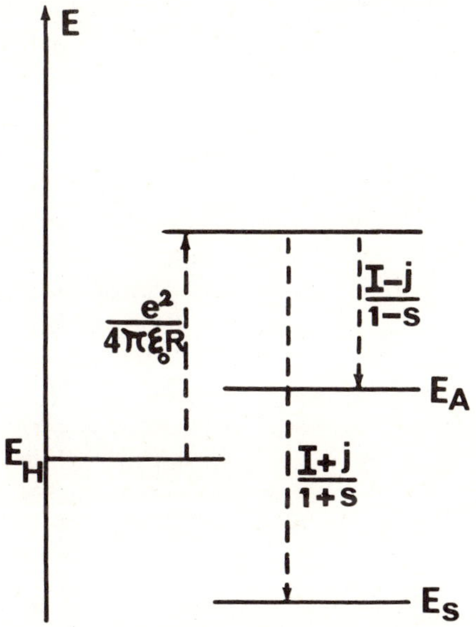

Figure 31. Diagram of the energy levels of H_2^+ for a given value of R.

Calculation of the integrals: S, J, and I. The calculation is easily done using elliptical coordinates: ξ, η, and χ (see Appendix IV). We recall that we have:

$$\xi = \frac{r_a + r_b}{R} \quad \text{varying from 1 to } +\infty$$

$$\eta = \frac{r_a - r_b}{R} \quad \text{varying from } -1 \text{ to } +1$$

$$\chi \text{ from 0 to } 2\pi \text{ with} \quad d\tau = \frac{R^3}{8}(\xi^2 - \eta^2)d\xi d\eta d\chi.$$

In all these cases it is a matter of calculating eulerian integrals of the form:

$$\int_x^\infty e^{-\lambda t}\, t^n dt = \frac{n\,!}{\lambda^{n+1}}\left(1 + \frac{\lambda x}{1} + \ldots + \frac{\lambda^n x^n}{n\,!}\right)e^{-\lambda x}.$$

We get then, by taking:

$$\frac{\alpha R}{2} = \frac{R}{a_0} = D, \quad \text{with} \quad a_0 = 4\pi\varepsilon_0 \frac{\hbar^2}{m_0 e^2}$$

$$S = \int \psi_a \psi_b d\tau = \frac{N^2 R^3}{8} \int_0^{2\pi} d\chi \int_{-1}^{+1} d\eta \int_1^{\infty} e^{-D\xi}(\xi^2 - \eta^2) d\xi$$

so that after integration:

$$S = \left(1 + D + \frac{D^2}{3}\right) e^{-D}. \tag{13}$$

Likewise, introducing the value α (§28), and the ionization energy of the hydrogen atom $E_0 = (1/(4\pi\varepsilon_0)^2)(m_0 e^4/2\hbar^2)$, we get:

$$I = \frac{2E_0}{D}[1 - (1 + D)e^{-2D}]$$

$$J = 2E_0(1 + D)e^{-D}.$$

The integral J is gotten immediately; the calculation of I is a little more difficult and can be made without using elliptical coordinates by taking the expression obtained (§27) in the calculation of the potential energy of an electron distribution. The expressions thus obtained require the following comments:

(*a*) The three integrals I, J, and S tend toward zero for large values of R. I has as its principal part $2E_0/D = e^2/(4\pi\varepsilon_0 R)$, which is the expression one gets when the electron cloud of H_a can be thought of as being at a uniform distance R from H_b^+.

(*b*) When R goes to zero, expansion of the three integrals gives:

$$I \sim \frac{2E_0}{D}[1 - (1 + D)(1 - 2D)] \sim 2E_0$$

$$J \sim 2E_0$$

$$S \sim 1.$$

These values will be used in the discussion which follows.

(*c*) The expressions for J and S are of a similar form. More precisely we consider the ratio:

$$\frac{J}{2E_0 S} = \frac{1 + D}{1 + D + \frac{D^2}{3}}.$$

This ratio, which varies slowly with $D = \alpha R/2$, can be thought of as practically constant for small variations in the distance.

Discussion of the Results

The preceding discussion developed the concept that the approach of a proton to within the distance R of a hydrogen atom in a 1*s* state leads to two possible states with energies, E_S and E_A respectively, governed by the molecular orbitals (MO) ψ_S and ψ_A, which are formed from the atomic orbitals ψ_a and ψ_b. The expressions for E_S and E_A are functions of the internuclear distance R, and are found from equations (13) and those following. The results obtained are represented in figure 32. We know that the electron state of energy E_S is always the most stable. The minimum which occurs for R = 1.32 Å corresponds to the formation of a stable molecular ion and thus to the existence of a *chemical bond* between the

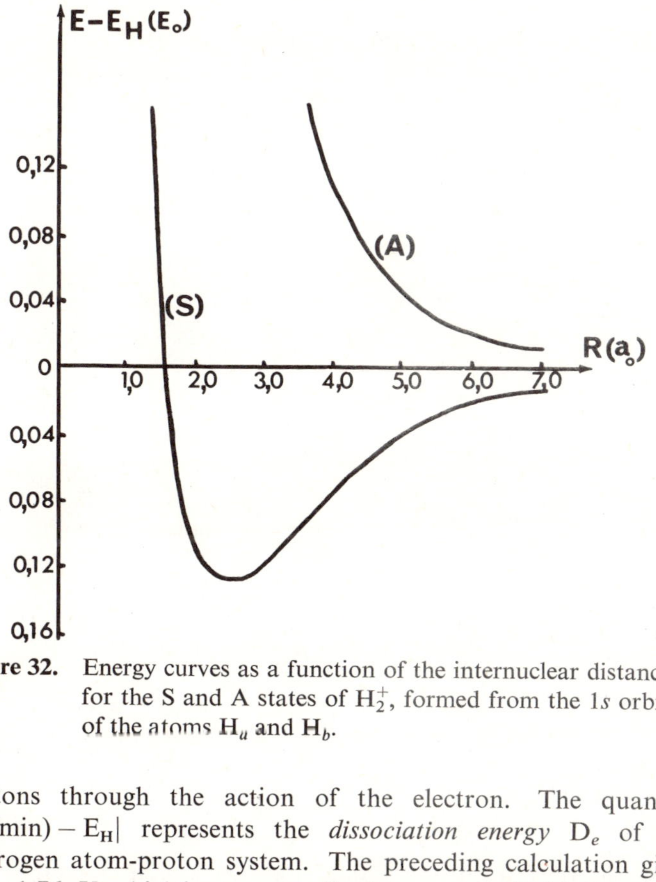

Figure 32. Energy curves as a function of the internuclear distance R for the S and A states of H_2^+, formed from the 1*s* orbitals of the atoms H_a and H_b.

protons through the action of the electron. The quantity $|E_S(\text{min}) - E_H|$ represents the *dissociation energy* D_e of the hydrogen atom-proton system. The preceding calculation gives $D_e = 1.76$ eV, which is a very small value compared with the energy of a hydrogen atom $E_0 = 13.60$ eV, and gives an intuitive justification for the formation of molecular orbitals from the bound states of hydrogen. We find good agreement with the order of magnitude of the experimental values ($r_e = 1.06$ Å; $D_e = 2.772$ eV). Such satisfactory agreement was hardly expected *a priori*, given that the approximation in the construction of the wave functions is rather bad as the distance R becomes small. At the limit, when R tends to zero, the function ψ_S tends toward the wave function of the hydrogen atom, where—ignoring the nuclear interaction—it should tend toward the 1*s* function of He^+, corresponding accordingly to $Z = 2$ instead of $Z = 1$. Another weakness characteristic of the approximation for small values of R appears in the evaluation of the limiting expression for E. Using the expansions given above for I, J, and S it becomes:

$$E_S \sim E_H + \frac{e^2}{4\pi\varepsilon_0 R} - 2E_0 = -3E_0 + \frac{e^2}{4\pi\varepsilon_0 R} \cdot$$

Besides the term representing the nuclear repulsion, which varies as R^{-1}, we find a second term $-3E_0$, which represents the limiting electronic energy of the system for R going to zero, which is nothing more than the ionized helium atom for which we should in fact

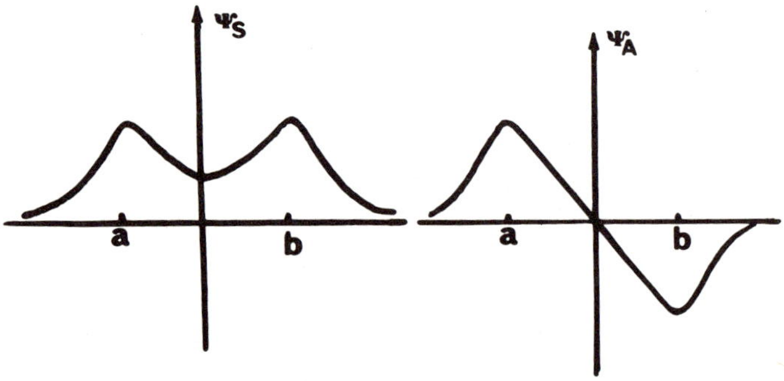

Figure 33. Diagram of the wave functions ψ_S and ψ_A along the internuclear line *ab* of H_2^+.

get $-2^2 E_0$, suggesting that the dissociation energy found in the preceding calculation must be too small.

For the time being, we will set aside the more elaborate techniques that require construction of the wave functions ψ_S with a variable parameter α so that we can get the relative value of hydrogen for large R and to that of He^+ for R tending to zero. On this point we can refer to J. C. Slater's complete treatment.†

Up to now we have pretty much limited our discussion to the symmetric state that leads to the formation of a stable molecular ion. By contrast, the antisymmetric state of the system corresponds (fig. 32) to a curve $E_A(R)$ which decreases monotonically. This cannot lead to any stable system, for if it were realized, it would necessarily evolve toward dissociation.

† J. C. Slater, *Quantum Theory of Molecules and Solids* (New York: McGraw-Hill, 1963), Vol. I, Sect. 1-1.

In conclusion, starting from the atomic orbitals ψ_a and ψ_b it is possible to construct a symmetric bonding orbital ψ_S and an antibonding antisymmetric orbital ψ_A. The bonding or antibonding character is apparent from the results on the variation in energy. This can be seen more directly by examining the curves representing the respective variations of ψ_S and ψ_A along the nuclear axis for a given value of R (fig. 33). The value of ψ_S is non-zero at the center of the system and one can intuitively see the possibility of getting a stable structure from the mutual attraction between the nuclei and the intermediate electron distribution. In the second case, however, the zero value of ψ_A at the center leads to a predominance in nuclear repulsion.

Classification of the Electron States of a Diatomic Molecule 31

The preceding discussion considers only the ground state of the diatomic molecule. A generalization is necessary to begin the discussion of the general case. We will proceed by studying the

Figure 34. The contour of a surface of constant potential energy for a symmetric diatomic molecule. The force $\vec{F}$ applied to an electron at a point on this surface of revolution intersects the line *ab* passing through the nuclei.

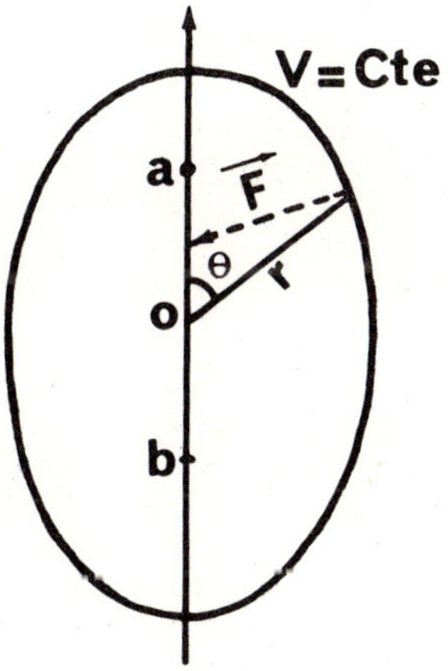

Schrödinger equation of a one-electron system formed (fig. 34) by two nuclei situated on the *z*-axis at a fixed distance R:

$$-\frac{\hbar^2}{2m_0}\Delta\psi + V\psi = E\psi.$$

We can write the potential energy V as arising from the action of the two nuclei *a* and *b* on the electron, but a more general discussion is possible allowing extension to the many-electron case by simply assuming a potential about the nuclear axis as a function $V(r, \theta)$ independent of χ (fig. 34).

The form of the laplacian (Appendix IV) implies that the separation of the variable χ in the equation above should be exploited by taking a solution of the form:

(14) $$\psi(r, \theta, \chi) = u(r, \theta) . v(\chi)$$

which leads to:

$$-\frac{r^2 \sin^2 \theta}{u}\left[\frac{1}{r^2}\frac{\partial}{\partial r}\left(r^2 \frac{\partial u}{\partial r}\right) + \frac{1}{r^2 \sin \theta}\frac{\partial}{\partial \theta}\left(\sin \theta \frac{\partial u}{\partial \theta}\right)\right]$$
$$+ \frac{2m_0}{\hbar^2}[V(r, \theta) - E]r^2 \sin^2 \theta = \frac{1}{v}\frac{d^2 v}{d\chi^2}.$$

The function of the variables r and θ on the left-hand side can only be equal to the function of χ on the right-hand side if each side is constant. Thus we can take:

$$\frac{1}{v}\frac{d^2 v}{d\chi^2} = \text{constant}$$

The function $v(\chi)$, the solution of a linear homogeneous equation of the second order is of the form:

$$v(\chi) = Ce^{m\chi} \quad \text{with} \quad \lambda = m^2.$$

The function ψ should be unique at each point, which requires $v(\chi) = v(\chi + 2\pi)$, hence:

$$e^{2\pi m} = 1.$$

The coefficient m must be purely imaginary so that $m = ki$ with integral k and $\lambda = -k^2$.

The energy E dependent on λ must be the same for the two values k and $-k$, from which we get twofold degeneracy of all the levels with the exception of that for $k = 0$. We saw earlier that when we deal with a degenerate energy level it is necessary to introduce the variables compatible with E for a complete determination of the state. As in the case of the atom, discussion of the classical mechanical problem defines the variables. The equations of motion for the electrons are written:

$$m_0\ddot{x} = X; \quad m_0\ddot{y} = Y; \quad m_0\ddot{z} = Z.$$

The force with components X, Y, and Z normal to the surfaces V = constant, touches the z-axis of the molecule, which requires $xY - yX = 0$, from which:

$$m_0(x\ddot{y} - y\ddot{x}) = \frac{d}{dt} m_0(x\dot{y} - y\dot{x}) = 0.$$

The component of the momentum along the molecular axis is therefore a constant, which suggests that we should use this same variable in the quantum discussion. We must in that case introduce the associated operator, which is given in spherical coordinates (Appendix VI-5) by:

$$\mathbf{L}_z = \frac{\hbar}{i}\frac{\partial}{\partial\chi}\,.$$

We see that for a wave function in the form of (14):

$$\mathbf{L}_z\psi = \frac{\hbar}{i}\frac{\partial\psi}{\partial\chi} = k\hbar\psi,$$

which shows that in the state defined by the function:

$$\psi_k = u(r,\,\theta)e^{ik\chi}$$

the component L_z is fixed and equal to $k\hbar$, and its value is likewise equal to $-k\hbar$ in the second base state corresponding to $-k$. From the classical point of view, the two values of the angular momentum correspond to two rotations in the opposite sense, the absence of degeneracy for $k = 0$ can be interpreted as a rest state.

The nomenclature of the levels based on the value of k is taken from the atomic levels, Greek letters being substituted for Latin. We get the following table:

$k = 0$	state σ (non-degenerate)
$k = 1$	state π (twofold degenerate)
$k = 2$	state δ (twofold degenerate)
$k = 3$	state φ (twofold degenerate).

Although the following observation is only interesting in the case of many-electron functions, we should note the action of a symmetry with regard to a plane, including the molecular axis which is a plane of symmetry of the molecule. If, for simplicity's sake, we take the xz-plane, the transformation is written in cartesian and spherical coordinates as:

$$x' = x\,; \qquad y' = -\,y\,; \qquad z' = z$$
$$r' = r\,; \qquad \theta' = \theta \quad ; \qquad \chi' = -\,\chi.$$

The immediate result is that the function ψ_k is transformed into ψ_{-k} and inversely. In other words, the reflection considered does not modify the energy value but inverts the value of L_z. This result is identical with that gotten classically by making the transformation on the component $L_z = m_0(x\dot{y} - y\dot{x})$. When we are dealing with a state σ for which $L_z = 0$, the state is transformed into itself, with the result that we must reproduce to within a constant the wave function itself. If this function is of the form $\psi = u(r,\theta)$, it necessarily remains invariant under the reflection. The discussion of the wave functions of many-electron states allows another possible behavior, corresponding to multiplication by (−1) under the reflection, so that occasionally we write σ^+ here, although it is superfluous because there are no other possibilities in the case of one-electron functions.

Symmetric Molecules A_2

Our discussion of the molecular ion H_2^+ showed that the construction of molecular orbitals from the 1*s* atomic orbitals leads to two

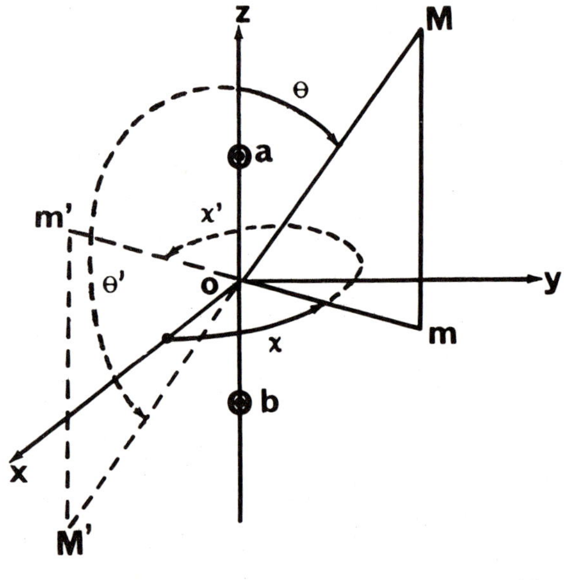

Figure 35. Action of the inversion (or the symmetry with respect to a point) on a point M defined by the spherical coordinates r, θ, and χ.

orbitals, symmetric and antisymmetric, relative to the permutation of the nuclei. We will generalize this result for the case of a general molecular orbital of a symmetric molecule. Such a molecule remains unchanged under an inversion *I*, that is, it has a symmetry with respect to the center of the molecule. This is given by the following formulas in cartesian coordinates and in spherical coordinates (fig. 35):

$$x' = -x; \qquad y' = -y; \qquad z' = -z$$
$$r' = r; \qquad \theta' = \pi - \theta; \qquad \chi' = \chi + \pi.$$

The operation leaves the Schrödinger equation invariant, so that if $\psi(r, \theta, \chi)$ is the solution for an eigenvalue E, it should also be one for $\psi(r, \pi - \theta, \chi + \pi)$. If we express ψ in the form (14) corresponding to the quantum number k we will have:

$$\psi_k(r, \theta, \chi) = u(r, \theta)e^{ik\chi}$$

$$\psi_k(r, \pi - \theta, \chi + \pi) = u(r, \pi - \theta)\cdot(-1)^k \cdot e^{ik\chi} = \varphi_k.$$

The transformed wave function (by inversion) has the same value of k as the original function. We denote this transformed function by the notation† $\varphi_k = \mathbf{I}\psi_k$. A second application of the inversion returns the molecule to the initial state, which is expressed by:

$$\mathbf{I}\varphi_k = \mathbf{I}^2\psi_k = \psi_k.$$

The simplest case is that of a system of values of E and k defining a single state; the functions ψ_k and φ_k must be identical to within a numerical factor λ which allows us to write:

$$\varphi_k = \mathbf{I}\psi_k = \lambda\psi_k$$

and by applying the operator **I** to these identical functions:

$$\mathbf{I}\varphi_k = \mathbf{I}^2\psi_k = \psi_k = \lambda\mathbf{I}\psi_k = \lambda\varphi_k$$

which implies:

$$\psi_k = \lambda^2\psi_k \quad \text{so that} \quad \lambda^2 = 1 \quad \text{and} \quad \lambda = \pm 1.$$

The wave functions are classified into two groups depending on their behavior under inversion. The existence of a degeneracy is not modified by this result since if the functions ψ_k and φ_k are distinct functions, that is, not proportional, it is sufficient to replace them by the linear combinations $\psi_k + \varphi_k$ and $\psi_k - \varphi_k$. We then get:

$$\mathbf{I}(\psi_k + \varphi_k) = \varphi_k + \psi_k; \qquad \mathbf{I}(\psi_k - \psi_k) = \varphi_k - \psi_k.$$

In this case we can always find symmetric or antisymmetric wave functions relative to inversion.

The usual terminology used to distinguish the levels of molecular symmetry are the words even (symmetric), and odd (antisymmetric). The nature of the level is usually marked by the indices g or u, derived from the corresponding terms in German (gerade = even and ungerade = odd).

† The action of the symmetry with respect to a point of inversion of the function is denoted by the action of an operator **I**. The calculation above shows that the eigenvalues of this operator are real (± 1), which suggests that it is a hermitian operator.

The fact that the state defined by a system of values E and k remains invariant under inversion, implies that the component L_z is not modified by this operation. This can be verified directly by considering the classical expression:

$$L_z = m_0(x\dot{y} - y\dot{x}),$$

which remains invariant in taking:

$$\mathbf{I}\vec{r} = -\vec{r} \quad \text{and} \quad \mathbf{I}\vec{v} = -\vec{v}.$$

In conclusion, the catalog of states of a symmetric diatomic molecule has the following terms, the multiplicity of the state being given in parentheses:

$$\sigma_g(1),\ \pi_g(2),\ \delta_g(2),\ \varphi_g(2),\ \ldots$$
$$\sigma_u(1),\ \pi_u(2),\ \delta_u(2),\ \varphi_u(2),\ \ldots$$

The even or odd character of the molecular orbitals implies that for the explicit approximation for H_2^+ which leads to the formation of these orbitals from the atomic orbitals of the constituent atoms, these atomic orbitals must be of the same form. If we identify these orbitals by ψ_a and ψ_b, the molecular orbitals must be of the form $c(\psi_a \pm \psi_b)$, and, depending on the nature of the atomic orbitals ψ_a and ψ_b, one of the molecular orbitals must eventually correspond to the even state and the other to the odd.

One further point, particularly important in the study of molecular structure, remains to be discussed, namely, whether the occupation of one of the molecular orbitals thus formed contributes to the bonding of the two atoms under study, that is, whether a bonding or an antibonding orbital results. A convenient method of dealing with this is to use the Mulliken correlation diagram, to which we now turn.

32 Correlation Diagrams

The classification of the molecular orbitals of a symmetric molecule according to the types cataloged in the last section is valid at any internuclear distance and not only if the molecule is in its equilibrium state. As this is a matter of a discrete characteristic which necessarily remains invariant under a distance variation, it can be fixed in the state where this distance is infinitely large, in which case the approximation developed for H_2^+ is rigorous. If ψ_a and ψ_b are the atomic orbitals of the two atoms a and b having the same set of quantum numbers n, l, and m, the possible orbitals have the form (§30):

$$\psi' = \frac{1}{\sqrt{2}}(\psi_a + \psi_b)\,; \qquad \psi'' = \frac{1}{\sqrt{2}}(\psi_a - \psi_b)$$

the normalization factor being $1/\sqrt{2}$ when the internuclear distance is large enough so that the overlap integral S can be ignored. The state of each of these orbitals must be characterized by a value of L_z. Applying the corresponding operator we get:

$$\mathbf{L}_z\psi' = \frac{1}{\sqrt{2}}(\mathbf{L}_z\psi_a + \mathbf{L}_z\psi_b).$$

We assume that for the two atoms a and b the usual z-axis has been selected along the internuclear line. Then:

$$\mathbf{L}_z\psi_a = m\hbar\psi_a\,; \qquad \mathbf{L}_z\psi_b = m\hbar\psi_b$$

where:

$$\mathbf{L}_z\psi' = m\hbar\psi'\,: \qquad \mathbf{L}_z\psi'' = m\hbar\psi''.$$

The value of the quantum number m of the orbitals ψ_a and ψ_b fixes the type of molecular orbital formed. Thus, starting with s atomic orbitals ($m = 0$) we get exclusively σ type molecular orbitals. If we start with p orbitals ($m = 0, \pm 1$) we can form σ ($k = 0$) orbitals and π ($k = \pm 1$) orbitals.

We now go on to the discussion of the parity of the state which is determined by the inversion operator so that:

$$\mathbf{I}\psi' = \frac{1}{\sqrt{2}}(\mathbf{I}\psi_a + \mathbf{I}\psi_b)\,; \qquad \mathbf{I}\psi'' = \frac{1}{\sqrt{2}}(\mathbf{I}\psi_a - \mathbf{I}\psi_b).$$

The function $\mathbf{I}\psi_a$ is an atomic orbital of atom b. We have introduced the spherical functions Y_l^m which occur in the expressions for the atomic orbitals as homogeneous polynomials of degree l. As a result, their behavior under an inversion is given by:

$$\mathbf{I}Y_l^m = (-1)^l Y_l^m \quad \text{where} \quad \mathbf{I}\psi_a = (-1)^l\psi_b, \quad \mathbf{I}\psi_b = (-1)^l\psi_a.$$

From this we can determine the action of the inversion on the preceding molecular orbitals:

$$\mathbf{I}\psi' = (-1)^l \frac{1}{\sqrt{2}}(\psi_b + \psi_a) = (-1)^l\psi'$$

$$\mathbf{I}\psi'' = (-1)^l \frac{1}{\sqrt{2}}(\psi_b - \psi_a) = -(-1)^l\psi''.$$

Thus, from an s ($l=0$) atomic orbital we form an even function $\psi'(g)$, and an odd function $\psi''(u)$ when the inversion acts on p ($l=1$) orbitals.

Hierarchy of Levels

When the internuclear separation R is infinite, the interaction vanishes, with the result that the states defined by ψ' and ψ'' have the same energy, namely, that of an isolated atom in the states ψ_a or ψ_b. This is no longer the case when R decreases, and it is important to know how to classify the levels corresponding to ψ' and ψ''. For this we can, as in equation (7-11) which gives the energy of H_2^+, consider this energy as arising from two sources: one, the term $e^2/2\pi\varepsilon_0 R$ due to nuclear repulsion which increases indefinitely as R tends toward zero; and the other part, the electronic energy which tends toward a finite limit for zero R. Mulliken's idea was to consider the limiting case in order to classify the energy levels, since the molecular state reduces in the limit to a set of atomic electrons (unitary or composite atom) and the hierarchy of states there is exactly known. The problem is to determine which state of the unitary atom one gets starting from the limiting state defined for infinite R. The answer requires two steps.

(*a*) The initial state is defined by a value of k and a parity characteristic. As R tends to zero, we must preserve from the continuity a state of the same kind. This requires classification of the electron states of the atom as if it were a matter of states along a molecular z-axis. The above remarks yield an immediate answer. A state (n, l, m from $-l$ to $+l$) can be considered as a unification of the states characterized by the values of k: 0, ± 1, ..., $\pm l$. The parity of these $2l+1$ states is the same, being $(-1)^l$. The 1s state of the unitary atom can be considered as a state σ_g, and the state 2p as the set σ_u and π_u.

(*b*) We know only the form of the initial states ($R \to \infty$) and the final states ($R \to 0$), and the problem is to establish the correlation, that is to say, to determine which of the states of the unitary atom we get in starting from a given state of the separated atoms. The result is obtained by assuming that all states of the dissociated molecule ($R \to \infty$) must lead to a state of the unitary atom by passing through a real molecular state, which, in turn, must be gotten from the unitary atom by the perturbation due to the progressive separation of the nuclei. There must be an unequivocal correspondence between the initial and the final states, which correspondence we can specify by using the rule called the "non-crossing rule." This rule can be stated as follows: The curves representing $E_I(R)$ and $E_{II}(R)$, which express the variation of energy of levels I and II as a function of the internuclear distance R, cannot have a common point when the levels have the same kind of symmetry. To show this property, we must note that the molecular orbitals formed above are the rigorous solutions of the problem for $R \to \infty$ but not for the general R. It should be possible, in principle, to construct the best solutions by hybridization of the limiting solutions, and we have shown (§19) that this process can only provide a good approximation when the integral H_{ab} is non-zero. With the notation used here:

$$H_{I,II} = \int \psi_I^*(\mathbf{H}\psi_{II})d\tau \neq 0.$$

We will show that this integral is necessarily zero if the levels I and II are of different types. We explain the reasoning by considering the parity character of the states. To do this, we make the following change of variables corresponding to an inversion which does not modify the value of the integral:

$$x' = -x; \quad y' = -y; \quad z' = -z.$$

We have written:

$$\int \psi_I^*(\mathbf{H}\psi_{II})d\tau = \int \mathbf{I}\psi_I^* \cdot \mathbf{I}(\mathbf{H}\psi_{II})d\tau.$$

The inversion leaves the molecule unchanged and does not modify the expression for $\mathbf{H}$, with the result that we have, calling l_1 and l_2 the parity indices of the two states:

$$\mathbf{I}\psi_I = (-1)^{l_1}\psi_I; \quad \mathbf{I}(\mathbf{H}\psi_{II}) = \mathbf{H}\mathbf{I}\psi_{II} = (-1)^{l_2}\mathbf{H}\psi_{II}$$

where:

$$\int \psi_I^*(\mathbf{H}\psi_{II})d\tau = (-1)^{l_1+l_2}\int \psi_I^*(\mathbf{H}\psi_{II})d\tau.$$

If the number $l_1 + l_2$ is odd, the integral is necessarily zero. We can only have coupling between levels of the same parity which then must be all of type *g* or type *u*. A similar argument, replacing $\mathbf{I}$ by $\mathbf{L}_z$, leads to the conclusion that the integral is also zero when the values k_1 and k_2 are different. In conclusion, the interaction between two arbitrary levels, which we take as having the same type of symmetry with equation (5-7), shows that under these conditions the real levels are distributed reciprocally on E_I and E_{II} with a maximum variation when $E_I = E_{II}$ (fig. 36). As a result, the curves giving the variation of the real energy levels do not intersect.

Finally, it is possible to construct a correlation diagram by starting with the orbitals formed from the ground state levels (1*s*) of the separate atoms, and by going progressively to higher and higher levels and satisfying the principle of non-crossing of levels of the same type, completely using all the levels of the unitary atom. The study of the structure of molecules A_2, where A is an atom from the first row of the periodic table, requires consideration of

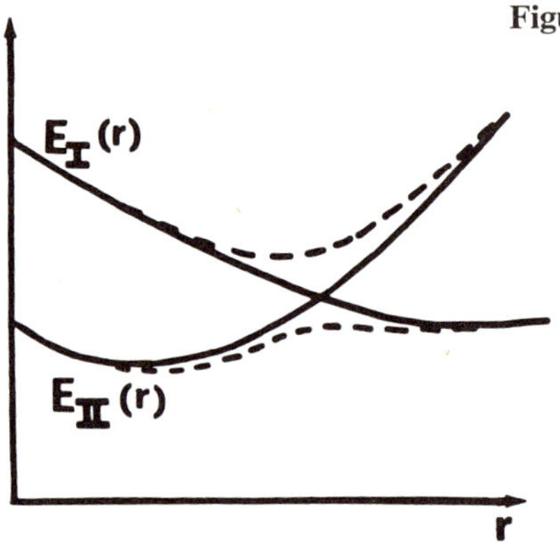

Figure 36. Variation of two energy levels of the same kind with the internuclear distance *r* (solid line). These curves can only represent the energy of the system at points removed from their point of intersection, in the neighborhood of which they must be replaced by the dotted curves (non-crossing rule).

the states 1*s*, 2*s*, and 2*p* of the isolated atoms, and we get the following table:

Table II

Type	Unitary Atom		Separated Atom	Type
$\sigma_g, \pi_g, \delta_g$	3 d	$\pi_g(A)$, $\sigma_u(A)$		
σ_u, π_u	3 p	$\sigma_g(L)$	2 p	$\sigma_g, \sigma_u, \pi_g, \pi_u$
σ_g	3 s	$\pi_u(L)$, $\sigma_u(A)$		
σ_u, π_u	2 p	$\sigma_g(L)$	2 s	σ_g, σ_u
σ_g	2 s	$\sigma_u(A)$		
σ_g	1 s	$\sigma_g(L)$	1 s	σ_g, σ_u

The study of H_2^+ suggested that of the two molecular orbitals constructed from $(1s)_a$ and $(1s)_b$ the symmetric one, which we call σ_g, is bonding, and the other, antisymmetric and called σ_u, is antibonding (A). This same characterization can be applied to each of the molecular orbitals constructed from different atomic orbitals of the separate atoms; the bonding orbitals correspond to the deepest level, which allows their immediate identification on the correlation diagram. This character can also be immediately identified by using the criterion developed in figure 33 for H_2^+, namely, the knowledge that bonding orbitals correspond to an accumulation of charge between the nuclei while the charge density goes to zero between the nuclei for non-bonding orbitals. Our previous discussion of the orbitals, now designated as $(\sigma_g 1s)$ and $(\sigma_u 1s)$, is easily extended by examining the diagram of the

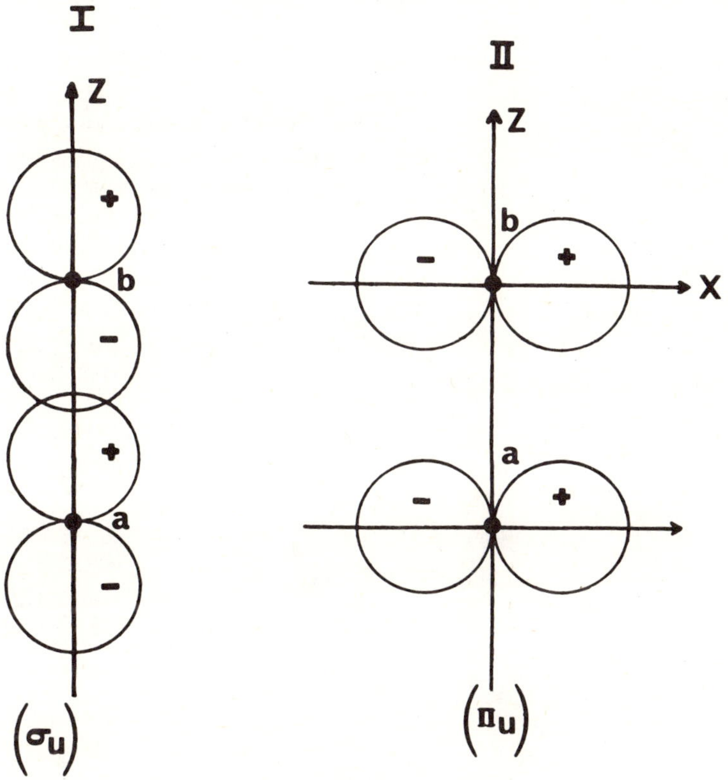

Figure 37. The relative disposition of the $2p$ orbitals of bound atoms for σ_u and for the components of π_u.

atomic orbitals of bound atoms (fig. 37). We see then that in the case of the orbital $(\sigma_u 2p)$ the functions representing $(2p_z)_a$ and $(2p_z)_b$ are the same and have opposite signs at all points in the median plane of the molecule. This illustrates the antibonding character of these orbitals. On the other hand, for $(\pi_u 2p)$ these functions are additive in the same plane, and thus we see the bonding character in the correlation diagram.

We will now indicate the electronic structure of a number of symmetric molecules from the first row of the periodic table. This structure is obtained in a manner similar to that operating in the case of many-electron atoms, that is by progressively filling the one-electron orbitals, taking into account their ordering as indicated in the correlation diagram.

The formation of H_2 corresponds to filling the level $(\sigma_g 1s)$ with two electrons, allowed by the Pauli principle because of two possible spin states. For the system involving one additional electron, as in the molecular ion He_2^+, this electron can only be placed in the next highest level $(\sigma_u 1s)$ which is repulsive and so lessens the bond energy. The situation found for two electrons in a bonding orbital and one electron in a corresponding antibonding orbital—comparable from the point of view of energy to the one-electron bond of H_2^+—is known as a *three-electron bond*, and is written $(He\ldots He)^+$. Starting from He, the orbitals $(\sigma_g 1s)$ and $(\sigma_u 1s)$ are filled; the bonding and non-bonding actions exactly compensate and the set of orbitals used cannot contribute to the bond between the atoms. The distribution of the four electrons is very similar to that of the isolated atoms corresponding to the saturation of each of their $n = 1$ shells and is designated by the symbol K.

Table III

Molecule	Electronic structure	n_L	n_A	$n_L - n_A$	D (eV)
H_2^+	$\sigma_g 1s$	1	0	1	2,65
H_2	$(\sigma_g 1s)^2$	2	0	2	4,48
He_2^+	$(\sigma_g 1s)^2(\sigma_u 1s)$	2	1	1	(2,5)
He_2	$(\sigma_g 1s)^2(\sigma_u 1s)^2 = KK$	2	2	0	0
N_2^+	$KK(\sigma_g 2s)^2(\sigma_u 2s)^2(\pi_u 2p)^4(\sigma_g 2p)$	7	2	5	6,35
N_2	$KK(\sigma_g 2s)^2(\sigma_u 2s)^2(\pi_u 2p)^4(\sigma_g 2p)^2$	8	2	6	7,38
O_2^+	$KK(\sigma_g 2s)^2(\sigma_u 2s)^2(\pi_u 2p)^4(\sigma_g 2p)^2(\pi_g 2p)$	8	3	5	6,48
O_2	$KK(\sigma_g 2s)^2(\sigma_u 2s)^2(\pi_u 2p)^4(\sigma_g 2p)^2(\pi_g 2p)^2$	8	4	4	5,08
F_2	$KK(\sigma_g 2s)^2(\sigma_u 2s)^2(\pi_u 2p)^4(\sigma_g 2p)^2(\pi_g 2p)^4$	8	6	2	(2,08)
Ne_2	$KK(\sigma_g 2s)^2(\sigma_u 2s)(\pi_u 2p)^4(\sigma_g 2p)^2(\pi_g 2p)^4(\sigma_u 2p)^2$	8	8	0	0

The structure of the following systems uses the molecular orbitals constructed from the $2s$ and $2p$ orbitals of the isolated atoms. For each case the table indicates the number n_L of electrons occupying the bonding orbitals, and n_A, the number occupying the non-bonding orbitals. The difference $n_L - n_A$ represents the effective number of electrons contributing to the bond and has close correlation with the molecular bond energy. By convention, the term single bond is applied to those bonds formed by a bonding orbital filled by two electrons. When two or three bonding orbitals are completely filled, the terms double bond and triple bond, respectively, are used. The triple bond is the highest that can be formed from the $2s$ and $2p$ orbitals and appears in the molecule N_2 which we write $\mathrm{N}\equiv\mathrm{N}$. The case of O_2 is a little more complex, since the representation we find by taking into account the number $n_L - n_A = 4$ should be a double bond, and does not allow an interpretation of the existence of the paramagnetism of molecular oxygen. We must therefore look more closely at the electronic structure of the molecule. There is compensation of the two bonding and non-bonding orbitals $(\sigma_g 2s)$ and $(\sigma_u 2s)$ and we have only to account for the terms $(\pi_u 2p)^4$, $(\sigma_g 2p)^2$ and $(\pi_g 2p)^2$. In this set we can see the group $(\sigma_g 2p)^2$ which corresponds to an ordinary single bond, and we further note that the two orbitals $(\pi_g 2p)$ are non-bonding orbitals corresponding to two bonding orbitals $(\pi_u 2p)$. The resulting situation involves two bonds with three electrons, and the structure of the oxygen molecule can then be written:

$$\mathrm{O} \overset{\cdots}{\underset{\cdots}{-}} \mathrm{O}$$

The other cases appearing in the table can also be represented in the same way. The complete list is:

$$(\mathrm{H}\cdot\mathrm{H})^+,\ \mathrm{H}-\mathrm{H},\ (\mathrm{He}\cdots\mathrm{He})^+,\ (\mathrm{N}=\mathrm{N})^+,$$

$$\mathrm{N}\equiv\mathrm{N},\ (\mathrm{O}\overset{\cdots}{=}\mathrm{O})^+,\ \mathrm{O}\overset{\cdots}{\underset{\cdots}{-}}\mathrm{O},\ \mathrm{F}-\mathrm{F}.$$

Different Types of Simple Covalent Bonds 33

The bond between two atoms of the same kind just discussed is known as the covalent bond, in contrast to the ionic bond, which arises when the electronegativity of the bound atoms is quite different. Our discussion, thus far limited to the case of MO's (molecular orbitals) formed from $2s$ and $2p$ orbitals, indicated two types of symmetry for bonding orbitals, σ_g and π_u; we now present a detailed discussion of these two types of orbitals which can be thought of as providing independent bonds called σ bonds and π bonds, the difference in symmetry forbidding their hybridization.

(a) σ Bonds

In the molecules treated in section 32, this bond can arise from $(\sigma_g 1s)$, $(\sigma_g 2s)$, or $(\sigma_g 2p)$. The corresponding MO's can hybridize when the energies are sufficiently close, as is the case for $(\sigma_g 2s)$ and $(\sigma_g 2p)$. This is similar to forming, for the isolated atoms, the hybrid AO's (atomic orbitals) constructed from $2s$ and $2p$, followed by formation of the MO by associating the hybrid AO's of the two bound atoms. This method offers the advantage of facilitating the discussion by showing how the hybridization can lead to a bond of greater energy. The normal approach, using the variational calculation of section 19, does not give simple results and we generally prefer to use a more simple criterion, due to Pauling, which suggests that the energy of the covalent bond formed from an AO is an increasing function of the modulus of this AO in the direction of the bond, the comparisons clearly being made at equal distances from the nucleus to which this AO pertains. The discussion is particularly simple in the case of AO's formed from $2s$ and $2p$, which, in the Slater approximation (§28), correspond to the same radial part of the wave function $\mathrm{N}r\exp[-\alpha r/2]$. We note that the approximation is legitimate here, since the bond affects the behavior of the orbitals in the exterior regions of the atom. We recall the form of the two basic AO's:

$$\psi_{2s} = f(r); \qquad \psi_{2p_z} = f(r)\sqrt{3}\,.\cos\theta$$

with:

$$f(r) = \mathrm{N}re^{-\frac{\alpha r}{2}} \quad \text{and} \quad 4\pi\int_0^\infty f^2r^2dr = 1.$$

The hybrid orbital is of the form:

$$\varphi = c_1\,\psi_{2s} + c_2\psi_{2p_z} = f(r).(c_1 + c_2\sqrt{3}\cos\theta).$$

This orbital must be normalized, which requires, since ψ_{2s} and ψ_{2p} are orthogonal:

$$\int\varphi^2\,d\tau = \int(c_1\psi_{2s} + c_2\psi_{2p_z})^2 d\tau = c_1^2 + c_2^2 = 1.$$

We can thus assume:

$$c_2 = \lambda; \qquad c_1 = \sqrt{1-\lambda^2}.$$

For a given value of r, the orbital ψ_{2s} has a value independent of the direction, whereas ψ_{2p_z} as well as φ depend on the angle θ with the z-axis, the maximum value of the modulus occurring for $\theta = 0$ or $\theta = \pi$. Since we have to compare the functions which are similarly dependent on the distance r from the nucleus, we introduce the idea of the *bonding force* F of an orbital, defined as equal to the modulus of AO in the direction being treated, the modulus of the $2s$ orbital being taken as unity. This definition of the bonding force is relative and independent of the radius r. For the orbitals ψ_{2s}, ψ_{2p_z}, and φ we have, respectively, the following values of F: 1, $\sqrt{3}$, and $\sqrt{1-\lambda^2} + \sqrt{3}$. The parameter λ which characterizes the hybridization varies from 0 to 1 and the variation of the bonding is given by the curve (fig. 38) which has a maximum $F = 2$ for $\lambda = \sqrt{3}/2$. The existence of such a maximum for F, with value obviously greater than that of the basis orbitals ψ_{2s}, ψ_{2p_z}, suggests that such hybridization must be the general rule for the establishment of covalent bonds between the elements in the first row of the periodic table. We must note that whatever the strength of

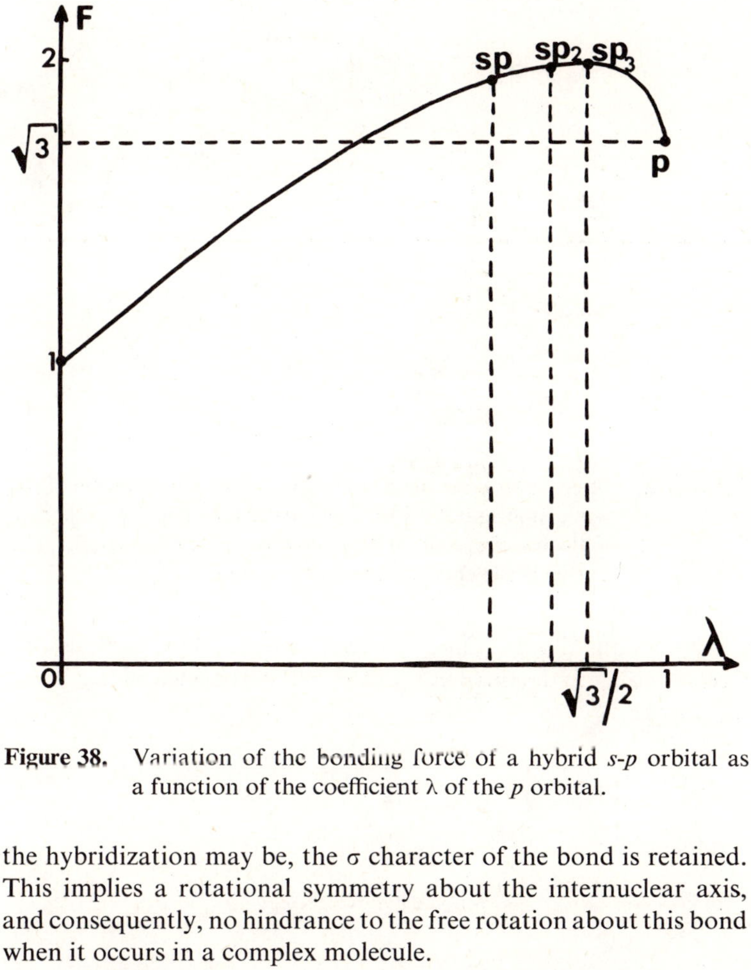

Figure 38. Variation of the bonding force of a hybrid *s-p* orbital as a function of the coefficient λ of the *p* orbital.

the hybridization may be, the σ character of the bond is retained. This implies a rotational symmetry about the internuclear axis, and consequently, no hindrance to the free rotation about this bond when it occurs in a complex molecule.

The value of the strength of hybridization defined by λ is variable, depending on the case, and generally corresponds to the particular geometry of the molecules. The most noteworthy values are:

Tetrahedral bond (sp^3): $\lambda = \sqrt{3}/2$; $F = 2$. This is the maximum bonding force. The maximum of F in one direction is accompanied by a diminution in the other directions, as demonstrated by the plot in figure 39*a* and should be compared to the interference phenomena of light waves, for example, for which the reinforcement of light in the bright fringes is tied to its diminution in the dark fringes, with compensation on the average.

We can easily verify that from among the four AO's (2*s* and 2*p*) it is possible to construct four orthogonal, independent tetrahedral orbitals. These orbitals are simply expressed by taking for the 2*p*

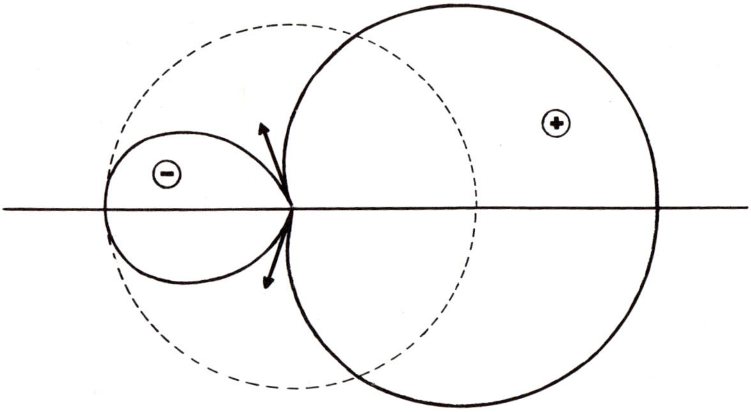

Figure 39a. Polar coordinate diagram of the amplitude variation of the wave function of an sp^3 (tetragonal) hybrid. The diagram indicates the sign of the function in each region. The dotted circle corresponds to the *s* state.

basis functions not $2p_1$, $2p_0$, and $2p_{-1}$, but the three orbitals $2p_0$ corresponding to the three axes of the reference trihedral (fig. 39*b*). We thus get:†

$$\varphi_A = \frac{1}{2}\psi_{2s} + \frac{\sqrt{3}}{2}\psi_{2p_{\overline{OA}}} = \frac{1}{2}(\psi_{2s} + \psi_{2p_x} + \psi_{2p_y} + \psi_{2p_z})$$

and likewise:

$$\varphi_B = \frac{1}{2}(\psi_{2s} - \psi_{2p_x} - \psi_{2p_y} + \psi_{2p_z})$$

$$\varphi_C = \frac{1}{2}(\psi_{2s} - \psi_{2p_x} + \psi_{2p_y} - \psi_{2p_z})$$

$$\varphi_D = \frac{1}{2}(\psi_{2s} + \psi_{2p_x} - \psi_{2p_y} - \psi_{2p_z}).$$

† $\varphi_{2p_{AO}}$ donates the $2p_0$ orbital corresponding to the *z*-axis directed along AO.

These orbitals, φ_A, φ_B, φ_C, and φ_D, allow us to see that the bonds are directed toward the apexes of a tetrahedron having *x*, *y*, and *z* as secondary axes.

Figure 39b. Directions of the four tetrahedral bonds formed in the sp^3 hybridization.

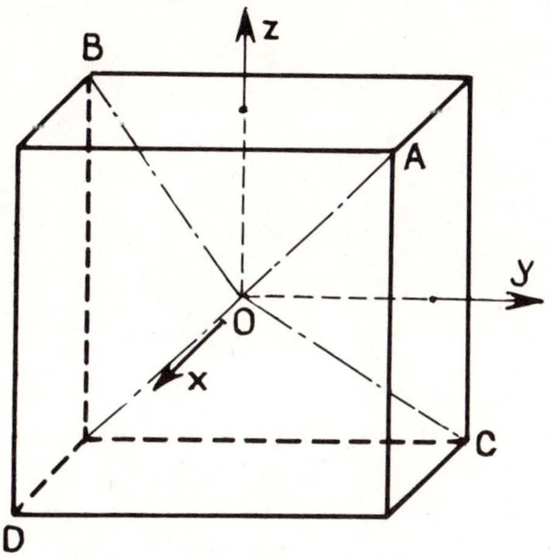

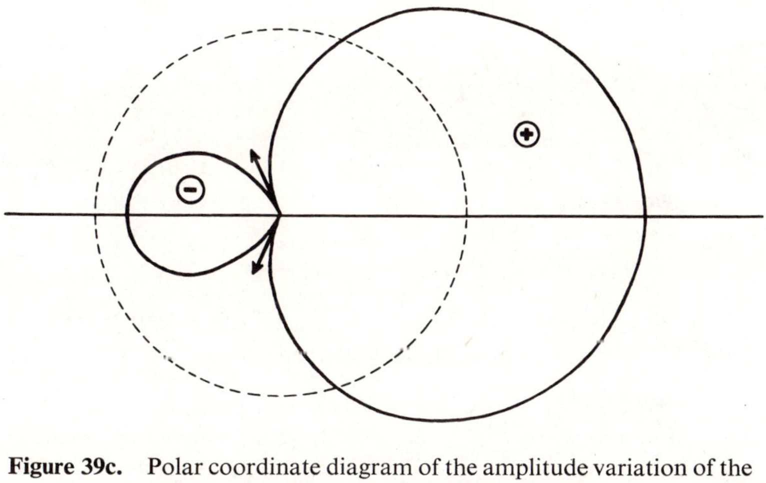

Figure 39c. Polar coordinate diagram of the amplitude variation of the wave function for the sp^2 (trigonal) hybrid.

Trigonal bond (sp^2): $\lambda = \sqrt{2/3}$; F = 1.991. The value of λ differs only slightly from the maximum, this being manifested by a small decrease in the value of the bonding force. The diagram (39*c*) shows the variation of this bonding force with the angle θ. It is possible to construct three orthogonal orbitals with the same force from the AO's $2s$, $2p_x$, and $2p_y$. The corresponding directions are coplanar with 120° angles between them. We write these orbitals:

$$\varphi_{0z} = \frac{1}{\sqrt{3}}(\psi_{2s} + \sqrt{2}\,\psi_{2p_x})$$

$$\varphi_{0z'} = \frac{1}{\sqrt{3}}\left(\psi_{2s} + \sqrt{\frac{3}{2}}\,\psi_{2p_y} - \frac{1}{\sqrt{2}}\,\psi_{2p_x}\right)$$

$$\psi_{0z''} = \frac{1}{\sqrt{3}}\left(\psi_{2s} - \sqrt{\frac{3}{2}}\,\psi_{2p_y} - \frac{1}{\sqrt{2}}\,\psi_{2p_x}\right).$$

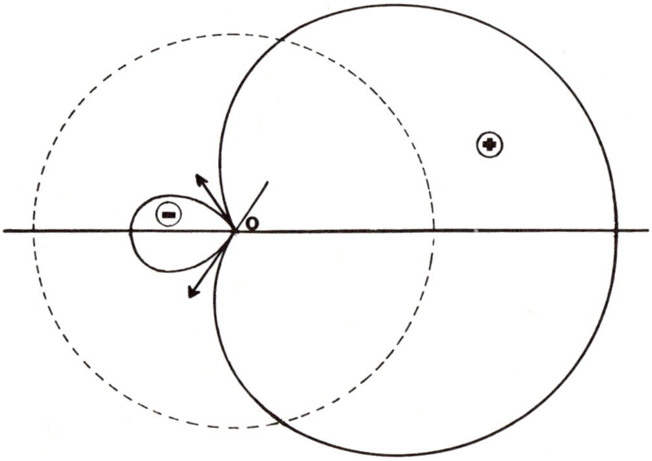

Figure 39d. Polar coordinate diagram of the amplitude variations of the *sp* hybrid wave function.

These orbitals appear in the interpretation of the electronic structure of ethylene and ethylene derivatives.

Triple bond (*sp*): $\lambda = 1/\sqrt{2}$; F = 1.933. We begin to depart from the maximum value of F = 2. The diagram shows the variation of the modulus of the orbital with direction for constant *r* (fig. 39*d*). Starting from the AO's $2s$ and $2p_z$ two such hybrid orbitals can be formed:

$$\varphi_{0z}^{+} = \frac{1}{\sqrt{2}}(\psi_{2s} + \psi_{2p_z})$$

$$\varphi_{0z}^{-} = \frac{1}{\sqrt{2}}(\psi_{2s} - \psi_{2p_z}).$$

The corresponding bonds are oppositely directed along the z-axis. This form of hybridization occurs in the representation of the acetylene bond.

(b) π Bonds

This type, arising from $(\pi_u 2p)$, corresponds to a binary level, with two orbitals developed respectively from the AO's $2p_1$ and $2p_{-1}$ of the bound atoms. An equivalent description can be gotten by replacing these AO's by $2p_x$ and $2p_y$, which are linear combinations and have the advantage of simple behavior (symmetric or antisymmetric) under the reflection on the xz- and yz-planes. As a result, when the bond appears in a more complex molecule but retains these symmetry elements (ethylene, for example), the degeneracy ceases. However, the MO thus constructed constitutes a first approximation, as shown in the treatment of perturbations for degenerate levels.

In figure 37 we gave the diagram of the two AO's $(2p_x)$ which contribute to the formation of the MO state $(\pi_u 2p)$. What is particularly interesting in the diagram is that none of these AO's has a preferential extension along the direction of the bond and therefore they become susceptible to the formation of a bond in any direction in the plane normal to the x-axis. As a result, as we will see, they have the possibility of hybridization with the adjacent ethylene bond, which is not equivalent to the σ bond, and which gives very unusual properties to molecules in which this occurs.

We note, finally, that when one of the components of the $(\pi_u 2p)$ bond exists, which is the case in an ethylenic bond, the absence of rotational symmetry of the MO corresponds to a generally elevated value of the potential barrier which prevents free rotation about the bond.

(c) Other Types of Covalent Bonds

We have limited ourselves here to the study of bonds constructed from the AO's: $1s$, $2s$, and $2p$. The discussion of elements with higher atomic number when they belong to the first row of the periodic table leads to the introduction of other orbitals, and consequently, to other kinds of bonds. This happens when we use the $3d$ AO which allows formations of the MO's $(\sigma 3d)$, $(\pi 3d)$, and $(\delta 3d)$, with of course, the character g or u. We will not examine the novel bonds which can be formed by the $(\delta 3d)$ orbital or the $(\pi 3d)$ MO case. Thus we restrict ourselves to the bonding MO $(\sigma_g 3d)$. The latter is of considerable interest since it is susceptible to hybridization with $(\sigma_g 4s)$ and $(\sigma_g 4p)$, whose energies are comparable. This leads to interpretation of the structure of the complexes of the first family of the transition elements.

Using the same method of discussion as in the preceding case, we begin by constructing the AO hybrids of $3d$, $4s$, and $4p$. We assume that these three orbitals correspond to the same radial function $f(r)$, and we will explain the directional variation of the wave function. We will limit ourselves to real hybridization coefficients, which implies the use not of the spherical functions (§17) but of the real functions which we get from them. We adopt the usual notation for these functions, indicating the expression by polynomials in x, y, and z having the same symmetry:

$$d_{(x^2-y^2)} = f(r) . \frac{\sqrt{15}}{2} \sin^2 \theta . \cos 2\chi$$

$$d_{xy} = f(r) . \frac{\sqrt{15}}{2} \sin^2 \theta . \sin 2\chi$$

$$d_{xz} = f(r) . \sqrt{15} \sin \theta \cos \theta \cos \chi$$

$$d_{yz} = f(r) . \sqrt{15} \sin \theta \cos \theta \sin \chi$$

$$d_{z^2} = f(r) . \frac{\sqrt{5}}{2} (3 \cos^2 \theta - 1)$$

using the same normalization condition as in the case of $2s$ and $2p$ studied previously, so that:

$$\int_0^\infty f^2 . 4\pi r^2 dr = 1.$$

By using the same methods as before, we can answer the following questions:

Finding the 3*d orbital with maximum force in a given direction.* This direction can always be taken as the z-axis. The only orbital of the d-type having a value different from zero in this direction $(\theta = 0)$, is d_{z^2} With the given definition of the bonding force taken equal to unity for an s orbital of the form $\psi_s = f(r)$; this leads to $F = \sqrt{5}$, compared to $\sqrt{3}$ for a p orbital.

Investigation of the d, s, p orbital with maximum force. Still considering the z direction, we need only combine the s, p, and d AO's so that:

$$\varphi = c_1 \psi_s + c_2 \psi_{p_z} + c_3 \psi_{d_{z^2}}$$

We assume the radial function $f(r)$ identical for these three orbitals, which is approximately true for $4s$, $4p$, and $3d$. The bonding force of this hybrid orbital is given by:

$$F = c_1 + c_2\sqrt{3} + c_3\sqrt{5}$$

from which we have to find the maximum, starting from the normalization condition:

$$N = c_1^2 + c_2^2 + c_3^2 = 1.$$

The problem is treated by introducing the Lagrange multiplier λ. The conditions:

$$\frac{\partial F}{\partial c_r} = \lambda \frac{\partial N}{\partial c_r}$$

lead to three equations:

$$1 = \lambda c_1; \qquad \sqrt{3} = \lambda c_2; \qquad \sqrt{5} = \lambda c_3$$

from which, by taking within the normalization condition:

$$\lambda^2 = 9, \quad \text{where} \quad F = 3.$$

The orbital is thus written:

$$\varphi = \frac{1}{3}(\psi_s + \sqrt{3}\,\psi_{p_z} + \sqrt{5}\,\psi_{d_{z^2}}).$$

Octahedral orbitals (d^2sp^3). We end by showing that by hybridization of the s, p, and d orbitals, it is possible to form an AO leading to the formation of bonds of the same force—close to the maximum force ($F = 3$)—lying in the six directions of the reference axes on either side of the origin, that is, in the direction of the apexes of a regular octahedron. The table of d-type orbitals presented above shows that they contribute a non-zero bonding force in various directions, which allows us to write along the x-axis:

$$\varphi_x^+ = c_1\psi_s + c_2\psi_{p_x} + c_3\psi_{d_{(x^2-y^2)}} + c_4\psi_{d_{z^2}}$$

$$\varphi_x^- = c_1'\psi_s + c_2'\psi_{p_x} + c_3'\psi_{d_{(x^2-y^2)}} + c_4'\psi_{d_{z^2}}.$$

We will get orbitals of the same force by taking:

$$c_1' = c_1; \qquad c_2' = -c_2; \qquad c_3' = c_3; \qquad c_4' = c_4.$$

We recall that this is a consequence of the fact that the parity character of the spherical functions of order l is given by $(-1)^l$.

For the hybrid orbital along the y-axis we will likewise have ($\chi = \pi/2$; $\cos 2\chi = -1$):

$$\varphi_y^+ = c_1\psi_s + c_2\psi_{p_y} - c_3\psi_{d_{(x^2-y^2)}} + c_4\psi_{d_{z^2}}$$

$$\varphi_y^- = c_1\psi_s - c_2\psi_{p_y} - c_3\psi_{d_{(x^2-y^2)}} + c_4\psi_{d_{z^2}}.$$

and, finally, along the z-axis:

$$\varphi_z^+ = c_1\psi_s + c_2\psi_{p_z} + c_5\psi_{d_{z^2}}$$

$$\varphi_z^- = c_1\psi_s - c_2\psi_{p_z} + c_5\psi_{d_{z^2}}.$$

The five coefficients c_r satisfy the equations obtained by making use of the fact that the six hybrid orbitals are on the one hand normalized, and on the other pairwise orthogonal, which leads to:

$$c_1^2 + c_2^2 + c_3^2 + c_4^2 = 1$$

$$c_1^2 + c_2^2 + c_5^2 = 1$$

$$c_1^2 - c_2^2 + c_3^2 + c_4^2 = 0$$

$$c_1^2 - c_2^2 + c_5^2 = 0$$

$$c_1^2 + c_4 c_5 = 0$$

$$c_1^2 - c_3^2 + c_4^2 = 0.$$

The solution of this system gives simply:†

$$c_1 = \frac{1}{\sqrt{6}}; \quad c_2 = \frac{1}{\sqrt{2}}; \quad c_3 = \frac{1}{2}; \quad c_4 = -\frac{1}{2\sqrt{3}}; \quad c_5 = \frac{1}{\sqrt{3}}.$$

The common value of the bonding force of the six orbitals is:

$$F = \frac{1}{\sqrt{6}} + \sqrt{\frac{3}{2}} + \sqrt{\frac{5}{3}} = \frac{1}{\sqrt{3}}(2\sqrt{2} + \sqrt{5}) = 2.924.$$

Such a bond is a little weaker than the maximal d, s, p bond, and this plays an important role in the interpretation of the structure of complexes.

---

† The calculation does not fix the signs of c_1, c_2, or c_3, and one of the coefficients c_4, and c_5, which should be chosen in such a way as to get the maximum bonding force for each of the directions considered.

8 THE MOLECULAR ORBITAL METHOD

The term molecular orbital (MO) was frequently used in the preceding chapter, principally with the object of explaining from a qualitative point of view, the nomenclature of the electron states of diatomic molecules and their energy levels. Here we present a quantitative, semi-empirical treatment. The behavior of each electron of the system is treated as a one-electron problem, the interaction with other electrons being introduced by a convenient choice of parameters which arise in this one-electron treatment. This treatment is frequently applied to molecules having conjugated ethylenic bonds and the principle of the method will be illustrated by first considering the case of an isolated ethylenic bond.

The Ethylenic Bond 34

The ethylenic bond, illustrated most simply by the ethylene molecule $CH_2{=}CH_2$, is the juxtaposition of a trigonal σ bond and a π bond, which we have already studied in detail.

The ethylene molecule (fig. 40) is planar in the most stable state and the three σ bonds, one joining the carbon atom C_a with C_b, and the others joined to two hydrogen atoms, are provided by the hybridization of the $2s$, $2p_x$ and $2p_z$ AO's.† The $2p_y$ orbital, which remains unused, couples with the orbital of the same kind on C_b, forming a first approximation to the bonding MO which we take *a priori* in the form:‡

(1) $$\varphi = c_a(\psi_{2p_y})_a + c_b(\psi_{2p_y})_b$$

or in the abridged notation:

$$\varphi = c_a\psi_a + c_b\psi_b.$$

† We can get *a priori* the same result by constructing the hybrid orbitals from the set $(2p_1, 2p_0, 2p_{-1})$, or from $(2p_x, 2p_y, 2p_z)$, since the functions in one of these sets are linear combinations of the other. The second system has the double advantage of being best suited to the symmetry of the molecule and of bringing into play real functions only.

‡ This method of approximating the formation of molecular orbitals by means of a linear combination of the atomic orbitals of the bound atoms is known as the LCAO (Linear Combination of Atomic Orbitals) approximation.

The treatment can be shortened considerably by using the results of the preceding chapter, which implies that this is a bonding MO. However, because of the subsequent generalization, it seems preferable to reproduce the energy calculations already developed for the case of H_2^+ (§30) in their entirety, since they can be applied to the case of more complex molecules.

To do this we form the expression (5-1) for the mean energy in the state defined by φ. It would be useless to distinguish between

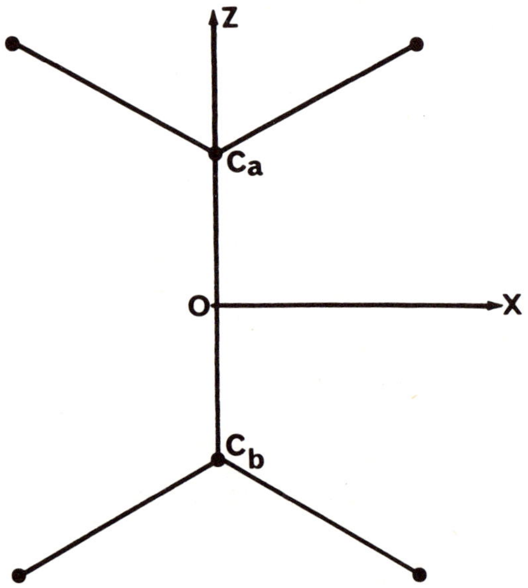

Figure 40. Diagram of the ethylene molecule indicating the reference axes.

a number and its complex conjugate here, since the atomic orbitals we will use are all real.† Thus we can write:

(2) $$E = \int \varphi \mathbf{H} \varphi d\tau.$$

The integral (2) extends over all space with $d\tau = dx\,dy\,dz$. Its evaluation requires a knowledge of the hamiltonian operator which we can take in the form:

(3) $$\mathbf{H} = -\frac{\hbar^2}{2m_0}\Delta + V_a + V_b + U$$

† As a result, it is possible to limit ourselves to cases where c_a and c_b are real.

where V_a and V_b represent, respectively, the potential energy of the electron in the presence of atoms C_a and C_b, taking into account the screening effect of their K shell, and $U(x,y,z)$, the mean potential energy of interaction with the nearest electrons, that is with the two electrons of the σ bond between the two carbon atoms and with the second electron whose behavior is governed by the same orbital φ actually under consideration. The form of this function $U(x,y,z)$ can only be gotten by use of self-consistent field methods (§46) which cannot be easily treated in the limit of the one-electron approximation. However, the calculation of this function U will eventually be unnecessary, and it suffices to assume that it takes the same form in each of the different bonds under study.

The usual method of determining c_a and c_b involves finding the extremum of (2), which we write:

$$E = Q(c_a^2 + c_b^2) + 2\gamma c_a c_b$$

with

$$Q = \int \psi_a \mathbf{H} \psi_a d\tau = \int \psi_b \mathbf{H} \psi_b d\tau$$

$$\gamma = \int \psi_a \mathbf{H} \psi_b d\tau = \int \psi_b \mathbf{H} \psi_a d\tau.$$

The coefficients c_a and c_b are related by the normalization condition of φ:

$$N = c_a^2 + c_b^2 + 2Sc_a c_b = 1$$

with

$$S = \int \psi_a \psi_b d\tau \quad \text{(overlap integral of } \psi_a \text{ and } \psi_b\text{).}$$

The extremum of E, taking into account the normalization, is written as in section 30 by introduction of the Lagrange multiplier (Appendix III):

$$Qc_a + \gamma c_b = \lambda(c_a + Sc_b)$$
$$\gamma c_a + Qc_b = \lambda(Sc_a + c_b).$$

The term by term addition of these two equations after multiplication by c_a and c_b, respectively, gives $\lambda = E$, with the result that the equations above take the form:

$$(Q - E)c_a + (\gamma - ES)c_b = 0$$
$$(\gamma - ES)c_a + (Q - E)c_b = 0.$$

The compatibility condition on these two linear homogeneous equations in c_a and c_b gives the allowed energy values so that:

$$(Q - E)^2 = (\gamma - ES)^2.$$

We get, as in section 30, two solutions, one symmetric, and one antisymmetric; in keeping with the general notation for the states of diatomic molecules we use the letters g and u† and we have:

$$(4) \quad \begin{cases} E_u = \dfrac{Q + \gamma}{1 + S}; \\ E_g = \dfrac{Q - \gamma}{1 - S}; \end{cases}$$

where

$$c_a = c_b = \frac{1}{\sqrt{2(1 + S)}} \quad \text{and} \quad \varphi_u = \frac{\psi_a + \psi_b}{\sqrt{2(1 + S)}};$$

$$c_a = -c_b = \frac{1}{\sqrt{2(1 - S)}} \qquad \varphi_g = \frac{\varphi_a - \psi_b}{\sqrt{2(1 - S)}}.$$

The results of this calculation are shown on the diagram (fig. 41), and we note that the two integrals Q and γ are negative,‡ which explains the relative disposition of the levels. To compare this with experiment it is necessary to note that the most stable state corresponds to two electrons filling the lowest energy state (E_u) in the normal state denoted by N, so that we get the value $2E_u$ for the electron energy in this state. The difficulty lies in finding the experimental significance of the total energy of the π electrons. We get this by observing that if the behavior of the electron is represented by ψ_a,

† The character of the p-functions involved should be noted (fig. 37) in that the symmetric combination of these functions is of the u type while the antisymmetric combination is of the g type.

‡ We get this result by writing the hamiltonian (3) in the form: $H = H_a + v_b + U$, where H_a refers to the isolated a atom so that: $H_a\psi_a = E\psi_a$, and E designates the (negative) energy of the electron fixed on this isolated a atom. We can then assume that the main part of Q is given by E, from the fact that V_b is weak near a, and that U is a corrective term, from which: $Q \sim E < 0$.

For γ, the defining equation leads us to write with this hamiltonian:

$$\gamma = ES + \int \psi_b V_b \psi_a d\tau + \int \psi_b U \psi_a d\tau.$$

If we ignore the correction U as usual, there remains, the potential V_b being negative:

$$\gamma - ES \sim \gamma - QS \sim \int \psi_b V_b \psi_a d\tau < 0,$$

which shows that not only γ, but also $\gamma - QS$, are negative quantities.

that is to say, in the absence of the bond, the energy calculated by means of (1) has as its value:

$$\int \psi_a \mathbf{H} \psi_a d\tau = Q.$$

The same is true for the other electron, from which, by subtraction, we find the expression for the contribution of the π electrons to the total bond energy to be:

$$2E_u - 2Q = \frac{2(\gamma - QS)}{1 + S}.$$

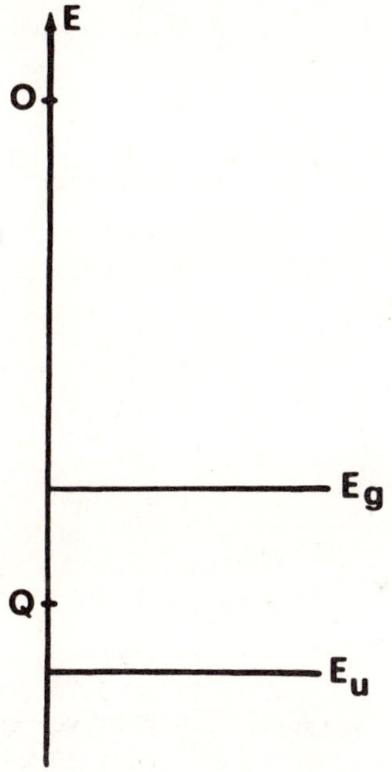

Figure 41. Disposition of the E_u and E_g levels of a π electron in the ethylene molecule.

This quantity is negative, which is natural, since the establishment of the π bond has the effect of stabilizing the molecule thereby making that level deepest. It is still convenient in applications to designate the bond energy by a positive number, this simply requiring a sign change. To avoid confusion we in this case use the letter W, which leads to:

$$W_\pi = -\frac{2(\gamma - QS)}{1 + S}. \quad (5)$$

In graphic representations, we take the W-axis descending vertically so as to get the same arrangement of levels as for E taken along the ascending vertical. We stress, additionally, that two representations are possible, one giving the one-electron levels and their population, the other denoting the total energy of the molecule. The transition between these is shown in Figure 42. This brings us to the problem of comparing these results with experiments which bear on bond energy and on transition energy.

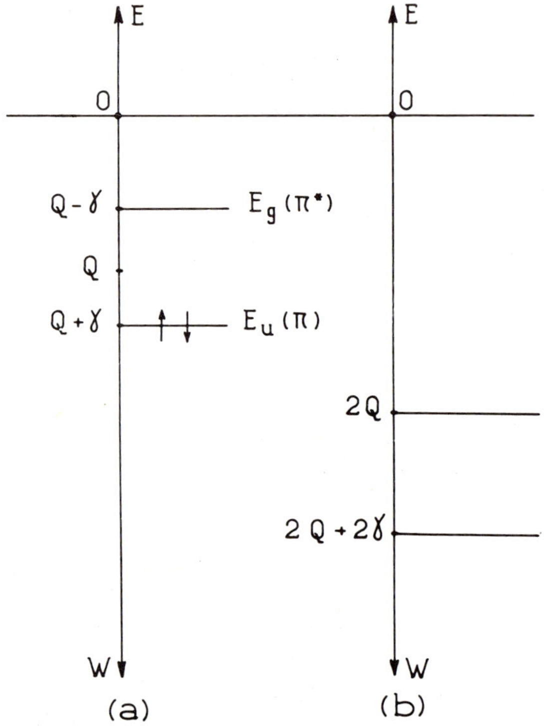

Figure 42. Comparative representation of the π electron states in ethylene: (*a*) indicates the one-electron levels and their occupation; (*b*) indicates the total energy for the two case π_2(N) and $\pi\pi^{\star}$(V).

(a) Bond Energy

First of all we give the commonly used energy, which assumes the value of 170 kcal/mole for the heat of sublimation of graphite:

Single Bond:

C—C $W_1 = 82.5$ kcal/mole $r_1 = 1.54$ Å

Double Bond:

C=C $W_2 = 145.2$ kcal/mole $r_2 = 1.34$ Å

where r_1 and r_2 indicate the internuclear distances for the molecule in equilibrium.

It is important to recall that these values are gotten by analyzing, in bond increments, the energy of formation in the standard state (298.1°K) starting from C and H atoms. This analysis assumes that the CH bond has the same energy in saturated derivatives and in ethylene, which is an approximation whose validity is difficult to assess.

The most simple application of these results is made by noting that the ethylenic bond is a juxtaposition of a σ bond and of a π bond, so that:

$$W_2 = W_\sigma + W_\pi \tag{6}$$

and by assuming that this σ bond is identical to that of energy W_1, which is found in saturated carbon compounds, so that:

$$W_\pi = W_2 - W_1 = 62.7 \text{ kcal/mole.}$$

The assumption that $W_\sigma = W_1$ is, however, subject to criticism for several reasons. First of all, the bonds are built up starting from AO's of different types, tetrahedral for the simple W_1 bond and trigonal for W_σ; in addition, the effective charge of the carbon atom is not the same in the two cases, since the introduction of a double bond leads to the disappearance of one C—H bond on each of the bound carbon atoms, and hence a modification of the effective charge on these atoms. Finally, it seems difficult to identify the energy of the σ bond of length $r_2 = 1.34$ Å in ethylene with W_1 which corresponds to a bond length of $r_1 = 1.54$ Å. However, a correction is possible as far as this last point is concerned, since the energy difference:

$$W_1(r_1) - W_1(r_2)$$

can be calculated starting from the bond force constant f_1 determined from the vibrational spectrum (Raman spectroscopy). The illustration (fig. 43) indicates the relative disposition of the levels, which is expressed by:

$$\begin{cases} W_\sigma = W_1 - \dfrac{1}{2} f_1 (r_1 - r_2)^2 \\ W_2 = W_\pi + W_1 - \dfrac{1}{2} f_1 (r_1 - r_2)^2. \end{cases} \tag{7}$$

The same calculation can be made for a value r of the internuclear distance, which differs from the equilibrium value r_2, giving:

$$W_2(r) = W_\pi(r) + W_1(r) .$$

Using the quadratic approximation of W(r) involving the force constant, this becomes:

$$\text{(8)} \qquad W_\pi(r) = W_2 - W_1 - \frac{1}{2} f_2(r - r_2)^2 + \frac{1}{2} f_1(r - r_1)^2.$$

We can imagine getting an expression closer to $W_\pi(r)$ by using the more elaborate forms of $W_2(r)$ and $W_1(r)$, but if we take as

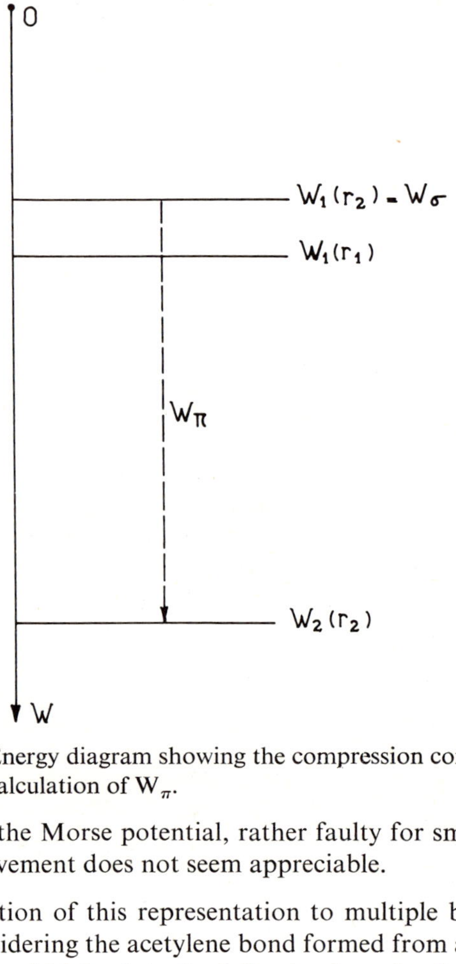

Figure 43. Energy diagram showing the compression correction in the calculation of W_π.

an example the Morse potential, rather faulty for small values of r, the improvement does not seem appreciable.

One application of this representation to multiple bonds can be seen by considering the acetylene bond formed from a σ bond and two π bonds, which leads to the following form for the total energy $W_3(r)$:

$$\text{(9)} \qquad W_3(r) = W_1(r) + 2W_\pi(r)$$

so that following expression (8) for $W_\pi(r)$:

$$W_3(r) = 2W_2 - W_1 - f_2(r - r_2)^2 + \frac{1}{2} f_1(r - r_1)^2.$$

The equilibrium position is defined by the minimum of $W_3(r)$, which leads to:

$$r_3 = \frac{2f_2r_2 - f_1r_1}{2f_2 - f_1}.$$

We can also deduce the force constant:

$$f_3 = -\frac{d^2W_3}{dr^2} = 2f_2 - f_1.$$

We now indicate the results obtained by taking $f_1 = 4.5 \times 10^5$ dynes/cm and $f_2 = 9.6 \times 10^5$ dynes/cm:

	Calculated	Experimental
r_3	1.28 Å	1.21 Å
f_3	14.7×10^5 dynes/cm	15.6×10^5 dynes/cm
W_3	224 kcal/mole	199 kcal/mole

The agreement is not too good, but the case being treated is *a priori* unfavorable since it involves an extrapolation. It is likely that equation (8) is a better representation of the facts when it is a question of bonds that are intermediate between single and double bonds; in the example of the triple bond, it is likely that the contribution of the repulsion to the total energy would be underestimated,† which would lead to an excessively high value of the bond energy.

The evaluation of W_π by means of (7) leads to a correction of 13 kcal/mole, hence a value of:

$$W_\pi = 62.7 + 13 = 75.5 \text{ kcal/mole}.$$

Equation (5) allows us to evaluate the quantity $\beta = \gamma - QS$, provided that we know the value of the overlap integral S which can be found directly using the Slater form of the orbitals $2p_y$. Another evaluation of β can be made by taking $S = 0$, from which $\beta \sim -38$ kcal. These two methods of dealing with the experimental data all lead to acceptable results as long as the same basic hypothesis is retained in the applications, namely, taking the overlap integral into account in the first case and ignoring it in the second.

† It may be desirable to separate in $W(r)$ the term of nuclear repulsion, or more precisely, the K shell core of the bound atoms, from the attractive force due to the energy of the bonding electrons, but such analysis seems difficult. One other factor of disagreement rests on the fact that the linear σ bond which occurs in C≡C is certainly of lower energy than the tetrahedral bond.

(b) Transition Energy

The excited state nearest to the normal state is described by the transition of an electron from the φ_u to the φ_g orbital. The transition can be effected by the absorption of a photon of frequency ν, and the conservation of energy imposes the Bohr condition:

$$h\nu = E_g - E_u = -\frac{2(\gamma - QS)}{1 - S^2} = \frac{W_\pi}{1 - S}.$$

The excited state thus obtained, described by the same AO's as the ground state, is denoted by V, so that we can remember that the bound atoms are in the same valence state as in the normal state N. The transition frequency N $\rightarrow$ V can then be calculated in principle from the energies given earlier. In fact, by taking $W_\pi = 75.7$ kcal/mole and $S = 0.25$, equation (9) leads to a value of 0.28 μ for the absorption wavelength much greater than the experimental value of 0.18 μ. The lack of agreement is essentially due to the fact that in state V the repulsion between the two electrons placed in different MO's is less than in state N, in which these two electrons occupy the same orbital, which leads to a different value of the parameter γ such that $|\gamma'| > |\gamma|$. The electronic energies of the two states have for their respective values:

$$W_N = -\frac{2(Q + \gamma)}{1 + S};$$

$$W_V = -\left[\frac{Q + \gamma'}{1 + S} + \frac{Q - \gamma'}{1 - S}\right] = -\frac{2(Q - \gamma' S)}{1 - S^2}$$

hence, the energy of transition:

$$(10) \qquad h\nu = W_N - W_V = -\frac{2}{1 - S^2}[\gamma - QS + (\gamma' - \gamma)S]$$

is higher than that calculated from (9).

Although the semi-empirical MO method is very effective for comparing the normal levels of molecules with π electrons or for the study of the excited levels of molecular analogs, it is difficult to use for relating the transition energy to the electronic energy in the ground state. We will see later (§37) that this result can be gotten more directly by using an approximation derived from a different point of view.

The Electronic Structure of Benzene 35

The physical properties of benzene can be interpreted by assuming that the molecule is planar with hexagonal symmetry (fig. 44). We are thus lead to conclude that each atom of carbon is bound to two adjoining carbon atoms, as well as to a hydrogen atom, by three bonds (trigonal AO's) similar to those which occur in ethylene. This accounts for a number of electrons equal to 3 × 6 or 18. The distribution of the remaining six valence electrons must be provided

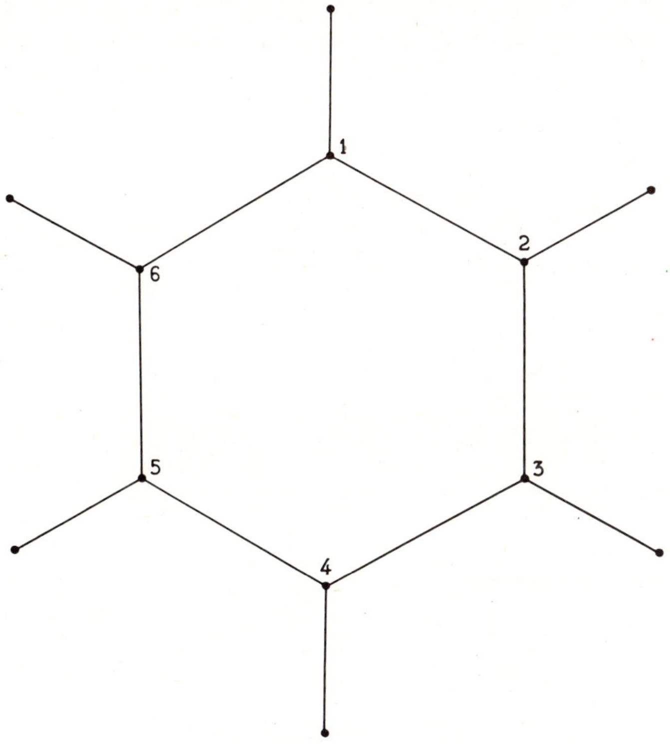

Figure 44. The benzene molecule.

for by the MO constructed from the 6 yet unused six $2p_y$ AO's. These MO's are of the form:

$$\varphi = \sum_r c_r \psi_r$$

where ψ_r represents the $2p_y$ orbital of the atom indexed by r. The determination of the coefficients c_r proceeds by the same variational method used in the study of the ethylenic bond. First of all the coefficients must satisfy the normalization condition:

$$\int |\varphi^2| \, d\tau = \sum_{r,s} c_r^* c_s \int \psi_r^* \psi_s d\tau = 1, \quad \text{with } r \text{ and } s \text{ from 1 to 6.}$$

The overlap integral decreases very rapidly with increasing internuclear distance, with the result that we limit our treatment to the overlap between adjacent atoms so that:

$$S = \int \psi_r^* \psi_s d\tau, \quad \text{with : } \quad |r - s| = 1.$$

The normalization condition is written:

(11) $$\sum_r c_r^2 + 2S \sum_r c_r c_{r+1} = 1$$

assuming the coefficients real, which is plausible, since the initial AO's are real functions.

It is also necessary to form the energy expression. We write this taking into account only the matrix elements $H_{r,s}$ relative to immediate neighbors. With the same notation as in the case of the ethylenic bond ($H_{r,r} = Q$; $H_{r,r+1} = \gamma$) we have:

(12) $$E = \int \varphi^* (\mathbf{H} \varphi) d\tau = Q \sum_r c_r^2 + 2\gamma \sum_r c_r c_{r+1}.$$

The extremum condition on the energy taken with the normalization condition is written by introducing the Lagrange multiplier. The six equations obtained have the same form, due to the symmetry of the molecule, so that:

$$Qc_r + \gamma(c_{r-1} + c_{r+1}) = \lambda[c_r + S(c_{r-1} + c_{r+1})].$$

The significance of the Lagrange multiplier is again found by adding the six conditions above term by term after multiplication by c_r. Taking these with (11) and (12) we get:

$$\lambda = E$$

with the result that the extremum condition is written:

(13) $$(Q - E)c_r + (\gamma - ES)(c_{r-1} + c_{r+1}) = 0.$$

The compatibility condition of this system of six linear homogeneous equations with six unknowns c_r can be written as the vanishing of a determinant of order six.† It is, however, quicker

† The equation in E we get is often called the *secular equation* as in celestial mechanics where equations of this type are often used.

to use the symmetry of equation (13) which implies the existence of particular solutions:†

$$c_r = c\theta^r$$

provided that it satisfies the condition gotten by substitution in (13), that is:

(14) $$Q - E + (\gamma - ES)(\theta + \theta^{-1}) = 0.$$

Equation (13) is useful for any value of r, in particular, for $r = 6$, which implies $c_7 = c_1$. The value of θ cannot be arbitrary and must satisfy:

(15) $$\theta^6 = 1.$$

where:

$$\theta = \exp\left(\frac{2k\pi i}{6}\right), \quad \text{with } k \text{ an integer.}$$

The energy calculated from (14) is given by:

(16) $$E = \frac{Q + 2\gamma \cos \frac{2k\pi}{6}}{1 + 2S \cos \frac{2k\pi}{6}}.$$

We get six distinct solutions (fig. 45) which can be characterized by the following values of k: 0, ± 1, ± 2, 3. The values k and $-k$ correspond to the same energy value, which is thus degenerate since $\theta \neq \theta^{-1}$, which then implies the most general form of c_r:

$$c_r = c\theta^r + c'\theta^{-r} = c \exp\left(\frac{2k\pi r i}{6}\right) + c' \exp\left(-\frac{2k\pi r i}{6}\right).$$

By hypothesis, we have assumed the coefficients c_r real, which leads us to replace the expression above by a linear combination

---

† This is a consequence of the symmetry of order six of the benzene molecule which implies that if $\varphi = \sum_r c_r \psi_r$ is a solution for a value E of energy, it must also be a solution for a function φ' deduced from φ by a rotation of order six, which shifts all the ψ_r by one number, hence: $\varphi' = \sum_r c_r \psi_{r-1} = \sum_r c_{r+1} \psi_r$. In the case of a non-degenerate level E, the two functions φ' and φ should be identical to within a factor θ, so that $\varphi' = \theta\varphi$, hence $c_{r+1} = \theta c_r$ for all values of r, which implies the equation given earlier $c_r = c\theta^r$.
This method, which should in principle give the non-degenerate levels, allows us to get all of them, each doublet level corresponding to two values of θ, namely, θ and θ^{-1}.

with the arbitrary real coefficients of the two solutions which we call, respectively, c'_r and c''_r:

$$c'_r = N' \cos \frac{2k\pi r}{6}; \qquad c''_r = N'' \sin \frac{2k\pi r}{6}, \tag{17}$$

where N′ and N″ are the normalization coefficients. Thus we get the following system of possible states, each characterized by the energy and the MO expression. We indicate, in addition, the

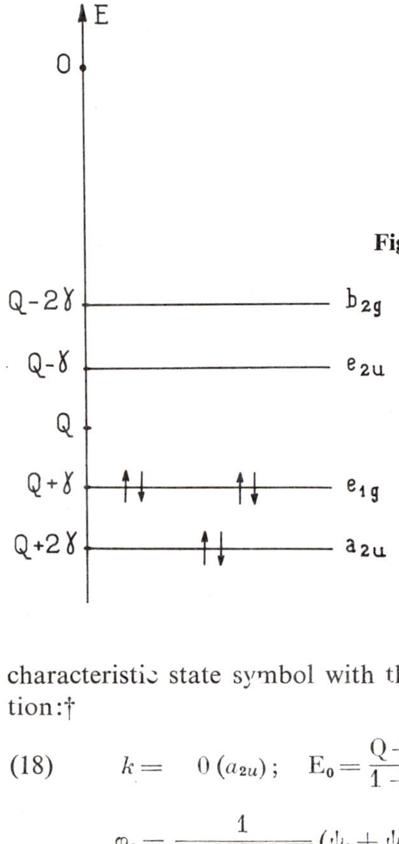

Figure 45. Disposition of the energy levels of the π electrons in benzene, and their occupation in the normal (ground) state.

characteristic state symbol with the usual group theoretic notation:†

$$k = 0\ (a_{2u}); \quad E_0 = \frac{Q + 2\gamma}{1 + 2S}; \tag{18}$$

$$\varphi_0 = \frac{1}{\sqrt{6(1 + 2S)}}(\psi_1 + \psi_2 + \psi_3 + \psi_4 + \psi_5 + \psi_6)$$

† We identify the index *g* or *u* with the behavior upon inversion. *a* and *b* designate the non-degenerate levels and *e* the degenerate levels of order 2. We use the notation recommended by Mulliken by taking the lower-case letters to designate the one-electron states, the capitals being reserved for spectroscopic terms.

$$k = \pm 1\ (e_{1g})\,; \quad E_1 = \frac{Q+\gamma}{1+S}\,;$$

$$\varphi_1' = \frac{1}{\sqrt{12(1+S)}}(\psi_1 - \psi_2 - 2\psi_3 - \psi_4 + \psi_5 + 2\psi_6)$$

$$\varphi_1'' = \frac{1}{\sqrt{4(1+S)}}(\psi_1 + \psi_2 - \psi_4 - \psi_5)$$

$$k = \pm 2\ (e_{2u})\,; \quad E_2 = \frac{Q-\gamma}{1-S}\,;$$

$$\varphi_2' = \frac{1}{\sqrt{12(1-S)}}(\psi_1 + \psi_2 - 2\psi_3 + \psi_4 + \psi_5 - 2\psi_6)$$

$$\varphi_2'' = \frac{1}{\sqrt{4(1-S)}}(\psi_1 - \psi_2 + \psi_4 - \psi_5)$$

$$k = \quad 3\ (b_{2g})\,; \quad E_3 = \frac{Q-2\gamma}{1-2S}\,;$$

$$\varphi_3 = \frac{1}{\sqrt{6(1-2S)}}(\psi_1 - \psi_2 + \psi_3 - \psi_4 + \psi_5 - \psi_6).$$

The ground state corresponds to filling the levels $k = 0$ and $k = \pm 1$ by the six π electrons, hence the electron energy value:

$$E = \frac{2(Q+2\gamma)}{1+2S} + \frac{4(Q+\gamma)}{1+S}. \tag{19}$$

To get the contribution E_π of the π electrons to the bond energy, it is necessary—as in the ethylene case—to make comparison with the reference state where each of the six electrons is localized on one of the carbon atoms, the electron energy of which is raised to 6Q. We then get:

$$E_\pi = E - 6Q = \frac{4(\gamma - QS)}{1+2S} + \frac{4(\gamma - QS)}{1+S}.$$

This quantity is negative, which, with the use of the inverse sign convention for defining the energy W_π of the π bond electrons, leads to:

$$W_\pi = -\frac{4(\gamma - QS)}{1+2S} - \frac{4(\gamma - QS)}{1+S}. \tag{20}$$

This expression is to be compared to the value of W_π calculated for the ethylene molecule, and given by equation (5). Some correction is necessary so as to take into account, through (8), the variation of the distance between the neighboring carbon atoms. We deduce the total electronic bond energy of the ring by adding the contributions of the σ bonds, so that by definition:

$$W_B = 6W_1(r) + W_\pi(r)$$

explicitly:

$$W_B = 6W_1 - 3f_1(r - r_1)^2$$
$$+ 2\left[W_2 - W_1 - \frac{1}{2}f_2(r - r_2)^2 + \frac{1}{2}f_1(r - r_1)^2\right]\left(1 + \frac{1+S}{1+2S}\right).$$

The most stable configuration corresponds to the minimum of W_B, so that:

$$\frac{dW_B}{dr} = 0.$$

We make the calculation taking $(1 + S)/(1 + 2S) = 1 - \lambda$ which leads to:

$$\frac{dW_B}{dr} = -2(1 + \lambda)f_1(r - r_1) - 2(2 - \lambda)f_2(r - r_2) = 0.$$

The second derivative allows us to derive the force constant f_B of the benzene bond, since the deformation by variation of r leaves the valence angles unchanged, and thus we can write:

$$W_B(r) = W_B(r_B) - 6 \cdot \frac{1}{2}f_B(r - r_B)^2$$

where:

$$\frac{d^2W_B}{dr^2} = -6f_B = -2(1 + \lambda)f_1 - 2(2 - \lambda)f_2$$

so that:

$$f_B = \frac{1}{3}[(1 + \lambda)f_1 + (2 - \lambda)f_2].$$

This calculation does not take into account the variation of S with the internuclear distance r, which does not have much significance here. We outline the numerical results calculated from the formulas above by taking $W_1 = 82.5$ kcal/mole, $W_2 = 145.2$ kcal/mole, $f_1 = 4.5 \times 10^5$ dynes/cm, and f_2 9.6×10^5 dynes/cm with two assumptions, either $S = 0$ and $\lambda = 0$, or $S = 0.25$ and $\lambda = \frac{1}{6}$.

	$S = 0$	$S = 0.25$	Experimental
r_B	1.375 Å	1.385	1.383
f_B	7.9×10^5 dyne/cm	7.61×10^5	7.6×10^5
W_B	725 kcal/mole	701	720

The determination of r_B and f_B leads to results in good agreement with the experimental values with little relation to the choice of S. The same is not true for W_B. The comparison with experimental results poses additional tricky problems which are far from resolved. However, detailed examination of these is far beyond the scope of this book.

We again point out that the energy value is often expressed by taking the hypothetical form of Kékulé as the reference state. If we take—and this is quite arbitrary—the energy in this form:

$$W_K = 3W_1 + 3W_2,$$

the resonance energy R is calculated by subtraction from W_B thus:

$$\begin{aligned} R &= W_B - W_K \\ &= (1 - 2\lambda)(W_2 - W_1) - (1 + \lambda)f_1(r_B - r_1)^2 - (2 - \lambda)f_2(r_B - r_2)^2. \end{aligned}$$

This definition of R leads to quite different values as a result of the hypothesis, so that $R = 42$ kcal/mole for $S = 0$ and $R = 18$ kcal/mole for $S = 0.25$.

The General Case: Bond Index 36

The method of calculation demonstrated in the case of benzene can be extended for the treatment of π electrons in any molecule. We form the MO by means of the linear combination of AO's ψ_r of the type $2p_y$ belonging to different carbon atoms with trigonal hybridization. For any of the MO's, denoted φ^α, we write:

(21) $$\varphi^\alpha = \sum_r c_r^\alpha \psi_r.$$

Throughout this paragraph, we neglect the value of the overlap integral because the preceding study indicated that this simplification has little effect on the precision with which we can determine the bond length. The normalization of φ^α takes the simple form:

(22) $$\sum_r (c_r^\alpha)^2 = 1$$

limited here to real orbitals, which is always possible. Using the notation of the preceding two paragraphs, the expression for the mean energy in the state defined is written:

(23) $$E^\alpha = Q\sum_r (c_r^\alpha)^2 + 2\sum_{r,s} \gamma_{rs} c_r^\alpha c_s^\alpha.$$

The coefficients c_r^α of φ^α are fixed by the extremum condition on E^α, so that:

$$Qc_r^\alpha + \sum_s \gamma_{rs} c_s^\alpha = \lambda c_r^\alpha.$$

As in the previous case, addition of these equations multiplied by c_r^α leads to $\lambda = E^\alpha$. Finally, the equations are written:

$$(24) \qquad (E^\alpha - Q)c_r^\alpha = \sum_s \gamma_{rs} c_s^\alpha.$$

The possible values of E^α are fixed by the compatibility conditions of the set of equations (24) gotten by the vanishing of the determinant of the coefficient of the c's. For each of the roots, the coefficients c_r^α can then be calculated, taking into account the normalization condition (22). It only remains to place the π electrons in the lowest levels so as to describe the ground state of the molecule. We will use a standard example to demonstrate the calculations for a particularly simple case.

Butadiene

$$(\underset{(1)}{CH_2} = \underset{(2)}{CH} - \underset{(3)}{CH} = \underset{(4)}{CH_2}).$$

We know that this molecule has an electronic structure close to the form foreseen in classical organic chemistry and behaves like two ethylenic bonds with a mesomeric form† with a central double bond having appreciably smaller energy. Thus we have a case which differs from that of benzene in which the bonds are all equal. Nevertheless, it is convenient to treat the calculations in the first approximation by adopting the hypothesis of aromaticity, i.e., assuming all bond lengths equal, as a consequence of which we introduce a single dynamical parameter $\gamma_{12} = \gamma_{23} = \gamma_{34} = \gamma$. We will repeat the previous calculation with expression (21) for the MO's so that:

$$\varphi = c_1\psi_1 + c_2\psi_2 + c_3\psi_3 + c_4\psi_4.$$

† Chemists understand this word as designating different formulas written with the same skeleton of the molecule, and differing only in the electron distribution and thus in the nature of the bond (here we have only $CH_2{=}CH{-}CH{=}CH_2$ and $\dot{C}H_2{-}CH{=}CH{-}\dot{C}H_2$). Such formulas will be introduced in section 52 and correspond to different bonding schemes. The idea of mesomeric forms is a little more restrictive, however, this designation being reserved for schemes involving a large number of possible bonds between adjacent atoms even if a single arbitrarily more important form occurs in the sketch of the bond scheme (see §52, **Exercise 1**).

Equations (22) and (23) are written, respectively:

$$c_1^2 + c_2^2 + c_3^2 + c_4^2 = 1.$$

$$E = Q(c_1^2 + c_2^2 + c_3^2 + c_4^2) + 2\gamma(c_1c_2 + c_2c_3 + c_3c_4).$$

The system of equations (34) is then written:

$$\begin{aligned}(E - Q)c_1 &= \gamma c_2\\(E - Q)c_2 &= \gamma(c_1 + c_3)\\(E - Q)c_3 &= \gamma(c_2 + c_4)\\(E - Q)c_4 &= \gamma c_3.\end{aligned}$$

The compatibility equation of this system is obtained by the vanishing of its determinant. The solution of the equation thus obtained can be seen *a priori*, as a consequence of the symmetry of the equations, to lead to two types of solutions:

Symmetric solutions: $c_4 = c_1$; $c_3 = c_2$. The system reduces to two equations:

$$\begin{aligned}(Q - E)c_1 + \gamma c_2 &= 0\\(Q - E)c_2 + \gamma(c_1 + c_2) &= 0\end{aligned}$$

from which the compatibility condition is written:

$$(Q - E)^2 + \gamma(Q - E) - \gamma^2 = 0$$

with the two roots:

$$E_s' = Q + \frac{1 + \sqrt{5}}{2}\gamma = Q + 1.618\gamma$$

$$E_s'' = Q + \frac{1 - \sqrt{5}}{2}\gamma = Q - 0.618\gamma.$$

Antisymmetric solutions: $c_4 = -c_1$; $c_3 = -c_2$. We likewise get:

$$\begin{aligned}(Q - E)c_1 + \gamma c_2 &= 0\\(Q - E)c_2 + \gamma(c_1 - c_2) &= 0.\end{aligned}$$

The roots of the compatibility equations are:

$$E_a' = Q - \frac{1 - \sqrt{5}}{2}\gamma = Q + 0.618\gamma$$

$$E_a'' = Q - \frac{1 + \sqrt{5}}{2}\gamma = Q - 1.618\gamma.$$

We find four roots of the general equation which are classified in the order indicated in the diagram (fig. 46). The normal state of the molecule corresponds to the occupation by four π electrons of the two lowest levels, E'_s and E'_a, which correspond to the orbitals whose coefficients are calculated from the equations above, along with the normalization, and which are written:

$$\varphi'_s = 0.373\psi_1 + 0.604\psi_2 + 0.604\psi_3 + 0.373\psi_4$$

$$\varphi'_a = 0.604\psi_1 + 0.373\psi_2 - 0.373\psi_3 - 0.604\psi_4.$$

In the state described, the contribution of the π electrons to the bond energy has the value:

$$-W_\pi = 2(E'_s - Q) + 2(E'_a - Q) = 2\sqrt{5}\gamma,$$

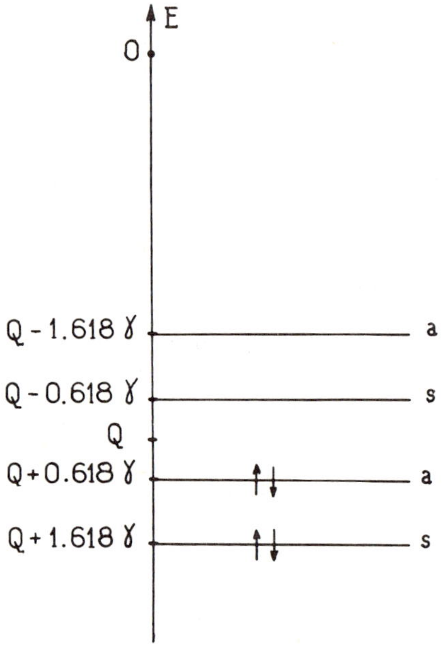

Figure 46. Disposition of the π electron energy levels in butadiene, and their occupation in the normal (ground) state.

the negative sign is introduced as before as a result of the opposing sign convention for E and W. This value is greater than that obtained by considering the two ethylenic bonds as independent, amounting to 4γ. The difference 0.472γ represents the delocalization energy of the electrons—known as the resonance energy—which, as in the case of benzene, is defined relative to some hypothetical form.

Bond Index

This idea is introduced as a generalization of the considerations of the compression energy developed in the particular cases of ethylene and benzene. We start with the total energy expression of the π bonds, gotten in the general case beginning from (23):

(25) $$W_\pi = -2\sum_{r,s,\alpha} \gamma_{r,s} c_r^\alpha c_s^\alpha$$

the index α numbering one of the occupied orbitals, and the summation being performed on the assembly of π electrons, with the result that each orbital occupied by two electrons is counted twice. The total energy is obtained by adding the σ-bond energy, which, taking into account the compressive correction, is written:

$$W_\sigma = \sum_l \left[W_1 - \frac{1}{2} f_1 (r_l - r_1)^2 \right].$$

the summation being extended to the different bonds. We carry out the calculations by neglecting the integrals $\gamma_{r,s}$ between atoms not united by a bond (labeled l). Each parameter $\gamma_{r,s} = \gamma_l$ is a function of the length r_l, and therefore each of the coefficients c_r depends on the set of distances r_l through the equations (24). The equilibrium state of the molecule is defined by the system of values c_r, which renders the total energy a minimum, thus:

$$\frac{\partial W}{\partial r_l} = -f_1(r_l - r_1) - 2\frac{d\gamma_l}{dr_l}\sum_\alpha c_r^\alpha c_s^\alpha - 2\sum_{r,s,\alpha} \gamma_{r,s}\frac{d}{dr_l}\left(c_r^\alpha c_s^\alpha\right) = 0.$$

The summation with respect to the indices r and s vanishes in the second term, since γ_l depends only on r_l. However, the extremum condition which fixes the coefficients c_r^α is written (23):

$$\sum_{r,s} \gamma_{rs} d\left(c_r^\alpha c_s^\alpha\right) = 0$$

with the result that the last term of the preceding sum is zero. This leads to the equation used for each of the bonds l:

(26) $$f_1(r_l - r_1) + 2\frac{d\gamma_l}{dr_l}\sum_\alpha c_r^\alpha c_s^\alpha = 0.$$

We then take:

(27) $$K_l = \sum_\alpha c_r^\alpha c_s^\alpha.$$

The number K_l defines the bond index l between the atoms r and s; remember that the summation is extended over all the occupied orbitals, each being counted as many times as the level's electron population. Thus for butadiene we get:

$$K_{1,2} = K_{3,4} = 0.88; \qquad K_{2,3} = 0.40.$$

The introduction of K_l allows us to write the π-bond energy as the sum of bond increments:

$$W_\pi = 2\sum_l |\gamma_l| K_l \tag{28}$$

with the result that K_l defines the contribution of bond l to the total energy of the π electrons. The index K_l is tied to the length

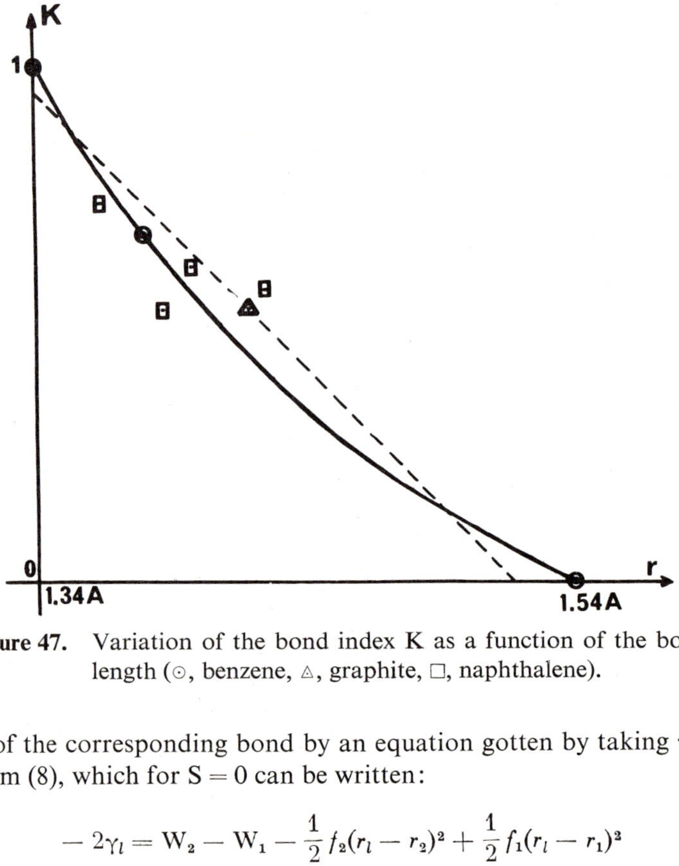

Figure 47. Variation of the bond index K as a function of the bond length (⊙, benzene, △, graphite, □, naphthalene).

r_l of the corresponding bond by an equation gotten by taking γ_l, from (8), which for $S = 0$ can be written:

$$-2\gamma_l = W_2 - W_1 - \frac{1}{2} f_2(r_l - r_2)^2 + \frac{1}{2} f_1(r_l - r_1)^2$$

where:

$$2\frac{d\gamma_l}{dr} = f_2(r_l - r_2) - f_1(r_l - r_1).$$

Equation (26) lets us write, suppressing the index l:

$$K = \frac{f_1(r_1 - r)}{f_1(r_1 - r) + f_2(r - r_2)} \simeq \frac{r_1 - r}{r_1 + r - 2r_2} \tag{29}$$

by taking $f_2 \simeq 2f_1$. Figure 47 represents $K(r)$ taking for r_1 and r_2 the measured values for the single and double bond, 1.54 Å and

1.34 Å. Various other points are given corresponding to the benzene bond ($K = 0.66$, $r = 1.38$ Å), graphite ($K = 0.525$, $r = 1.42$ Å) and to various other bonds from the naphthalene molecule. The representation is not too good, reflecting on the one hand a lack of precise knowledge of the internuclear distance, and on the other, the simplified character of the theory. If we note that the majority of the bonds examined correspond to the upper part of the curve, we can see the possibility of getting a somewhat better representation by means of the straight line:

$$r = 1.52 - 0.19\,K \quad (r \text{ in Å})$$

which corresponds to the dotted line in the figure.†

† See also, Coulson, C. A., *Proc. Roy. Soc.* (1939), **160**, 413.

9 THE FREE ELECTRON MODEL

Approximation of the Transition Energy for Ethylene 37

The treatment (§34) of the ethylenic bond by the MO method in its simplest form does not allow us to connect the energy of the $\pi_u \rightarrow \pi_g$ transition to the energy of the π bond in a satisfactory way. We can obtain better results by replacing the function φ_g which defines the π_g level, by a function giving almost the same electronic distribution but better adapted to the energy calculation.

The method is based upon the fact that the radiation absorption for a molecule in the $\pi_g \rightarrow \pi_u$ transition is governed by:†

$$\vec{r}_{12} = \int \vec{r} \varphi_g^* \varphi_u d\tau.$$

The transition is forbidden when the three components of this vector are zero. In the case of the ethylene molecule we can easily imagine that the major component of this vector for the $\pi_u \rightarrow \pi_g$ transition should be directed along the z-axis of the molecule. The matrix element $z_{1,2}$ appears in the expansion of the function $z\varphi_u$ on an orthonormal system of functions containing the two functions φ_u and φ_g which are orthogonal as a result of being two solutions of the state equation. We can thus take *a priori* a series of the form:

$$z\varphi_u = z_{1,1}\varphi_u + z_{1,2}\varphi_g + z_{1,3}\varphi_3 + \ldots$$

In the example of ethylene considered here, the symmetry of the level π_u has as a consequence:

$$z_{1,1} = \int z\varphi_u^* \varphi_u d\tau = 0.$$

The possibility of an approximate treatment becomes evident when one of the possible transitions is predominant over the others, which reduces the preceding series to a single term. We will verify

† This equation, where $\vec{r}$ represents the radius vector from the origin taken at the center of the molecule to the electron, is gotten from the theory of electron transitions which is not treated here. A simple approach to this involves seeking *a priori* the form of the coordinate function in x, y, and z with which we must treat φ_u to obtain a function having the symmetry characteristics of φ_g. The factor z is suitable since it reverses the parity of the function, and furthermore, has bias toward the common antisymmetric character of φ_u and φ_g during reflection in the plane of the ethylene molecule, which leads to the test of (1) as the first approximation for representation of φ_g.

that this eventuality arises in the study of the $\pi_u \to \pi_g$ transition for the ethylene molecule, which we are going to treat in the first approximation:

(1) $$z\varphi_u = z_{1,2}\varphi_g,$$

where the matrix element $z_{1,2}$ is found from:

$$z_{1,2} = \int z\varphi_g^*\varphi_u d\tau.$$

We treat the calculation by taking the MO expressions for φ_u and φ_g obtained previously (§34), so that:

$$\varphi_u = \frac{\psi_a + \psi_b}{\sqrt{2(1+S)}}, \qquad \varphi_g = \frac{\psi_a - \psi_b}{\sqrt{2(1-S)}}$$

where ψ_a and ψ_b designate the $2p_y$ orbitals of the bound atoms. Thus we get:

(2) $$z_{1,2} = \frac{1}{2\sqrt{1-S^2}} \int z(\psi_a^2 - \psi_b^2)d\tau.$$

We have the difference of two integrals which are calculated by taking the nuclei a and b, respectively, as origins. This leads to:

$$z = -\frac{R}{2} + z_a = \frac{R}{2} + z_b.$$

Thus (2) reduces to:

(3) $$z_{1,2} = -\frac{R}{2\sqrt{1-S^2}}.$$

It now remains to verify that the expansion of $z\varphi_u$ reduces to the single terms $z_{1,2}\varphi_g$, which is shown by the equation:†

$$\int (z\varphi_u - z_{1,2}\varphi_g)^2 d\tau = 0,$$

indicating that the square of the difference between these two functions is zero.

By expanding and taking into account the value of $z_{1,2}$ we see that this condition takes the form:

(4) $$\int z^2\varphi_u^2 d\tau = (z_{1,2})^2.$$

The integral becomes explicitly:

$$\int z^2\varphi_u^2 d\tau = \frac{1}{2(1+S)} \int z^2(\psi_a^2 + 2\psi_a\psi_b + \psi_b^2)d\tau.$$

---

† The sign of the complex conjugate is suppressed since we are dealing with real MO's.

In a first approximation we can neglect the contribution of the term in $\psi_a\psi_b$ which is maximum for $z = 0$. The preceding change in variables leads to the expression:

$$\int z^2\varphi_u^2 d\tau = \frac{1}{4(1+S)}\left(R^2 + 4\overline{z_a^2}\right).$$

We now have to evaluate the integral:

$$\overline{z_a^2} = \overline{z_b^2} = \int z_a^2\psi_a^2 d\tau = \int z_b^2\psi_b^2 d\tau.$$

The calculation is easily performed by using the Slater form of the $(2p_y)_a$ orbital and leads to:

$$\overline{z_a^2} = \overline{z_b^2} = \frac{24a_0^2}{Z^{\star 2}} = \frac{\rho^2}{4},$$

a_0 designating the radius of the first Bohr orbit. With this notation the desired integral is written:

(5) $$\int z^2\varphi_u^2 d\tau = \frac{R^2 + \rho^2}{4(1+S)}.$$

Condition (4) then takes the form:

(6) $$1 - S = \frac{R^2}{R^2 + \rho^2}.$$

The calculation is made taking the usual value $Z^\star = 3.25$ (§28) and $R = 1.34$ Å, leading to $S = 0.22$ which is an entirely acceptable value for the overlap integral. The basic approximation of the method can thus be considered as well justified.

As we shall see, this approximation permits the calculation of the transition energy without our having to write the hamiltonian explicitly. We take the identity:

(7) $$\mathbf{H}z\varphi - z\mathbf{H}\varphi = -\frac{\hbar^2}{m_0}\frac{\partial\varphi}{\partial z},$$

which is immediately verified by examining the operator $\mathbf{H}$:

$$\mathbf{H} = -\frac{\hbar^2}{2m_0}\Delta + V,$$

and so (7) is true regardless of the form of the potential energy V.

By multiplying both sides of (7) by $z\varphi$ and by integrating over all space, we get:

$$\int z\varphi\mathbf{H}z\varphi d\tau - \int z^2\varphi\mathbf{H}\varphi d\tau = -\frac{\hbar^2}{m_0}\int z\varphi\frac{\partial\varphi}{\partial z}d\tau.$$

The right-hand side is transformed by integration by parts and for a normalized function φ gives as a result:

$$-\frac{\hbar^2}{2m_0}\int z\frac{\partial\varphi^2}{\partial z}d\tau = \frac{\hbar^2}{2m_0}\int\varphi^2 d\tau = \frac{\hbar^2}{2m_0}.$$

If we apply this relation to φ_u we get, taking account of (1):

$$\int z\varphi_u \mathbf{H} z\varphi_u d\tau = z_{1,2}^2\int\varphi_g \mathbf{H}\varphi_g d\tau = z_{1,2}^2 \mathrm{E}_g$$

$$\int z^2\varphi_u \mathbf{H}\varphi_u d\tau = \mathrm{E}_u\int z^2\varphi_u^2 d\tau = z_{1,2}^2 \mathrm{E}_u$$

hence the well-defined result:

$$\mathrm{E}_g - \mathrm{E}_u = \frac{\hbar^2}{2m_0 z_{1,2}^2} = \frac{2\hbar^2(1-\mathrm{S}^2)}{m_0\mathrm{R}^2}. \tag{8}$$

We have gotten the value of the transition energy starting from the geometry of the system without explicit recourse to the dynamical parameter γ (§34). In reality, it is quite evident that the internuclear distance R is defined by the dynamics of the system which thus enters indirectly. This does not mean that the replacement of R by γ is of interest, since what is involved is a quantity (R) directly accessible from experiment, while the discussion in section 34 shows the difficulty of determining γ.

Equation (8) leads to quite satisfactory results in the case of the $\pi_u \rightarrow \pi_g$ transition in ethylene. By taking R = 1.34 Å and S = 0.22 we get as the value of the absorption wavelength:

$$\lambda = \frac{2\pi^2 mc\mathrm{R}^2}{h(1-\mathrm{S}^2)} = 1.520 \times 10^{-8} \text{ cm}$$

while the observed wavelength is 1.630 Å.

Exercise 1. The preceding method is not applicable to all cases. We will verify this for the case of the $1s \rightarrow 2p_z$ transition in the helium atom by using the Slater form of the AO's, calculated with the usual screening coefficients. To do this calculate $z_{1,2}$ and verify that the square of the difference between $z_{1,2}\psi_{2p_z}$ and $z\psi_{1s}$ differs noticeably from zero. As a result, the transition energy determined from (8) is much too large. This brings us back to the statement that there is more than one possible transition originating from the $1s$ state.

The Free Electron Model 38

The preceding discussion—which allowed us to establish a relationship between the transition energy and the bond length—can be generalized to more complex molecules, but because this raises various difficulties we will not pursue this further.†

We will assume that the electron transition energy is quite insensitive to the choice of potential V as long as this energy corresponds to a fixed geometry of the molecule. We will attempt to establish a connection between the method treated above and the so-called *free electron* (or metallic model of color) to which we now turn.

Once again the treatment is via the one-electron model, that is to say, each electron is treated as isolated. The states are calculated as if the electron were free on the length L of the bond, with the result that the quantization is written in the simple form (2-26):

$$E = \frac{k^2 h^2}{8 m_0 L^2}, \qquad \text{with :} \qquad k = 1, 2, 3, \ldots$$

This model has achieved some success in the treatment of electron transitions in molecules having a sequence of alternating single and double bonds. One frequently cited example (Bayliss) is that of diethylene diamine, whose structure in the ionic state is interpreted by hybridization of the two mesomeric forms:

$$(CH_3)_2\overset{\oplus}{N}{=}CH-CH{=}CH-\overline{N}(CH_3)_2 \leftrightarrow$$

$$(CH_3)_2\overline{N}-CH{=}CH-CH{=}\overset{\oplus}{N}(CH_3)_2.$$

The equivalence of these mesomeric forms implies the equality of the bonds between the CH groups. For a chain involving n double bonds (here $n = 2$) the number of mobile electrons is $N = 2n + 2$ if we include the pair localized on one of the nitrogen atoms in each form. The total length of the chain on which one of the π electrons can be displaced is thus $L = 2nR = (N - 2)R$ by taking the same length R for the N—C and C—C bonds.

---

† J. Barriol and S. Regnier, *J. Chim. Phys.* (1956), 929; (1957), 231.
N. S. Bayliss, *J. Chem. Phys.* (1948), **16**, 287.
H. Kuhn, *Helv. Chem. Acta* (1948), **31**, 1441.
S. Nikitine, *J. Chem. Phys.* (1953), **50**, 282.
W. T. Simpson, *J. Chem. Phys.* (1948), **16**, 1124.
J. R. Platt, *J. Chem. Phys.* (1949), **17**, 484.

To make use of the equation above, it is necessary to define the ground state which corresponds to the filling of the first N/2 levels, all non-degenerate, by the N π electrons. The lowest transition corresponds to passing from $k = N/2$ to $k + 1$ giving the value of the transition energy:

$$\Delta E = \frac{(2k+1)h^2}{8m_0L^2} = \frac{(N+1)h^2}{8m_0(N-2)^2R^2}.$$

To make the comparison with the experimental results it is important to note the simplistic character of the infinite potential well model which led to the writing of the above equation. The model is improved by introducing an "edge correction" whose size l is connected to the depth V_0 of the potential well (2-27) by:

$$l = \frac{2}{\rho} = \frac{2\hbar}{\sqrt{2m_0V_0}}.$$

By taking $V_0 = 10$ eV (the ionization potential of ethylene), we get $l = 1.24$ Å, which gives an idea of the size of the correction term. The preceding formula is then written:

$$\Delta E = \frac{h^2}{8m_0R^2} \cdot \frac{(N+1)}{\left[N - 2 + \frac{l}{R}\right]^2} = \frac{hc}{\lambda}$$

hence the expression for the transition wavelength:

$$\lambda = \frac{8m_0c}{h} R^2 \frac{\left[(N+1) + \left(\frac{l}{R} - 3\right)\right]^2}{(N+1)}$$

$$= \frac{8m_0c}{h} R^2 \left[N + \left(\frac{2l}{R} - 5\right) + \frac{\left(\frac{l}{R} - 3\right)^2}{N+1}\right].$$

The final term is generally of relatively minor importance which leads to an approximately linear relationship between λ and the number of electrons.

The structure of the series which we happen to consider has been studied by W. T. Simpson (*loc. cit.*) who obtained a good description of the following absorption wavelengths:

$n = 2$	$\lambda = 309$ mμ
$n = 3$	$\lambda = 409$ mμ
$n = 4$	$\lambda = 511$ mμ.

It is necessary to take as the expression for the length of the chain L:

$$L = 2nR + l,$$

with:

$$R = 1.24 \text{ Å}, \qquad l = 1.55 \text{ R}.$$

These values obviously deviate from the experimental results, noticeably in the value of R which should be of the order of 1.39 Å.

The same method can be used for a closed chain system such as benzene. The quantization condition is that of the plane rotor (§15) and leads to the energy expression:

$$E = \frac{k^2h^2}{2m_0L^2} \; (k = 0, 1, 2, \ldots).$$

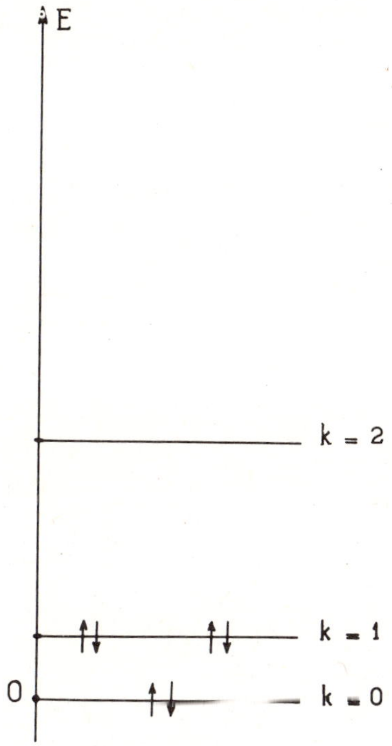

Figure 48. Energy levels of the π electrons of benzene in the free electron model.

L designating the length of the carbon chain in benzene which is here treated like a circle.

We recall that aside from the lowest level ($k = 0$), which is non-degenerate, all of the others are degenerate of order two, corresponding to the two possible senses of motion of the electron on the circle. In the case of benzene the six π electrons are situated on the levels $k = 0$ and 1, with the result that the first transition (fig. 48) corresponds to the energy:

$$\Delta E = \frac{3h^2}{2m_0L^2} = \frac{hc}{\lambda}$$

and the absorption wavelength is:

$$\lambda = \frac{2m_0cL^2}{3h}.$$

To find the observed value ($\lambda = 255$ mμ) it is necessary to take L = 8.86 Å, so that R = 1.48 Å, assuming that the path of the π electrons is effectively the perimeter of the hexagon formed by the C—C bonds. We again find the order of magnitude of the observed distance, which, given the simple character of the model, is quite satisfactory.

In conclusion, the free electron model leads to equations analogous to those obtained by means of the treatment in section 34, which we found from fundamental principles. It is clear, however, that this model, interesting because of its simplicity, cannot claim to furnish a valid representation of the electron distribution in the molecule, particularly in the case of small molecules. This is easily demonstrated if we attempt to treat the ethylene molecule by this method, as we then obtain a value for the transition energy which is much too high, even with an edge correction. However, as a simple model, the free electron model presents a convenient way of discussing the spectral properties of complex molecules.

10 WAVE FUNCTIONS ON A TRANSLATIONAL LATTICE

General Properties of Crystalline Lattices 39

The study of a quantum mechanical system rapidly becomes complicated when the number of constituent atoms becomes moderately large, unless the symmetry properties of the system introduce simplifications. A particularly important case in practice is that of a perfect crystalline lattice for which the invariance of the translational group of the lattice significantly simplifies the treatment, without need to resort to the formalism of group theory. We will first examine various general properties of such systems, beginning with the case of the *linear crystal*, which is made up of a series of identical atoms equidistantly spaced on a line. The ideas introduced in the study of this simple case are easily generalized to the three-dimensional case.

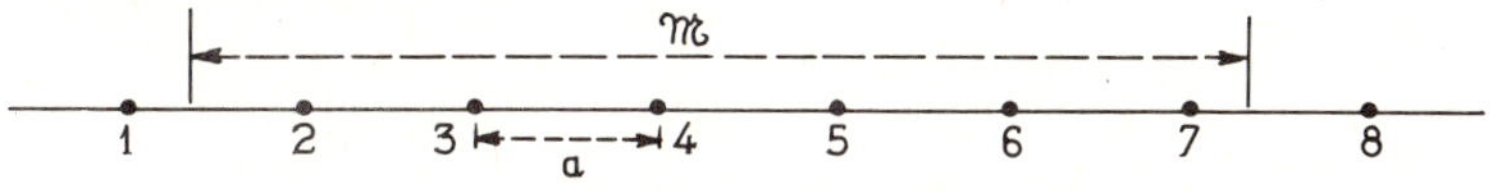

Figure 49. Diagram of a monoatomic linear lattice, with a lattice multiple of six unit cells.

One of the tricky problems in the study of crystals, linear or otherwise, rests in the choice of boundary conditions which we know must be somewhat arbitrary, since most of the properties of the crystal are *bulk properties*, insensitive to the physical state of the surface. In a precise way we can observe that on the one hand, examination of a sample of finite size does not permit rigorous use of the periodicity; while on the other hand, analysis of the states and levels presents difficulties when we deal with an infinite crystal. These difficulties have been eliminated by M. Born through the introduction of cyclic conditions for the functions of interest in the crystal. These are then taken to be periodic on a lattice multiple M, grouping an arbitrarily large number N of elementary lattice distances. We see immediately that the Born conditions allow us to consider a finite number of constituent particles while preserving the rigorous periodicity of the crystal. We can also say that the description of the crystal states becomes sufficiently precise when we examine a set of adjacent lattice sites of sufficiently large size. We will now examine the consequences of the use of the Born conditions by taking the linear crystal formed of atoms distributed along $x'x$ with equal spacing a (fig. 49).

The function $f(x)$ satisfies cyclic conditions if it is periodic on a lattice multiple of length $L = Na$, which permits representation by a Fourier series:

$$f(x) = \sum_{n'=-\infty}^{+\infty} c_{n'}.\exp\left(\frac{2\pi i n' x}{Na}\right), \quad \text{with } n' \text{ an integer.}$$

In this sum we collect the corresponding terms of the various values n' of the form $n' = n + kN$, where n is fixed and where the integer k runs from $-\infty$ to $+\infty$. We get N families of terms characterized respectively by $n = 0, 1, 2, \ldots, N-1$. The terms of each family have in common the factor $\exp\{2\pi inx/Na\}$, with the result that the function $f(x)$ can be written:

$$f(x) = \sum_{n=0}^{N-1} \exp\left(\frac{2\pi inx}{Na}\right) \sum_{k=-\infty}^{+\infty} c_{n,k}\cdot\exp\left(\frac{2\pi ikx}{a}\right).$$

The second sum on the values of k represents a function $u(x)$ periodic on the *elementary cell*, that is, on the period a. We are then able to put $f(x)$ in the form:

(1)
$$f(x) = \sum_{n=0}^{N-1} u_n(x)\cdot\exp\left(\frac{2\pi inx}{Na}\right).$$

$f(x)$ is thus a linear combination of functions of the form:

(2)
$$f_n(x) = u_n(x)\exp\left(\frac{2\pi inx}{Na}\right)$$

where $u_n(x+a) = u_n(x)$ designates a function periodic on the lattice. The functions $f_n(x)$ are called Bloch functions and are characterized by their behavior under a translation a:

(3)
$$\begin{aligned} f_n(x+a) &= u_n(x+a)\cdot\exp\left(\frac{2\pi in(x+a)}{Na}\right) \\ &= f_n(x)\cdot\exp\left(\frac{2\pi in}{N}\right). \end{aligned}$$

Conversely, if a function $f(x)$ behaves like (3) under a lattice translation, the function $f(x)\exp\{-2\pi inx/Na\}$ remains invariant, and as a result, represents a certain function $u_n(x)$ periodic on the lattice, the function $f(x)$ can then be put in the form (2).

Wave Functions for the Linear Lattice 40

We repeat the discussion in section 35, but this time for an indefinite line of equidistant atoms. We will keep the same notation, that is, we will form the MO's describing the behavior of an electron on the lattice as a linear combination of atomic orbitals of the same kind belonging to different atoms so that:

$$\varphi = \sum_r c_r \psi_r.$$

The problem is quite similar to the treatment of the benzene molecule (§35), the only difference being in the extension to an infinite number of atoms. We are again led to take the coefficients c_r of the MO in the form $c_r = c\theta^r$, but whereas in the case of benzene the existence of a closed chain implies $c_{r+6} = c_r$, so that $\theta = \exp\{2\pi in/6\}$, we in this case have only some similar dependence. A generalization of this calculation can only occur if we assume that wave functions of the lattice satisfy the Born cyclic conditions, so that $c_{r+N} = c_r$, and we get:

$$\theta = \exp\left(\frac{2\pi in}{N}\right), \qquad \text{with:} \quad n = 0, 1, 2, \ldots, N - 1.$$

The procedure for the calculations is thus identical with that in section 35, the number of levels being equal to the number N of distinct Bloch functions.

The treatment sketched here cannot be considered as absolutely general, however, since it assumes that the MO solutions of the stationary states are capable of being expressed as linear combinations of atomic orbitals (LCAO approximation). Because of this we will take the most general approach† without any assumptions about the structure of the solutions of the Schrödinger equation, which we now rewrite:

(4) $$\mathbf{H}\varphi = E\varphi$$

where the hamiltonian $\mathbf{H}$ is of the form:

$$\mathbf{H} = -\frac{\hbar^2}{2m_0}\Delta + V(x, y, z)$$

with the lattice periodicity expressed by the conditions:

(5) $$V(x + a, y, z) = V(x, y, z).$$

† This is treated thoroughly in L. Brillouin, *Wave Propagation in Periodic Structures* (New York: Dover, 1953).

We assume that we know a solution of (3), that is, an energy E and a wave function $\varphi(x,y,z)$. We now consider a new function defined by:

$$\varphi'(x, y, z) = \varphi(x + a, y, z) \tag{6}$$

where:

$$\varphi'(x - a, y, z) = \varphi(x, y, z).$$

Substitution in (4) lets us write:

$$\mathbf{H}.\varphi'(x - a, y, z) = \mathrm{E}.\varphi'(x - a, y, z).$$

We now carry out the change in variables:

$$x' = x - a; \qquad y' = y; \qquad z' = z$$

which leads to:

$$\mathbf{H}'.\varphi'(x', y', z') = \mathrm{E}.\varphi'(x', y', z').$$

But the hamiltonian $\mathbf{H}'$, gotten from $\mathbf{H}$ by the translation, is identical to $\mathbf{H}$, with the result that we rigorously reproduce (4), the only difference being the fact that the variables are here called x', y', and z'. Finally, if $\varphi(x,y,z)$ is a solution of the state equation belonging to the energy E, E is also the solution for the function $\varphi(x + a,y,z)$.

If the level E is non-degenerate, this result shows that the two wave functions differ only by a proportionality factor λ, so that:

$$\varphi(x + a, y, z) = \lambda.\varphi(x, y, z). \tag{7}$$

The property extends to the case of a degenerate energy level and we outline the proof assuming the degeneracy twofold. The level is defined by two orthonormal functions φ_1 and φ_2. The functions derived by the translation a have the same energy, which implies that they should be of the form:

$$\varphi_1(x + a, y, z) = \lambda_{11}\varphi_1 + \lambda_{12}\varphi_2$$
$$\varphi_2(x + a, y, z) = \lambda_{21}\varphi_1 + \lambda_{22}\varphi_2.$$

We will look for a wave function from this family that is of the form:

$$\theta = c_1\varphi_1 + c_2\varphi_2$$

which transforms according to the same rule (7) as the non-degenerate case. This is written:

$$c_1(\lambda_{11}\varphi_1 + \lambda_{12}\varphi_2) + c_2(\lambda_{21}\varphi_1 + \lambda_{22}\varphi_2) = \lambda(c_1\varphi_1 + c_2\varphi_2)$$

which means, since the system φ_1, φ_2 is by hypothesis orthonormal:

$$\lambda_{11}c_1 + \lambda_{21}c_2 = \lambda c_1$$
$$\lambda_{12}c_1 + \lambda_{22}c_2 = \lambda c_2.$$

The possible values of λ are determined as roots of the compatibility equation:

$$\begin{vmatrix} \lambda_{11} - \lambda & \lambda_{21} \\ \lambda_{12} & \lambda_{22} - \lambda \end{vmatrix} = 0.$$

We can show that the two functions θ obtained here are orthogonal, so that they can replace the functions φ_1 and φ_2. We can thus conclude that it is possible to find a complete set of wave functions on a translational lattice transforming according to (7).

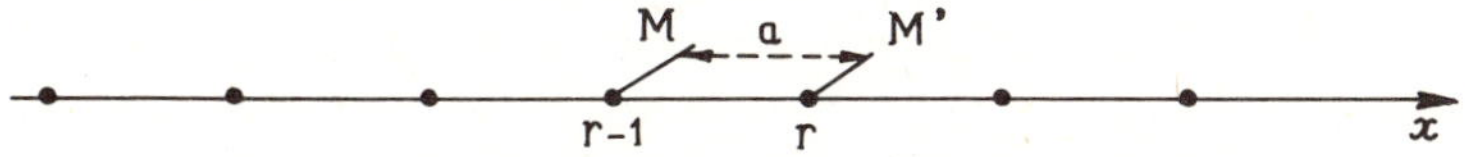

Figure 50. Action of a translation on the wave function of the lattice: the diagram shows immediately that $\psi_{r-1}(\mathrm{M}) = \psi_r(\mathrm{M}')$.

We will now demonstrate the application of the cyclic conditions which require that the functions φ be periodic on the lattice multiple $\mathrm{L} = \mathrm{N}a$. Equation (7) lets us write:

$$\varphi(x + 2a, y, z) = \lambda\varphi(x + a, y, z) = \lambda^2\varphi(x, y, z).$$

and term by term:

$$\varphi(x + \mathrm{N}a, y, z) = \lambda^{\mathrm{N}}\varphi(x, y, z).$$

The cyclic conditions are satisfied by taking:

(8) $\lambda^N = 1$; so that $\lambda = \exp\left(\frac{2\pi in}{\mathrm{N}}\right)$, with n an integer,

hence, N possible levels each characterized by a value of n.

Now that the general treatment of the wave functions representing the stationary lattice states has been given, we return to the particular case of the LCAO approximation by taking the MO in the form:

$$\varphi = \sum_r c_r\psi_r.$$

We will show that this function is completely determined by virtue of being a Bloch function. Diagram 50 shows that if M and M′ are homologous points in the lattice translation a, we can write:

$$\psi_r(x + a, y, z) = \psi_r(\mathrm{M}') = \psi_{r-1}(\mathrm{M}).$$

We should thus have:

$$\varphi(x+a, y, z) = \sum_r c_r \psi_{r-1}(x, y, z) = \lambda \sum_r c_r \psi_r(x, y, z),$$

which implies:

$$c_r = \lambda c_{r-1} = \lambda^2 c_{r-2} = \dots$$

This result, obvious if the various functions ψ_r can be considered as orthogonal, is easily demonstrated† if we take into account the overlap integral between nearest neighbors. In conclusion, we can show, using (8):

$$c_r = c\lambda^r = c \exp\left(\frac{2\pi i n r}{N}\right).$$

The determination of c assumes a convention for writing the normalization condition on the domain of integration and that will naturally be the Born lattice multiple, since the Bloch functions (2) are orthogonal in this domain. In the integration, the coordinates y and z are taken on the domain $-\infty$ to $+\infty$, this choice is unimportant since the functions $\psi_r(x,y,z)$ are practically negligible when y and z appreciably exceed the order of magnitude of the atomic dimensions. We get:

$$\int_0^L dx \int_{-\infty}^{+\infty} dy \int_{-\infty}^{+\infty} |\varphi^2|\, dz$$

$$= c^2\left[\sum_r \int |\psi_r^2|\, d\tau + \sum_{r,s} \exp\left(\frac{2\pi i n}{N}(r-s)\right) \int \psi_s^* \psi_r\, d\tau\right] = 1.$$

We only take into account the overlap integral between nearest neighbors ($r - s = \pm 1$), from which, assuming the integral real:

$$S = \int \psi_r^* \psi_{r\pm 1}\, d\tau:$$

$$\text{(9)} \qquad Nc^2\left[1 + 2S \cos\left(\frac{2\pi n}{N}\right)\right] = 1.$$

The function φ so determined should be the eigenfunction of the hamiltonian, hence:

$$\sum_r \exp\left(\frac{2\pi i n r}{N}\right) \mathbf{H} \psi_r = E \sum_s \exp\left(\frac{2\pi i n s}{N}\right) \psi_s.$$

† It is sufficient that the base functions be independent so that the expansion can be made only one way.

We will calculate the eigenvalue E by integration, following multiplication by $\psi_r^\star$. The only integrals involved are:

$$\int \psi_r^\star \psi_r d\tau = 1 \qquad \int \psi_r^\star \psi_{r\pm 1} d\tau = S$$

$$\int \psi_r^\star \mathbf{H} \psi_r d\tau = Q \qquad \int \psi_r^\star \mathbf{H} \psi_{r\pm 1} d\tau = \gamma$$

and we get:

(10) $$E\left(1 + 2S \cos \frac{2\pi n}{N}\right) = Q + 2\gamma \cos \frac{2\pi n}{N} .$$

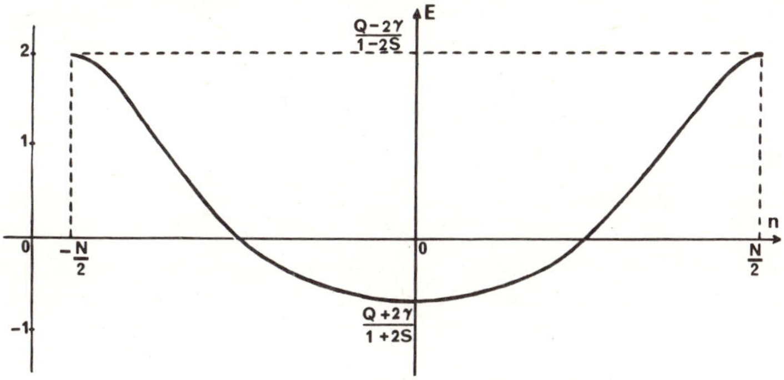

Figure 51. Diagram of the eigenvalues of the energy E(n) on the inside of the first Brillouin zone for a linear lattice.

We then get N levels falling between the two limits:

$$(Q + 2\gamma)/(1 + 2S), \qquad \text{for } n = 0,$$

and

$$(Q - 2\gamma)/(1 - 2S), \qquad \text{for } n = N/2.$$

The N levels are arranged between these two limits, giving a quite dense distribution, since the arbitrary number N of atoms forming the lattice multiple is generally quite large. Such a distribution of levels, characteristic of crystals, is known as an *electron band* in contrast to the discrete levels in separate atoms.

The graphical representation of E(n) (see fig. 51) tends toward a continuous function, which we represent by taking as the abscissa n/N so as to suppress the arbitrary nature of the choice of N. As related to the domain of variation of n we note that the wave functions are identical for n and $n' = n + N$, so that it is sufficient to vary n/N on the interval equal to unity but with an arbitrary origin. Current convention assigns the smallest possible value which is achieved by taking n/N on the interval $-\frac{1}{2}$ to $+\frac{1}{2}$, which constitutes what is called the *first Brillouin zone*. We note that the values n and $-n$ correspond to distinct wave functions but to the same energy, with the result that the energy axis fixed by this choice is the axis of symmetry of the curve $\mathrm{E}(n)$.

The evaluation of the number of levels, N, does not take the spin into account. If we consider the two possibilities corresponding to the two spin components, we see that the total number of states in the band is 2N, being two electrons per bond between the adjacent atoms. Saturation of the band corresponds to filling the set of bonds and this forms a model for an isolated crystal.

41 Localization of an Electron on the Linear Lattice

The problem which we now discuss is similar to that of the free particle (§4) where, as a result of the incompatible nature of the variables p and x, localization of the particle implied an uncertainty in the energy. The reasoning will be developed in a parallel manner by taking the lattice wave functions in the Bloch form (2), so that:

$$\varphi_n = u_n(x, y, z) \exp\left(\frac{2\pi i n x}{\mathrm{N}a}\right)$$

u_n being a function periodic on the lattice. We note that the function is analogous to that describing the state of a free particle:

$$\varphi_n' = \exp\left(\frac{ipx}{\hbar}\right).$$

The similarity of these two functions leads us to associate with a lattice electron a variable p comparable to momentum, given by:

$$p = \frac{2\pi n\hbar}{\mathrm{N}a} = \frac{nh}{\mathrm{N}a}. \tag{11}$$

We state explicitly that p does not represent the momentum of the electron on the lattice, which should be gotten using the associated operator $\mathbf{P} = (\hbar/i)(\partial/\partial x)$. This then becomes:

$$\mathbf{P}\varphi_n = \left[\frac{\hbar}{i}\frac{\partial u_n}{\partial x} + \frac{nh}{\mathrm{N}a}u_n\right] \exp\left(\frac{2\pi i n x}{\mathrm{N}a}\right).$$

φ_n is not an eigenfunction of $\mathbf{P}$, with the result that the momentum does not have a precisely defined value in the state φ_n, and thus cannot be identified with p. The importance of this quantity appears in reconsidering the problem of the localization of the electron. We treat the question exactly as in section 4, by examining a superposition state of two waves, defined by the neighboring values n and $n' = n + \delta n$, so that:

$$(12) \qquad \theta = \frac{1}{\sqrt{2}}(\varphi_n - \varphi_{n'})$$

$$= \frac{1}{\sqrt{2}}\left[u_n - u_{n'} \exp\left(\frac{2\pi i \delta n x}{Na}\right)\right] \exp\left(\frac{2\pi i n x}{Na}\right)$$

the normalization coefficient being equal to $1/\sqrt{2}$ as a result of the orthogonality of two Bloch functions belonging to different n values.

The electron density distribution is given by:

$$\theta^*\theta = \frac{1}{2}\left[\overset{*}{u}_n - \overset{*}{u}_{n'} \exp\left(-\frac{2\pi i \delta n x}{Na}\right)\right]\left[u_n - u_{n'} \exp\left(\frac{2\pi i \delta n x}{Na}\right)\right]$$

so that by taking $u_n'/u_n = \rho e^{i\alpha}$ to simplify the expression:

$$\theta^*\theta = \frac{1}{2}\overset{*}{u}_n u_n\left[1 + \rho^2 - 2\rho \cos\left(\frac{2\pi \delta n x}{Na} + \alpha\right)\right].$$

The same relation is found for any time t, the evolution of the function φ_n over time being given by:

$$\varphi_n(t) = \varphi_n(0) \exp\left(-\frac{iE_n t}{\hbar}\right).$$

We get, finally:

$$(13) \quad \theta^*\theta = \frac{1}{2}\overset{*}{u}_n u_n\left[1 + \rho^2 - 2\rho \cos\left(\frac{2\pi \delta n x}{Na} - \frac{\delta E_n t}{\hbar} + \alpha\right)\right].$$

The quantity $\theta^*\theta$ is in approximately the same form as we get in the free electron case (§4). We find a sinusoidal function with an argument:

$$(14) \qquad \frac{2\pi \delta n x}{Na} - \frac{\delta E_n t}{\hbar} = 2\pi\left(\frac{x \delta n}{Na} - \frac{t \delta E_n}{h}\right)$$

but with amplitude and phase defined by u_n, ρ, and α which are here periodic functions of the lattice and not constants, so that the results deduced in the case of the free electron are not directly carried over. We can only assume that (13) represents the behavior of a wave modulated on a lattice period that is very short compared to the wavelength defined by (14), namely, $\lambda = Na/\delta n$. As a result, the mean of $\theta^*\theta$ on the elementary lattice length should behave exactly as in the case of a traveling wave whose argument is given by (14), corresponding to the velocity:

(15) $$v = \frac{\delta E_n}{h} \frac{Na}{\delta n} = \frac{\delta E_n}{\delta p} \rightarrow \frac{dE_n}{dp}$$

introducing the quantity p already defined in (11), which plays the role of momentum. The localization of the electron which we thus find is on a segment of the order of magnitude of the wavelength λ, and always corresponds to a large number of lattice lengths, given that we have $N \gg \delta n$.

The intuitive method that we follow for getting (15), similar to that used in section 4 for the free particle, has been treated in detail by several authors† by using a method which we will outline.

We start with the fact that the linear combination pictured in (12) to represent a state where the electron is localized, gives $\partial\varphi/\partial p$ when the parameter $\delta p = h\delta n/Na$ tends to zero. If we write the equation satisfied by φ in the form:

$$\left(-\frac{\hbar^2}{2m_0}\Delta + V - E\right)\varphi = 0,$$

we introduce $\partial\varphi/\partial p$ by differentiating relative to p, so that:

(16) $$\frac{\partial}{\partial p}\left[\left(-\frac{\hbar^2}{2m_0}\Delta + V - E\right)\varphi\right] = \left(-\frac{\hbar^2}{2m_0}\Delta + V - E\right)\frac{\partial\varphi}{\partial p} - \varphi\frac{dE}{dp} = 0,$$

noting that the only quantities dependent on p are E and φ.

We multiply by $\varphi^\star$ and integrate on the volume C of the unit cell, which in the particular case of the linear crystal will be a parallelepiped whose edge along $x'x$ will be taken equal to a. The other two dimensions are chosen large enough so that φ, which decreases rapidly when we move away from the axis, can be taken as negligible. Condensing the results of this operation, we write:

$$\frac{dE}{dp}\int_C \varphi^\star\varphi\, d\tau = \int_C \varphi^\star(\mathbf{H} - E)\frac{\partial\varphi}{\partial p}\, d\tau.$$

It seems useful to emphasize that we cannot write:

$$\int_C \varphi^\star(\mathbf{H} - E)\frac{\partial\varphi}{\partial p}\, d\tau = \int_C \frac{\partial\varphi}{\partial p}(\mathbf{H} - E)\varphi^\star d\tau = 0,$$

since the hermiticity condition is only applicable when the integration is carried out on the Born lattice multiple, which is not the

† W. Shockley, *Electrons and Holes in Semiconductors* (New York: D. Van Nostrand, 1956), 15-4.

case here. It is, however, possible to develop the second term of this latter equation by adding and subtracting the integral:

$$\frac{\partial\varphi}{\partial p}\left(-\frac{\hbar^2}{2m_0}\Delta\varphi^\star\right)d\tau,$$

which leads to:

$$\int_c \varphi^\star\left[-\frac{\hbar^2}{2m_0}\Delta+V-E\right]\frac{\partial\varphi}{\partial p}d\tau$$
$$=\int_c\frac{\partial\varphi}{\partial p}\left[-\frac{\hbar^2}{2m_0}\Delta+V-E\right]\varphi^\star d\tau-\frac{\hbar^2}{2m_0}\int_c\left(\varphi^\star\Delta\left(\frac{\partial\varphi}{\partial p}\right)-\frac{\partial\varphi}{\partial p}\Delta\varphi^\star\right)d\tau.$$

The first integral is zero, since the operator H is real, $H\varphi = E\varphi$, implying $H\varphi^\star = E\varphi^\star$. As a result, equation (16) can be put in the form:

$$(17)\qquad \frac{dE}{dp}\int_c\varphi^\star\varphi d\tau=-\frac{\hbar^2}{2m_0}\int_c\left(\varphi^\star\Delta\left(\frac{\partial\varphi}{\partial p}\right)-\frac{\partial\varphi}{\partial p}\Delta\varphi^\star\right)d\tau.$$

One frequently used method of calculation (§4) allows us to transform the volume integral into a surface integral, which puts (17) in the form:

$$(18)\quad \frac{dE}{dp}\int_c\varphi^\star\varphi d\tau=-\frac{\hbar^2}{2m_0}\int_c\left(\varphi^\star\,\overrightarrow{\text{grad}}\left(\frac{\partial\varphi}{\partial p}\right)-\frac{\partial\varphi}{\partial p}\overrightarrow{\text{grad}}\,\varphi^\star\right)\cdot\vec{n}dS$$

the integration being taken over the surface of the unit cell defined above. This integral goes to zero with φ, with the result that the contribution of the faces of the cell parallel to $x'x$ disappears. Therefore, we need only take into account the two faces normal to $x'x$ at a distance a, which leads to:

$$(19)\qquad \int\left(\varphi^\star\frac{\partial}{\partial x}\left(\frac{\partial\varphi}{\partial p}\right)-\frac{\partial\varphi}{\partial p}\frac{\partial\varphi^\star}{\partial x}\right)dydz\Big|_x^{x+a}.$$

Expressing the quantity under the integral sign by writing the Bloch function in the form:

$$\varphi = ue^{ipx/\hbar}.$$

We then get

$$\varphi^\star\frac{\partial}{\partial x}\left(\frac{\partial\varphi}{\partial p}\right)-\frac{\partial\varphi}{\partial p}\frac{\partial\varphi^\star}{\partial x}=\frac{ix}{\hbar}\left(\varphi^\star\frac{\partial\varphi}{\partial x}-\varphi\frac{\partial\varphi^\star}{\partial x}\right)$$
$$+\left[\frac{i}{\hbar}\varphi^\star\varphi+\varphi^\star\frac{\partial}{\partial x}\left(\frac{\partial u}{\partial p}e^{\frac{ipx}{\hbar}}\right)-\frac{\partial u}{\partial p}\frac{\partial\varphi^\star}{\partial x}e^{\frac{ipx}{\hbar}}\right].$$

The quantity between brackets is a periodic function of x, which has the same value for x and $x + a$, with the result that its contribution vanishes in (19). The term:

$$\varphi^\star\frac{\partial\varphi}{\partial x}-\varphi\frac{\partial\varphi^\star}{\partial x}$$

is also periodic in x, the difference of the integrals for x and $x + a$ can be put (19) in the form:

$$\frac{ia}{\hbar}\int\left(\varphi^\star\frac{\partial\varphi}{\partial x} - \varphi\frac{\partial\varphi^\star}{\partial x}\right)dydz$$

hence the final form of (18):

$$\frac{dE}{dp}\int_C \varphi^\star\varphi d\tau = \frac{\hbar a}{2im_0}\int\left(\varphi^\star\frac{\partial\varphi}{\partial x} - \varphi\frac{\partial\varphi^\star}{\partial x}\right)dydz.$$

This calculation involves the expression:

$$J_x = \frac{\hbar}{2im_0}\left(\varphi^\star\frac{\partial\varphi}{\partial x} - \varphi\frac{\partial\varphi^\star}{\partial x}\right)$$

which represents the component of the particle current density along $x'x$ (1-17). This component occurs in the integral which defines the resultant value J_x^0 of the current crossing one of the end faces of C, so that:

$$J_x^0 = \frac{\hbar}{2im_0}\int\left(\varphi^\star\frac{\partial\varphi}{\partial x} - \varphi\frac{\partial\varphi^\star}{\partial x}\right)dydz.$$

The equation which we get is written:

$$\frac{dE}{dp}\int_C \varphi^\star\varphi d\tau = aJ_x^0.$$

We can note that on the one hand: $\int_C \varphi^\star\varphi d\tau = \int u^\star u d\tau$ does not vary on a unit cell translation, which allows us to define the mean particle density $\bar{\rho}$ by:

$$\bar{\rho} = \frac{1}{a}\int_C \varphi^\star\varphi d\tau$$

and finally:

$$J_x^0 = \bar{\rho}\frac{dE}{dp},$$

which is in the form: $\bar{J}_x = \bar{\rho}v$ which leads to the value of the mean velocity:

$$v = \frac{dE}{dp}$$

previously gotten (15) by intuitive arguments.

Effective Mass Concept

This concept is frequently introduced with the *a priori* assumption that the variation of the energy of an electron with apparent momentum p has the form of kinetic energy, the velocity being the mean velocity defined in (15). We write then by definition:

$$dE = m^\star v dv. \tag{21}$$

By taking

$$v = \frac{dE}{dp}, \qquad \frac{d^2 E}{dp} dp = dv$$

we get:

$$m^\star = \frac{dE}{vdv} = \left(\frac{d^2 E}{dp^2}\right)^{-1}. \tag{22}$$

The effective mass $m^\star$ thus defined is introduced when the variation of the energy is due to the action of a force F acting on the electron. The conservation of energy implies:

$$F\, dx = m^\star v dv \qquad \text{with} \qquad dx = vdt$$

hence:

$$F = m^\star \frac{dv}{dt} = m^\star \gamma.$$

Generalization to the Case of a Three-Dimensional Lattice 42

We will briefly indicate how the ideas introduced in the study of the linear lattice are generalized to the three-dimensional case. We examine a general lattice defined by the elementary translations $\vec{a}_1$, $\vec{a}_2$ and $\vec{a}_3$. The cyclic Born conditions are generalized in this case by taking the periodicity on a lattice multiple $n_1\vec{a}_1$, $n_2\vec{a}_2$ and $n_3\vec{a}_3$ having $N = n_1 n_2 n_3$ elementary cells. The three-dimensional Bloch functions have the following generalized form from (2):

$$f_{\vec{k}}(\vec{r}) = u_{\vec{k}}(\vec{r}) e^{i\vec{k}.\vec{r}} \tag{23}$$

where $u_{\vec{k}}(\vec{r})$ is a periodic function on the unit cell and $\vec{k}$ a vector such that the periodicity on the multiple lattice is assured, which implies:

$$\vec{k}.\vec{a}_1 = 2\pi n_1'/n_1 ; \qquad \vec{k}.\vec{a}_2 = 2\pi n_2'/n_2 ; \qquad \vec{k}.\vec{a}_3 = 2\pi n_3'/n_3,$$

with n_1', n_2', and n_3' integers.

The vector $\vec{k}$ so defined cannot be simply expressed on the reference system formed by the vectors $\vec{a}_i$. It is preferable to introduce the vectors $\vec{b}_1$, $\vec{b}_2$, and $\vec{b}_3$ which are tied to the translations a_i by the set of equations:

$$\vec{a}_i.\vec{b}_j = 2\pi\delta_{ij} \qquad (\delta_{ij} = \text{Kronecker delta}) \tag{24}$$

These equations, which show that the vector $\vec{b}_1$, for example, is normal to $\vec{b}_2$ and $\vec{b}_3$ when its modulus is given by $\vec{b}_1 \cdot \vec{a}_1 = 2\pi$, completely determine the vectors $\vec{b}_i$ (fig. 52). We are thus able to express the vector $\vec{k}$ defining the Bloch wave in this system so that:

$$\vec{k} = \lambda_1 \vec{b}_1 + \lambda_2 \vec{b}_2 + \lambda_3 \vec{b}_3.$$

The foregoing equations are then written:

$$\vec{k} . \vec{a}_1 = 2\pi \frac{n_1'}{n_1} = 2\pi\lambda_1, \quad \text{where} \quad \lambda_1 = \frac{n_1'}{n_1}.$$

The Bloch function (23) is thus characterized by the vector:

(25) $$\vec{k} = \frac{n_1'}{n_1} \vec{b}_1 + \frac{n_2'}{n_2} \vec{b}_2 + \frac{n_3'}{n_3} \vec{b}_3.$$

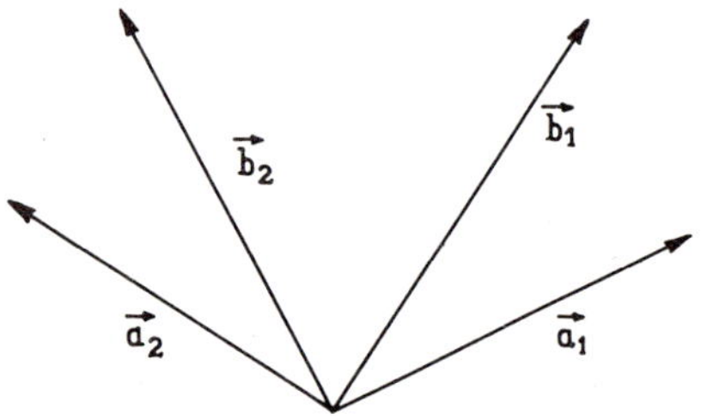

Figure 52. Relative disposition of the vectors $\vec{a}_1$ and $\vec{a}_2$ defining the elementary translations of a planar lattice, and the vectors $\vec{b}_1$ and $\vec{b}_2$ defining the reciprocal lattice (for graphite).

The behavior of the Bloch function during a lattice translation then describes the generalization of equation (3) gotten for the one-dimensional case. We get for example:

(26) $$f_{\vec{k}}(\vec{r} + \vec{a}_1) = f_{\vec{k}}(\vec{r}) \exp\left(2\pi i \frac{n_1'}{n_1}\right).$$

We have shown in the one-dimensional case that the solution of the equations of state are Bloch functions. This remains true on a three-dimensional lattice and can be shown in an analogous manner. As a result we can associate with each energy level a value of the vector $\vec{k}$ of the form (25), hence our interest in describing the possible distinct values of $\vec{k}$ and their distribution.

The possible values of $\vec{k}$. Two Bloch functions characterized respectively by the sets (n_1', n_2', n_3') and (n_1'', n_2'', n_3'') are the same kind if their behavior (26) under a general lattice transformation is the same for both functions. It is sufficient to consider the action of three elementary lattice translations, and the equality of the exponential factors for each of these leads to the conditions:

$$n_1'' = n_1' + r_1 n_1 ; \qquad n_2'' = n_2' + r_2 n_2 ; \qquad n_3'' = n_3' + r_3 n_3$$

with r_1, r_2, and r_3 integers. Designating by $\vec{k}'$ and $\vec{k}''$ the corresponding vectors given by (25) we must then have:

$$\vec{k}'' = \vec{k}' + r_1\vec{b}_1 + r_2\vec{b}_2 + r_3\vec{b}_3.$$

The vector $r_1\vec{b}_1 + r_2\vec{b}_2 + r_3\vec{b}_3$ represents a translation on the translational lattice formed from $\vec{b}_1$, $\vec{b}_2$, and $\vec{b}_3$, and is called the

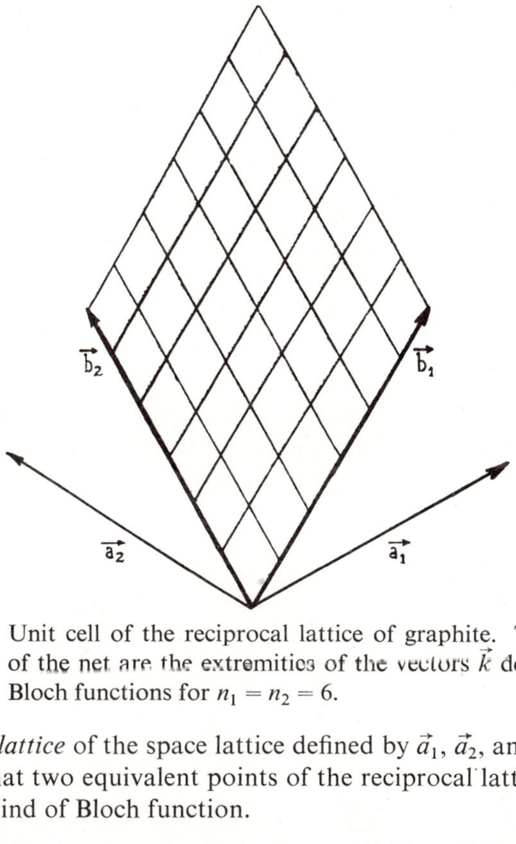

Figure 53. Unit cell of the reciprocal lattice of graphite. The nodes of the net are the extremities of the vectors $\vec{k}$ defining the Bloch functions for $n_1 = n_2 = 6$.

reciprocal lattice of the space lattice defined by $\vec{a}_1$, $\vec{a}_2$, and $\vec{a}_3$. We see then that two equivalent points of the reciprocal lattice define the same kind of Bloch function.

The result thus gotten involves the possibility of limiting our consideration to the vectors $\vec{k}$ in the form (25), whose extremities are constrained within an elementary cell of the reciprocal lattice. We take as an example (fig. 53) the diagram relative to the two-dimensional graphite lattice by taking $n_1 = n_2 = 6$.

Brillouin zones. The preceding study establishes the possibility of representing all of the types of Bloch functions, limiting ourselves to the use of a volume of the reciprocal lattice equal to that of an elementary cell defined by $\vec{b}_1$, $\vec{b}_2$, and $\vec{b}_3$, but arbitrarily positioned. This volume can be chosen in particular so that the vectors $\vec{k}$ are as short as possible and thus form the *first Brillouin zone* already defined in the one-dimensional case. The construction of the zone limits is carried out quite simply in the general case.

We take for this (fig. 54) a point M of the reciprocal lattice (not the origin O) and we connect it to the point M′ symmetric with respect to O. The lattice translations always permit us to join the extremity of an arbitrary vector $\vec{k}$ to a point situated at the interior

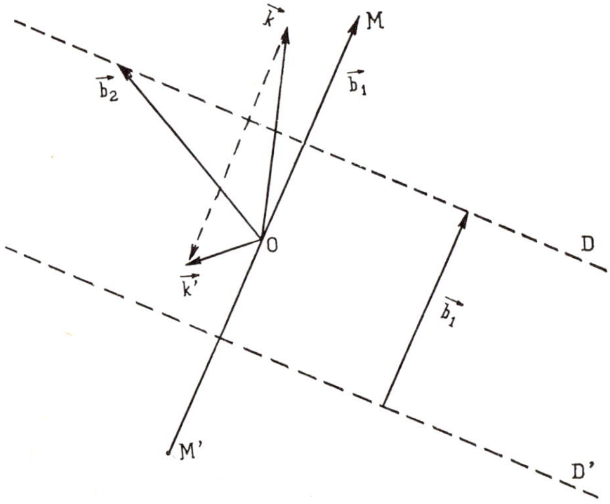

Figure 54. Reduction of a vector in the reciprocal lattice for the purpose of constructing the first Brillouin zone.

of a region fixed by the two median planes of the translations $OM = \vec{b}_1$ and $OM' = -\vec{b}_1$, which are separated by a translational vector $\vec{b}_1$. The diagram (fig. 54) gives one such reduction in the case of a two-dimensional lattice, used so as to give the clearest picture. We pass from the vector $\vec{k}$ to the equivalent vector $\vec{k}'$, whose tip lies between the two median planes D and D′. The construction proceeds by taking another node M_2 of the lattice such that it does not coincide with O and joining it to M_2, and so forth. Figure 55 represents the contour of the Brillouin zone as delimited in the case of graphite. We can see that the number of distinct ends of the vectors $\vec{k}$ given by (25) is the same (here 36) as in the unit cell (fig. 53), taking into account the fact that the points on the edges of the zone are not distinct from those on the opposite edges, which differ by a translation of the reciprocal lattice. For the same reason, the three apexes of the zone define an equilateral triangle corresponding to the same value of $\vec{k}$.

In a general way, by returning to the case of a three-dimensional lattice, the ends of the vectors $\vec{k}$ of the form (25) define $N = n_1 n_2 n_3$

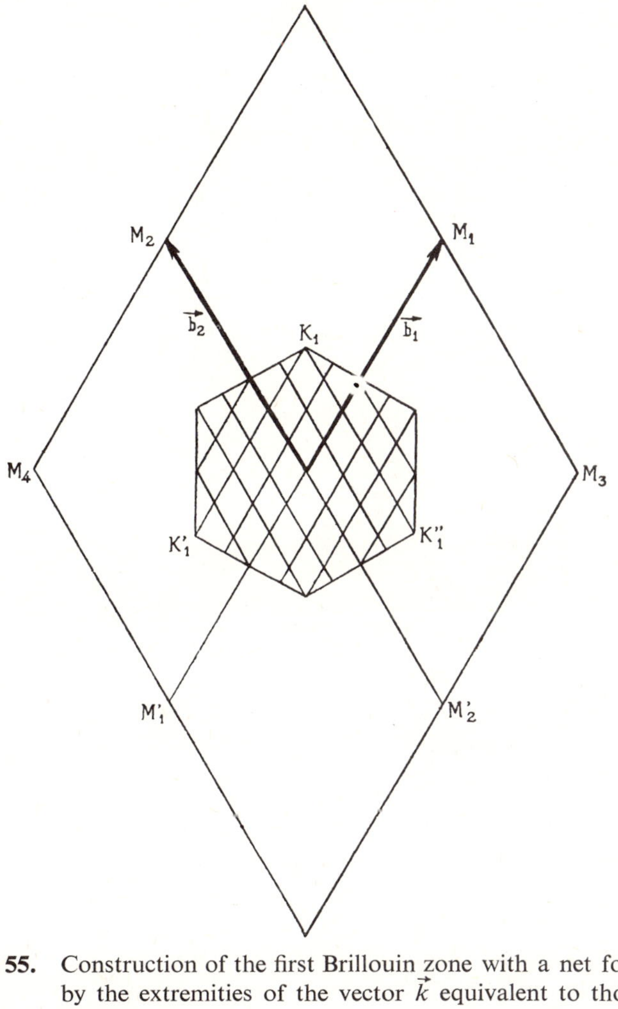

Figure 55. Construction of the first Brillouin zone with a net formed by the extremities of the vector $\vec{k}$ equivalent to those in figure 53.

points distributed with a uniform density on the zone, this density being large because the Born lattice multiple is taken large.

Exercise. Consider a face-centered cubic lattice (*fcc*) and determine the three elementary lattice translations (fig. 56) as a function of the edge of the cubic cell *a*. Calculate the corresponding three vectors $\vec{b}_1$, $\vec{b}_2$, and $\vec{b}_3$, and show that the reciprocal lattice thus obtained is a body-centered lattice. Show in perspective (fig. 57) the construction of the first Brillouin zone.

Finally, among the uses of the first Brillouin zone we note the possibility of representing the electron energy levels by tracing the constant energy surfaces. Figure 58 gives an example of one such representation in the case of graphite taken as a two-dimensional lattice.

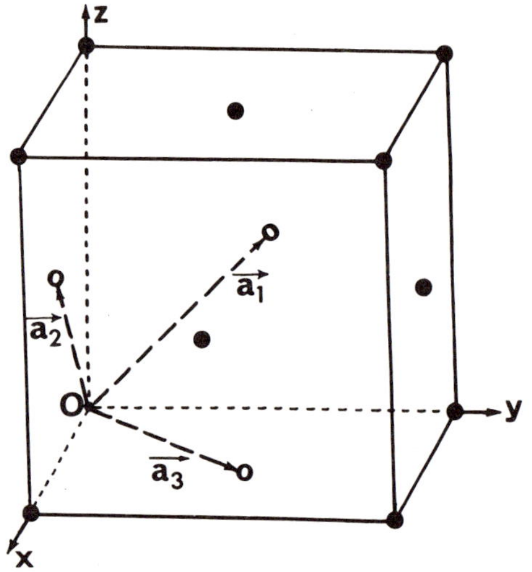

Figure 56. Perspective representation of the disposition of the atoms in a face-centered cubic lattice showing the three elementary translations.

Localization of an Electron on the Lattice

The calculation of the one-dimensional case is immediately transposed. We introduce a vector playing the role of momentum, analogous to (11), here defined by:

$$\vec{p} = \hbar\vec{k}. \tag{27}$$

The introduction of the electron velocity on the lattice leads to the general result (15) so that:

$$\vec{v} = \overrightarrow{\text{grad}}_p \, E = \frac{1}{\hbar} \overrightarrow{\text{grad}}_k \, E. \tag{28}$$

The generalization of the effective mass is a little more difficult. We always take the same basis by assuming that the variation of energy dE is in the kinetic energy, so that:

$$dE = m^{\star} \, v dv. \tag{29}$$

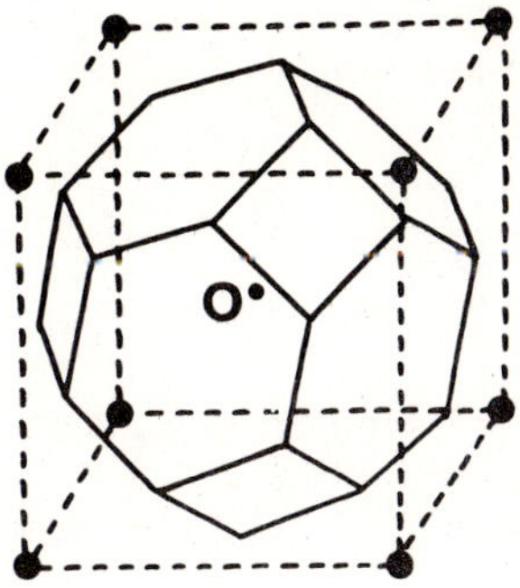

Figure 57. Contour of the first Brillouin zone of a face-centered cubic lattice.

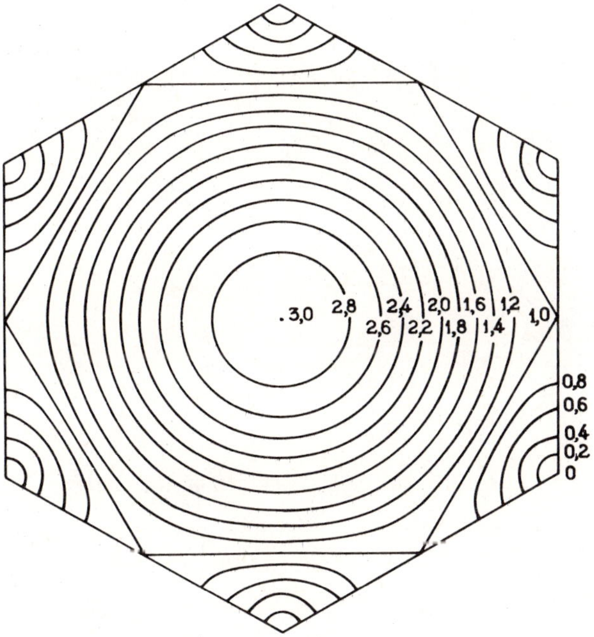

Figure 58. Constant energy curves in the first Brillouin zone (from the planar model of graphite). The diagram is constructed by taking $S = 0$ and by giving the values of $|E - Q|/\gamma$ for each of the curves traced.

The energy E is determined by the vector $\vec{k}$ with components $\vec{k}_1$, $\vec{k}_2$, and $\vec{k}_3$ in cartesian coordinates. We can write

$$v^2 = \frac{1}{\hbar^2}\sum_r \left(\frac{\partial E}{\partial k_r}\right)^2, \qquad \text{hence} \qquad v\,dv = \frac{1}{\hbar^2}\sum_{r,s} \frac{\partial E}{\partial k_r} \cdot \frac{\partial^2 E}{\partial k_r \partial k_s}\,dk_s$$

so that by substituting in (29):

$$\frac{\hbar^2}{m^*}\sum_s \frac{\partial E}{\partial k_s}\,dk_s = \sum_s \frac{\partial E}{\partial k_s}\sum_r \frac{\partial^2 E}{\partial k_r \partial k_s}\,dk_r.$$

The simplification which allows us to relate equation (22) to the effective mass no longer occurs here. The preceding equation becomes an identity if we have:

$$\frac{\hbar^2}{m^*}\,dk_s = \sum_r \frac{\partial^2 E}{\partial k_r \partial k_s}\,dk_r$$

for all systems of variation dk, which is obviously impossible. This is no longer the case however if $\hbar^2/m^*$ is no longer considered as a scalar but as a second order tensor, that is to say, such that:

$$\left(\frac{\hbar^2}{m^*}\right)_{rs} = \frac{\partial^2 E}{\partial k_r \partial k_s}$$

equation (22) then represents a special case.

PART THREE

GENERAL THEORY OF MANY-ELECTRON SYSTEMS

11 DESCRIPTION OF THE STATES OF A SYSTEM OF MANY PARTICLES

Principles of a Many-Electron Treatment 43

We have already considered many-electron systems in the study of the behavior of an isolated particle in the field produced by other particles, but whatever our interest in such methods may be, they have a fundamentally inconvenient attribute. They show the behavior of only one particle, and this is an unrealistic treatment in the case of a system containing several identical particles. For this reason it is necessary to outline the more complete treatment.

The most direct approach is to turn our attention to the variables that we can use to define the state of such a system. From the classical point of view it is necessary to give at every point in time the coordinates and the velocity components of each particle. In the treatment of an isolated particle, we saw that the incompatibility of x_r and p_r allows us to retain only three of the six variables to define the state, and this is true for each particle treated. By contrast, when the particles were independent, it appeared natural to assume that the various variables x_r and p_s relative to different particles should be compatible. This broad postulate has immediate consequences which we will now elucidate.

(*a*) If we consider a system of many particles defined by the positions $(x_1,y_1,z_1),(x_2,y_2,z_2),\ldots$, the state of the system will be defined by a wave function of amplitude $\psi(x_1,y_1,z_1,x_2,y_2,z_2,\ldots)$ such that $|\psi^2|dx_1\,dy_1\,dz_1\,dx_2\,dy_2\,dz_2\ldots$ or in shortened form $|\psi^2|d\tau_1\,d\tau_2\ldots$, represents the probability of simultaneously having particle 1 in the volume element $d\tau_1$ at point x_1, y_1, z_1, particle 2 in $d\tau_2$ at x_2, y_2, z_2, and so forth. This generalization of the meaning of the one-particle wave function implies the normalization condition on ψ:

$$\int |\psi^2|\, d\tau_1 d\tau_2 \ldots = 1. \tag{1}$$

In the case where the particles are electrons, it is necessary to add the spin components to each, so that integral (1) on the space variables would be completed by a summation on the different values of the spin components of the various electrons.

(*b*) The compatibility of the coordinates x_r, y_r, z_r, of the various electrons allows us to define the mutual separation of these electrons, and consequently, to retain the classical potential energy expression of the system which depends only on the distance between the particles. Therefore, for the helium atom we can write:

$$V = \frac{1}{4\pi\varepsilon_0}\left(-\frac{2e^2}{r_1} - \frac{2e^2}{r_2} + \frac{e^2}{r_{12}}\right).$$

To get the total energy expression it is necessary to add the kinetic energy terms for the various electrons. We have shown (§13) that the sum of variables can always be defined whether or not they are compatible. By assuming that this principle is general, we can immediately form the energy expression. In the case of the helium atom already mentioned, we have:

$$E = \frac{1}{2m_0}\left(p_{x_1}^2 + p_{y_1}^2 + p_{z_1}^2 + p_{x_2}^2 + p_{y_2}^2 + p_{z_2}^2\right)$$
$$+ \frac{1}{4\pi\varepsilon_0}\left(-\frac{2e^2}{r_1} - \frac{2e^2}{r_2} + \frac{e^2}{r_{12}}\right).$$

The corresponding operator can be formed by adopting the same rules as in the one-electron case, that is by replacing p_r by $(\hbar/i)(\partial/\partial x_r)$. We then get for the associated energy operator, which we again call the hamiltonian operator:

$$(2) \qquad \mathcal{H} = -\frac{\hbar^2}{2m_0}(\Delta_1 + \Delta_2) + \frac{1}{4\pi\varepsilon_0}\left(-\frac{2e^2}{r_1} - \frac{2e^2}{r_2} + \frac{e^2}{r_{12}}\right).$$

We see from this example that the ideas of wave function and equation of state are generalized without difficulty to systems containing more than one electron, the complication being essentially mathematical in that the treatment requires more approximations than in the more simple cases. Likewise, the properties indicated in section 13 are extended without difficulty to the many-electron case, notably the hermitian character of the operators associated with the physical variables and all of the consequences of this property, namely, the real nature of the eigenvalues of these operators whose eigenfunctions form an orthonormal set.

44 The Pauli Principle

The statement we made (§26) in the one-electron treatment of atoms and molecules cannot be simply introduced here; we therefore present a more general statement of the principle. We consider a system of n electrons where $\psi(q_1,q_2,\ldots,q_n)$ is a solution of the equation of state, where for simplicity we write q_r in place of the set of four coordinates x_r, y_r, z_r, and σ_r. A general permutation of the electrons must not modify the equation, so that if we have by hypothesis:

$$\mathcal{H}\psi(q_1, q_2, \cdots, q_n) = E\psi(q_1, q_2, \cdots, q_n),$$

then we must have also:

$$\mathcal{H}\psi(q_n, q_2, \cdots, q_1) = E\psi(q_n, q_2, \cdots, q_1). \tag{3}$$

The Pauli principle is then stated: the representative state function of a many-electron system must be *antisymmetric* relative to the permutation of any two electrons. We give an example for a system of three electrons: if $\psi(q_1,q_2,q_3)$ is a state function, the Pauli principle shows that the following identities must be satisfied:

$$\psi(q_2, q_1, q_3) = -\psi(q_1, q_2, q_3)$$
$$\psi(q_1, q_3, q_2) = -\psi(q_1, q_2, q_3)$$
$$\psi(q_3, q_2, q_1) = -\psi(q_1, q_2, q_3).$$

These three equations permit us to describe the action of a general perturbation on the function $\psi(q_1,q_2,q_3)$. We get:

$$\psi(q_2, q_3, q_1) = -\psi(q_2, q_1, q_3) = -\psi(q_1, q_2, q_3)$$
$$\psi(q_3, q_1, q_2) = -\psi(q_1, q_3, q_2) = -\psi(q_1, q_2, q_3).$$

We will show later that this statement of the Pauli principle reduces to that of section 26 as a special case.

This form of the Pauli principle can be presented by general consideration of the structure of the permutation group which presents in an interesting way a practical method for performing the calculations. To simplify the writing we limit our treatment to the case of three electrons.

The number of possible permutations of n objects is $n!$ We write our expressions for $n = 3$, indicating on the upper line the number of each of the three objects taken in any order. To avoid confusion we will take them in the natural order. The lower line gives the number of the object which replaces that above it within the specific permutation. We get the following list including in the series the permutation T_0, which is the identity element and leaves the objects unchanged:

$$T_0\begin{pmatrix}1, 2, 3\\1, 2, 3\end{pmatrix} \quad T_1\begin{pmatrix}1, 2, 3\\1, 3, 2\end{pmatrix} \quad T_2\begin{pmatrix}1, 2, 3\\3, 2, 1\end{pmatrix}$$

$$T_3\begin{pmatrix}1, 2, 3\\2, 1, 3\end{pmatrix} \quad T_4\begin{pmatrix}1, 2, 3\\3, 1, 2\end{pmatrix} \quad T_5\begin{pmatrix}1, 2, 3\\2, 3, 1\end{pmatrix}.$$

We denote the action of a perturbation on the wave function as follows:

(4) $$T_4\psi(q_1, q_2, q_3) = \psi(q_3, q_1, q_2).$$

The arguments which follow are based on the distinction between *even and odd permutations*, which we introduce by examination of the following functions of arbitrary variables $\xi_1, \xi_2, \ldots, \xi_3$, whose number equals the number of objects in the set, and which we write in the general form as well as for the case $n = 3$:

$$f = \prod_{r>s} (\xi_r - \xi_s) = (\xi_3 - \xi_2)(\xi_3 - \xi_1)(\xi_2 - \xi_1).$$

All of the differences $\xi_r - \xi_s$ affect the sign of the product f once, and only once. As a result of a permutation, we find all of the same factors $\xi_r - \xi_s$, except for their sign and ordering, and we thus get as a result $+f$ (even permutation) or $-f$ (odd permutation). As an example:

$$T_4(\xi_3 - \xi_2)(\xi_3 - \xi_1)(\xi_2 - \xi_1) = (\xi_2 - \xi_1)(\xi_2 - \xi_3)(\xi_1 - \xi_3) = f.$$

T_4 is even, as is T_5, and obviously, T_0. The permutations T_1, T_2, and T_3, are odd. Study of the function f gives evidence of the following properties:

(*a*) Successive action, or the *product*, of two permutations of the same kind is an even permutation.

(*b*) The product of two permutations of different kinds is an odd permutation.

(*c*) The number of distinct even permutations is equal to the number of odd permutations, $n!/2$.

This last property is demonstrated by use of a method very characteristic of group theory. We examine the set $\mathscr{E}_p(T_{p_1}, T_{p_2}, \ldots, T_{p_r})$ of even permutations and a general odd permutation T_i. All of the permutations $(T_iT_{p_1}, T_iT_{p_2}, \ldots, T_iT_{p_r})$ are odd, and all are distinct, since $T_iT_{p_1} = T_iT_{p_2}$ is only possible if $T_{p_1} = T_{p_2}$, which we can easily see by multiplying on the left by $\mathbf{T}_i^{-1}$ (the existence of which is assured since the permutations form a group). The second point of the proof is to see that we get all of the odd permutations. Let us take T_i'. The operation T_i^{-1} is odd and as a result $T_i^{-1}T_i'$ is even and appears in the set $\mathscr{E}_p$. The left hand product with T_i then forms the desired element T_i'.

We now introduce the antisymmetrization operator defined by:†

(5) $$\mathcal{A} = \sum_p T_p - \sum_i T_i$$

† The first sum is taken over all the even permutations, and the second over all the odd, where each sum has the same number of terms $n!/2$.

which satisfies the following equations as a result of the properties shown above:

$$\mathcal{A}\boldsymbol{T}_p = \boldsymbol{T}_p\mathcal{A} = \mathcal{A}\,; \qquad \mathcal{A}\boldsymbol{T}_i = \boldsymbol{T}_i\mathcal{A} = -\,\mathcal{A}$$

where:

(6) $$\mathcal{A}^2 = \mathcal{A}\left(\sum_p \boldsymbol{T}_p - \sum_i \boldsymbol{T}_i\right) = n\,!\;\mathcal{A}.$$

The property stated in (3) is written using the notation (4) for the permutation operators:

$$\mathcal{H}\boldsymbol{T}\psi = \mathrm{E}\boldsymbol{T}\psi = \boldsymbol{T}(\mathrm{E}\psi)$$

so that, since by hypothesis $\mathcal{H}\psi = \mathrm{E}\psi$:

$$\mathcal{H}\boldsymbol{T}\psi = \boldsymbol{T}\mathcal{H}\psi.$$

This equation shows that any eigenfunction ψ of $\mathcal{H}$ is capable of generalization to some function which can be expanded on the orthonormal system of eigenfunctions of $\mathcal{H}$. This property, which is true for a general permutation $\boldsymbol{T}$, remains valid for any linear combination, hence:

(7) $$\mathcal{H}\mathcal{A} = \mathcal{A}\mathcal{H}.$$

We know that when two hermitian operators commute it is possible to construct common eigenfunctions, which suggests that we can find eigenfunctions of $\mathcal{H}$ such that we can say:

(8) $$\mathcal{A}\psi = a\psi.$$

This result assumes the two operators $\mathcal{H}$ and $\mathcal{A}$ hermitian; for $\mathcal{H}$ this is a consequence of the fact that $\mathcal{H}$ is associated with a physical variable; the case of $\mathcal{A}$ is a little more tricky. A direct demonstration of the hermiticity of $\mathcal{A}$ can be given, but here we will satisfy ourselves by showing that the eigenvalues of this operator are real, this being a necessary (but not sufficient) condition for the operator to be hermitian. We begin with (8); application of $\mathcal{A}$ to the both sides of the equation gives:

$$\mathcal{A}^2\psi = \mathcal{A}(a\psi) = a\mathcal{A}\psi = a^2\psi$$

so that, since $\mathcal{A}^2 = n!\mathcal{A}$:

$$a^2\psi = n\,!\;a\psi$$

which requires for a function ψ not identically zero:

$$a' = n\,! \qquad \text{or} \qquad a'' = 0.$$

As a result, it is possible to separate the complete system of eigenfunctions of $\mathscr{H}$ into two families ψ'_r and ψ''_s, such that we have:

$$\mathscr{A}\psi'_r = n!\,\psi'_r, \qquad \mathscr{A}\psi''_s = 0.$$

We can also show that the matrix elements representing a general physical variable and cross-products of the states of the two families, are necessarily zero. If $\boldsymbol{F}$ represents the operator associated with one such variable, the matrix element to be considered will be of the form:†

$$f_{r,s} = (\psi'_r,\ \boldsymbol{F}\psi''_s).$$

Equation (8) allows us to replace ψ'_r by $(1/n!)\mathscr{A}\psi'_r$, and as a result, since these operators are hermitian:

$$f_{r,s} = \frac{1}{n!}(\mathscr{A}\psi'_r, \boldsymbol{F}\psi''_s) = \frac{1}{n!}(\psi'_r,\ \mathscr{A}\boldsymbol{F}\psi''_s).$$

Now a general quantity defined on a many-electron system must have its value unchanged by a permutation of the electrons. The proof given for $\mathscr{H}$ can be generalized, and as a result of the fact that $\mathscr{A}$ and $\boldsymbol{F}$ commute:

$$\mathscr{A}\boldsymbol{F}\psi''_s = \boldsymbol{F}\mathscr{A}\psi''_s = 0.$$

The matrix element $f_{r,s}$ is then necessarily zero.

It follows that if the system is represented at a given time by a state of the type ψ'_r, any perturbation whose calculation (§20) uses the matrix elements of the perturbation energy will not be capable of effecting a transformation to a state of the form ψ''_s, and reciprocally. Thus, there is complete separation between the two families of states which define two autonomous systems. It is therefore entirely normal that only one of them be found in the physical world. For electrons we have only the first of these, and this choice constitutes the most general statement of the Pauli principle: *the wave functions representing the state of a many-electron system belong to the non-zero eigenvalue of the anti-symmetrization operator* $\mathscr{A}$.

† We use the following current shorthand notation to represent the expression:

$$f_{r,s} = \sum_{\sigma_1, \ldots, \sigma_n} \int \psi_r'^{*}\boldsymbol{F}\psi''_s d\tau_1 \ldots d\tau_n$$

with the integral being taken over the spatial coordinates and the summation over the spin for the n electrons of the system.

The statement at the beginning of this paragraph is easily proven, so that for an eigenfunction of $\boldsymbol{H}$ which satisfies the statement above we have the identity:

$$\mathcal{A}\psi = n!\,\psi.$$

We examine the effect on ψ of a permutation of two electrons, the others remaining fixed. Examination of the function f shows that this permutation is of the odd type $\boldsymbol{T}_i$. This property is evident if the two electrons have consecutive indices r and $r+1$. The permutation of two electrons of general index can always be reduced to an odd number of permutations of electrons of adjacent index. We can then write:

$$\boldsymbol{T}\mathcal{A}\psi = n!\,\boldsymbol{T}\psi.$$

We have $\boldsymbol{T}_i\mathcal{A} = -\mathcal{A}$, hence:

$$n!\,\boldsymbol{T}_i\psi = -\,\mathcal{A}\psi = -\,n!\,\psi, \qquad \text{thus} \qquad \boldsymbol{T}_i\psi = -\,\psi$$

which is indeed the statement of the principle given earlier.

The Wave Functions of a Many-Electron Atom 45

We begin with the example of helium for which we again give equation (2) for the hamiltonian:

$$\mathcal{H} = -\frac{\hbar^2}{2m_0}(\Delta_1 + \Delta_2) + \frac{e^2}{4\pi\varepsilon_0}\left(-\frac{2}{r_1} - \frac{2}{r_2} + \frac{1}{r_{12}}\right). \tag{9}$$

The operator $\mathcal{H}$ is shown to be the sum of two one-electron hamiltonians relative to the behavior of each of the electrons in the field of the nucleus, and of the energy of repulsion for the two electrons:

$$\mathcal{H} = \boldsymbol{H}_1 + \boldsymbol{H}_2 + \frac{e^2}{4\pi\varepsilon_0 r_{12}}.$$

The treatment in section 26 involved taking the repulsion as a perturbation, which we disregard in the first approximation† by taking:

$$\mathcal{H}^0 = \boldsymbol{H}_1 + \boldsymbol{H}_2. \tag{10}$$

† The use of the perturbation method here is not without difficulty in principle. In effect, when the perturbation term comes from an external field, it tends to zero with the field intensity, with the result that the perturbation method is useful for fields which are sufficiently small. In this example, the magnitude of the perturbation energy is given *a priori* and we are not certain that the expansions in the perturbation method will converge.

This simplified form of the hamiltonian has separable variables and it is then possible to look for solutions of the Schrödinger equation in the form:

$$\psi = \varphi(q_1)\theta(q_2),$$

by using the shorthand notation q for the set of coordinates x, y, z, and σ_z of each electron. We then get:

$$\mathcal{H}^0\psi = (\boldsymbol{H}_1\varphi)\theta + \varphi(\boldsymbol{H}_2\theta) = \mathrm{E}\varphi\theta.$$

We get a solution of this equation by taking for φ and θ the eigenfunctions of $\boldsymbol{H}_1$ and $\boldsymbol{H}_z$, respectively, so that:†

$$\boldsymbol{H}_1\varphi = \varepsilon_1^0\varphi\,; \qquad \boldsymbol{H}_2\theta = \varepsilon_2^0\theta$$

which leads to:

$$\mathcal{H}^0\varphi\theta = (\varepsilon_1^0 + \varepsilon_2^0)\varphi\theta.$$

The one-electron hamiltonians $\boldsymbol{H}_1$ and $\boldsymbol{H}_2$ have the same form and differ only in the designation of the variables, with the result that φ and θ belong to the same family of eigenfunctions and are therefore called φ_r and φ_s hereafter We see then that it is possible to form an approximate solution to the many-electron equation starting from the one-electron solution. We will now examine the special case corresponding to the ground state level.

Configuration $(1s)^2$. There is only one wave function without spin u_{1s} which must be combined with two spin functions α and β, from which we get two wave functions $u_{1s}\alpha$ and $u_{1s}\beta$. The method that we will use then allows us to form the following four solutions which we write as:

$$\begin{aligned}
\psi_a &= u_{1s}(1)\ \alpha(1)\ u_{1s}(2)\ \alpha(2)\\
\psi_b &= u_{1s}(1)\ \alpha(1)\ u_{1s}(2)\ \beta(2)\\
\psi_c &= u_{1s}(1)\ \beta(1)\ u_{1s}(2)\ \alpha(2)\\
\psi_d &= u_{1s}(1)\ \beta(1)\ u_{1s}(2)\ \beta(2).
\end{aligned}$$

The four functions ψ_r belong to the same eigenvalue $2\varepsilon_{1s}^0$ of $\mathcal{H}_0$, with the result that the most general solution of:

$$\mathcal{H}^0\psi = \mathrm{E}\psi \quad \text{with} \quad \mathrm{E} = 2\varepsilon_{1s}^0$$

should be of the form:

$$\psi = a\psi_a + b\psi_b + c\psi_c + d\psi_d.$$

† A current notation adopted here uses the letter ε for the one-electron eigenvalues, E being reserved to designate the total energy of the system.

In section 44 we showed that the Pauli principle requires that the wave function should be antisymmetric under a permutation of the electrons. The condition $\psi(2,1) = -\psi(1,2)$ is written:

$$a\psi_a + b\psi_b + c\psi_c + d\psi_d = -(a\psi_a + b\psi_c + c\psi_b + d\psi_d).$$

We get an identity by taking $a = d = 0$, $b + c = 0$, so that we get only one acceptable solution with the form:

$$(11) \quad \psi = c(\psi_c - \psi_b) = cu_{1s}(1)\, u_{1s}(2)\, [\alpha(2)\, \beta(1) - \alpha(1)\, \beta(2)].$$

The factor c is determined by the normalization condition which leads, since ψ_b and ψ_c are orthonormal, to $c = 1/\sqrt{2}$.

We can note that the function (11) represents a hybrid state of the two possible states with opposite spins, which is in agreement with our first statement of the Pauli principle (§26).

The state equation (11) can be gotten by forming *a priori* the antisymmetric combinations from the action of the antisymmetrization operator $\mathcal{A}$ on the various functions ψ_r. In effect, φ being a general function, we consider $\mathcal{A}\varphi$ and $\mathcal{A}^2\varphi$. Equation (6) lets us write:

$$\mathcal{A}^2\varphi = n!\ \mathcal{A}\varphi, \quad \text{so that} \quad \mathcal{A}(\mathcal{A}\varphi) = n!\ \mathcal{A}\varphi,$$

which shows that whatever the function φ is, $\mathcal{A}\varphi$ is a function belonging to the eigenvalue $n!$ and therefore satisfies the Pauli principle if it is not identically zero. In our example, we get:

$$\begin{aligned}
\mathcal{A}\psi_a &= \psi_a(1,2) - \psi_a(2,1) = 0 \\
\mathcal{A}\psi_b &= u_{1s}(1)\, u_{1s}(2)\, [\alpha(1)\, \beta(2) - \alpha(2)\, \beta(1)] \\
\mathcal{A}\psi_c &= u_{1s}(1)\, u_{1s}(2)\, [\beta(1)\, \alpha(2) - \beta(2)\, \alpha(1)] = -\mathcal{A}\psi_b. \\
\mathcal{A}\psi_d &= 0.
\end{aligned}$$

We get the same result as in the previous calculation (11) to within the normalization factor. The state $(1s)^2$ is therefore non-degenerate, and the fact that it corresponds to only one possible spin configuration leads to the name *singlet state*.

Configuration $(1s)^1(2s)^1$. The calculation of the wave functions evolves as in the preceding example. The four initial functions are written:

$$\begin{aligned}
\psi_a &= u_{1s}(1)\ \alpha(1)\ u_{2s}(2)\ \alpha(2) \\
\psi_b &= u_{1s}(1)\ \alpha(1)\ u_{2s}(2)\ \beta(2) \\
\psi_c &= u_{1s}(1)\ \beta(1)\ u_{2s}(2)\ \alpha(2) \\
\psi_d &= u_{1s}(1)\ \beta(1)\ u_{2}\ (2)\ \beta(2).
\end{aligned}$$

We construct the functions obeying the Pauli principle by application of the antisymmetrization operator and we then get:

$$\mathcal{A}\psi_a = [u_{1s}(1)\ u_{2s}(2) - u_{1s}(2)\ u_{2s}(1)]\ \alpha(1)\ \alpha(2)$$
$$\mathcal{A}\psi_b = u_{1s}(1)\ \alpha(1)\ u_{2s}(2)\ \beta(2) - u_{1s}(2)\ \alpha(2)\ u_{2s}(1)\ \beta(1)$$
$$\mathcal{A}\psi_c = u_{1s}(1)\ \beta(1)\ u_{2s}(2)\ \alpha(2) - u_{1s}(2)\ \beta(2)\ u_{2s}(1)\ \alpha(1)$$
$$\mathcal{A}\psi_d = [u_{1s}(1)\ u_{2s}(2) - u_{1s}(2)\ u_{2s}(1)]\ \beta(1)\beta(2).$$

Contrary to the preceding case, we get four non-zero functions, for which we can verify the pairwise orthogonality. The energy degeneracy is thus of order four. We will show later that we are lead to consider it as composed of a singlet level and a triplet level.

Configuration $(1s)^1(2p)^1$. The argument in this case is again developed using the same principle of writing the initial functions for the three cases $(1s)(2p_1)$, $(1s)(2p_0)$, and $(1s)(2p_{-1})$, and by forming the antisymmetric combinations. This time the energy level is degenerate of order twelve. Its structure will be studied later.

The method set forth for this particular case is generalized without difficulty to the case of a general many-electron system, atomic or molecular, for which we will form the wave functions in the first approximation by ignoring the interactions between electrons. The hamiltonian $\mathcal{H}^0$ of the system divides into the sum of one-electron hamiltonians of the same form relative to each of the n electrons of the system, so that:

$$\mathcal{H}^0 = \boldsymbol{H}_1 + \boldsymbol{H}_2 + \dots + \boldsymbol{H}_n.$$

An eigenfunction of $\mathcal{H}^0$ can thus be made up of the product of eigenfunctions of the one-electron hamiltonians and we will write:

$$\psi = \varphi_\alpha(1)\ \varphi_\beta(2)\ \dots\ \varphi_\nu(n),$$

where the eigenfunction φ_λ depends on the space and spin variables, it not being profitable to give their detailed structure as in the example of the helium atom. The function ψ so formed belongs to an eigenvalue of $\mathcal{H}^0$: $E = \varepsilon_\alpha + \varepsilon_\beta + \dots + \varepsilon_\nu$. There are, in general, numerous possible combinations of $\varphi_\alpha, \varphi_\beta, \dots, \varphi_\nu$, corresponding to the same value of E, and, as a consequence, there is degeneracy in this level. We will pursue the calculation, retaining only the eigenfunctions permitted by the Pauli principle. These will be formed as before by applying the antisymmetrization operator, so that:

$$\mathcal{A}\psi = \mathcal{A}\varphi_\alpha(1)\ \varphi_\beta(2)\ \dots\ \varphi_\nu(n). \tag{12}$$

The expansion thus obtained by application of $\mathscr{A}$ corresponds to the definition of a determinant (Slater determinant) as shown from the expansion developed for the three-electron case:

$$\mathscr{A}\psi = \varphi_\alpha(1)\ \varphi_\beta(2)\ \varphi_\gamma(3) + \varphi_\alpha(2)\ \varphi_\beta(3)\ \varphi_\gamma(1) + \varphi_\alpha(3)\ \varphi_\beta(1)\ \varphi_\gamma(2) - \varphi_\alpha(1)\ \varphi_\beta(3)\ \varphi_\gamma(2) - \varphi_\alpha(3)\ \varphi_\beta(2)\ \varphi_\gamma(1) - \varphi_\alpha(2)\ \varphi_\beta(1)\ \varphi_\gamma(3)$$

which can be written:

$$\mathscr{A}\psi = \begin{vmatrix} \varphi_\alpha(1) & \varphi_\alpha(2) & \varphi_\alpha(3) \\ \varphi_\beta(1) & \varphi_\beta(2) & \varphi_\beta(3) \\ \varphi_\gamma(1) & \varphi_\gamma(2) & \varphi_\gamma(3) \end{vmatrix} \tag{13}$$

The eigenfunctions of the hamiltonian $\varphi_\alpha, \varphi_\beta, \ldots, \varphi_\nu$, are pairwise orthogonal, with the result that the integral of all of the rectangular terms disappears in the calculation of the normalization integral of $\mathscr{A}\psi$. There remain only the square terms of the number $n!$, and these are all equal to unity if the functions φ_γ are normalized. As a result, the normalized function formed from the one-electron orbitals φ_λ is written:

$$\frac{1}{\sqrt{n!}}\mathscr{A}\psi = \frac{1}{\sqrt{n!}}\mathscr{A}\varphi_\alpha(1)\ \varphi_\beta(2)\ \ldots\ \varphi_\nu(n). \tag{14}$$

We note, finally, that the form (13) of the wave function implies, from a classical property of determinants, that this function is identically zero if two of the functions φ_λ are the same. We thus see the basis of our first statement of the Pauli principle (§26) which is related to this approximate form of the many-electron wave function.

Self-Consistent Field Methods 46

We have shown that suppression of the electron-electron interaction terms must necessarily lead to a somewhat poor approximation, useful principally in the classification of electron states of the system. We get perceptibly better results, keeping the advantages of section 45, by looking for a system of functions φ_γ such that the wave function:

$$\psi = \varphi_\alpha(1)\ \varphi_\beta(2)\ \ldots\ \varphi_\nu(n) \tag{15}$$

is the best possible approximation, that is to say, that it corresponds to the extremum of the energy integral (§19):

$$\mathrm{E} = \int \psi^* \mathcal{H} \psi d\tau_1\ \ldots\ d\tau_n. \tag{16}$$

It must be emphasized that this is an approximation method, for the state functions certainly do not have such a simple form, which especially does not satisfy the asymmetry condition required by the Pauli principle. Given this reservation, we will treat the calculation expressing the hamiltonian according to the method indicated in section 43 for the case of the helium atom:

$$\mathcal{H} = \boldsymbol{H}_1 + \ldots + \boldsymbol{H}_n + \frac{e^2}{4\pi\varepsilon_0}\sum_{i>j}\frac{1}{r_{ij}} \tag{17}$$

with:

$$\boldsymbol{H}_i = -\frac{\hbar^2}{2m_0}\Delta_i - \frac{Ze^2}{4\pi\varepsilon_0 r_i}.$$

We stress that in the electron interaction term, the sum must be taken in such a way as to compute the interaction of pairs *i*, *j* of the electrons one time and only one time. This leads to our writing:

$$\frac{e^2}{4\pi\varepsilon_0}\sum_{i>j}\frac{1}{r_{ij}} = \frac{1}{2}\frac{e^2}{(4\pi\varepsilon_0)}\sum_{i\neq j}\frac{1}{r_{ij}}.$$

The evaluation of the energy by (16) leads to two series of terms, of which the first is written:

$$\int \psi^\star \boldsymbol{H}_i \psi d\tau_1 \ldots d\tau_n$$

$$= \int \overset{\star}{\varphi}_\alpha(1)\ldots\overset{\star}{\varphi}_\lambda(i)\ldots\overset{\star}{\varphi}_\nu(n)\boldsymbol{H}_i\varphi_\alpha(1)\ldots\varphi_\lambda(i)\ldots\varphi_\nu(n)d\tau_1\ldots d\tau_n$$

so that, since the φ_η are orthogonal:

$$\int \psi^\star \boldsymbol{H}_i \psi d\tau_2 \ldots d\tau_n = \int \overset{\star}{\varphi}_\lambda(i)\boldsymbol{H}_i\varphi_\lambda(i)d\tau_i = \varepsilon_\lambda^0.$$

The term representing the electron interactions is the sum of contributions of the form:

$$\frac{e^2}{4\pi\varepsilon_0}\int \overset{\star}{\varphi}_\alpha(1)\ldots\overset{\star}{\varphi}_\varkappa(i)\ldots\overset{\star}{\varphi}_\lambda(j)\ldots\overset{\star}{\varphi}_\nu(n)\frac{1}{r_{ij}}$$

$$\times\ \varphi_\alpha(1)\ldots\varphi_\varkappa(i)\ldots\varphi_\lambda(j)\ldots\varphi_\nu(n)d\tau_1\ldots d\tau_n$$

$$= \frac{e^2}{4\pi\varepsilon_0}\int\frac{1}{r_{ij}}|\varphi_\varkappa(i)|^2|\varphi_\lambda(j)|^2 d\tau_i d\tau_j.$$

The result obtained by separating in E all of the terms involving one wave function φ_η is:

$$\mathrm{E}_\eta = \varepsilon_\eta^0 + \frac{e^2}{4\pi\varepsilon_0}\sum_\lambda\int|\varphi_\eta(i)|^2\frac{1}{r_{ij}}|\varphi_\lambda(j)|^2 d\tau_i d\tau_j, \quad \text{with} \quad \lambda \neq \eta. \tag{18}$$

To write the extremum condition on E relative to variations of φ_η it is sufficient to limit considerations to E_η, since the other terms are independent of φ_η. To do this we take the interaction terms appearing in (18) in the equivalent form:

$$\frac{e^2}{4\pi\varepsilon_0}\int |\varphi_\eta(i)|^2 \frac{1}{r_{ij}} |\varphi_\lambda(j)|^2 d\tau_i d\tau_j$$

$$= \frac{e^2}{4\pi\varepsilon_0}\int \varphi_\eta^\star(i)\left[\int \frac{1}{r_{ij}} |\varphi_\lambda(j)|^2 \varphi_\eta(i) d\tau_j\right] d\tau_i.$$

The integral between brackets defines an operator acting on the function $\varphi_\eta(i)$, and can be denoted as the resultant of the action of an operator:

$$\text{(19)} \qquad \boldsymbol{J}_{\lambda\eta}\varphi_\eta(i) = \frac{e^2}{4\pi\varepsilon_0}\int \frac{1}{r_{ij}} |\varphi_\lambda(j)|^2 \varphi_\eta(i) d\tau_j.$$

Using the operators $\boldsymbol{J}_{\lambda\eta}$, which are linear and hermitian, equation (18) takes the form:

$$\text{(20)} \qquad E_\eta = \int \varphi_\eta^\star\left(\boldsymbol{H}_i + \sum_\lambda \boldsymbol{J}_{\lambda\eta}\right)\varphi_\eta d\tau_i, \quad \text{with} \quad \lambda \neq \eta.$$

We have thus returned to the case of the one-electron problem and have shown that there is an equivalence between the extremum condition on $\int \psi^\star \boldsymbol{H} \psi d\tau$, where ψ is normalized, and the desired solutions of $\boldsymbol{H}\psi = E\psi$. We have said that the function φ_η should be an eigenfunction of a specific hamiltonian, so that:

$$\text{(21)} \qquad \left(\boldsymbol{H}_i + \sum_\lambda \boldsymbol{J}_{\lambda\eta}\right)\varphi_\eta = \varepsilon_\eta\varphi_\eta, \quad \text{with } \lambda \neq \eta.$$

By multiplying by $\varphi_\eta^\star$ and integrating with respect to i, we see that the quantity E_η defined by (20) is an eigenvalue of the hamiltonian defined by (21) which is called the *self-consistent field* hamiltonian.

This same calculation done for each of the n electrons of the system leads to a set of n integro-differential equations of the form (21), where we should note that the action of the operator $\boldsymbol{J}_{\lambda\eta}$ on φ_η assumes a knowledge of the function φ_λ, with the result that the simultaneous solution of the n equations (21) seems *a priori* to be an unsolvable problem. This formulation lends itself to a successful method of approximation treated by first taking a set of n functions φ_η^1, as an approximate solution to the problem. These functions allow us to calculate the action of the operator $\boldsymbol{J}_{\lambda\eta}$ on any function giving a precise meaning to equations (21), from which the solution, only possible by numerical methods, leads to a new set of functions φ_η^2 generally different from the initial set φ_η^1. The process is repeated with this new set and continued until the wave functions φ_η^r and the values ε_η^r do not differ from those obtained in the previous iteration. The field acting on one electron arising from the others in the set is said to be self-consistent. This explains the nomenclature of the hamiltonian in (21), also called the Hartree hamiltonian to distinguish it from others of the same type which differ in the method of writing the many-electron wave function. We will now stress several important points concerning this method.

(a) The meaning of the eigenvalues ε_η. We form from (21):

$$\sum_\eta \varepsilon_\eta = \sum_\eta \int \varphi_\eta^* \Big(\boldsymbol{H}_i + \sum_\lambda \boldsymbol{J}_{\lambda\eta} \Big) \varphi_\eta d\tau_i, \quad \text{with} \quad \lambda \neq \eta.$$

On the one hand we see the terms ε_η^0, and on the other, the pairwise interaction energy between the electrons, each taken twice. We can then write:

$$\sum_\eta \varepsilon_\eta = \sum_\eta \varepsilon_\eta^0 + 2 \cdot \frac{e^2}{4\pi\varepsilon_0} \sum_{\lambda>\eta} \int \frac{1}{r_{ij}} \,|\, \varphi_\lambda(i) \,|^2 \,|\, \varphi_\eta(j) \,|^2 d\tau_i d\tau_j.$$

We see that the sum of the eigenvalues ε_η does not give the total energy as defined in (16). To give an interpretation to these values ε_η we compare E to the energy E′ of the system found by the removal of an electron occupying one of the orbitals, φ_ν for example. The theorem which we will state uses the extremum property of E′. This can be calculated in a first approximation by using the same orbitals as in the nonionized system. In particular, it is possible to use the same nomenclature for the orbitals in both cases. We can then write:

$$\mathrm{E} = \sum_\eta \varepsilon_\eta^0 + \frac{e^2}{4\pi\varepsilon_0} \sum_{\lambda>\eta} \int \frac{1}{r_{ij}} \,|\, \varphi_\lambda(i) \,|^2 \,|\, \varphi_\eta(j) \,|^2 d\tau_i d\tau_j$$

$$\mathrm{E}' = \sum_\eta{}' \varepsilon_\eta^0 + \frac{e^2}{4\pi\varepsilon_0} \sum_{\lambda>\eta}{}' \int \frac{1}{r_{ij}} \,|\, \varphi_\lambda(i) \,|^2 \,|\, \varphi_\eta(j) \,|^2 d\tau_i d\tau_j.$$

The notation $\sum'$ means that one of the values of the summation index (here ν) is excluded. If we form the difference of these two energies, which by definition represents the ionization energy for an electron occupying the orbital φ_ν, we get:

$$(22) \quad I_\nu = E' - E = -\varepsilon_\nu^0 - \frac{e^2}{4\pi\varepsilon_0} \sum_\lambda \int \frac{1}{r_{ij}} |\varphi_\lambda(i)|^2 |\varphi_\nu(j)|^2 d\tau_i d\tau_j$$

$$= -\varepsilon_\nu \text{ with } \lambda \neq \nu.$$

We get, except for the sign, the value of ε_ν defined by (21), hence the very important significance of the eigenvalues of the self-consistent field hamiltonians—they represent the ionization energy of the system for the corresponding state.

(b) Choice of the initial atomic orbitals. We note first that the preceding calculations did not require the use of spin variables, with the result that it is possible to exclude them and retain only the space variable part of φ_η. A frequently used approximation involves taking the Slater functions for the initial functions, with the screening coefficients of each level calculated as in (§28). We must note, however, that the Slater functions of the same type ($1s$ and $2s$, for example) are not orthogonal and must be replaced by an orthogonal set obtained by the Schmidt method (§22).

(c) The helium atom. We now repeat the example of the helium atom previously treated (§27) in the one-electron approximation. The hamiltonian that has been introduced contains a potential function $V(r)$, representing the interaction with the second electron. As a result, the working out of the calculations can be considered as an example of the self-consistent field method, with the reservation that the state functions are taken to be of the Slater type. This permits us to do all of the algebra but does not give the best possible orbitals. We can identify (1) with the self-consistent field hamiltonian whose eigenvalue ε_{1s} must represent, except for the sign, the ionization energy of the $1s$ level, so that:

$$I_{He} = -\varepsilon_{1s} = E_0(Z^{\star 2} - 2sZ^\star).$$

In fact, we have indicated one other value of this ionization energy as:

$$I_{He} = E(He^+) - E(He) = E_0(2Z^{\star 2} - Z^2),$$

which as a function of $Z^\star = Z - s$ is written:

$$I_{He} = E_0(Z^{\star 2} - 2sZ^\star - s^2).$$

The two values are very close and give the small value $s^2 = 0.096$.

(d) Other self-consistent field methods. The Hartree method has the disadvantage of leading to non-antisymmetric state functions. This is not the case with the Hartree-Fock self-consistent field method,† which involves determining the one-electron orbitals $\varphi_\alpha, \varphi_\beta, \ldots, \varphi_\nu$, which give an extremum of:

$$E = \int \psi^* \mathcal{H} \psi d\tau_1 \ldots d\tau_n$$

by taking ψ in the form (14) so that:

$$\psi = \frac{1}{\sqrt{n!}} \mathcal{A} \varphi_\alpha(1) \varphi_\beta(2) \ldots \varphi_\nu(n).$$

The calculations which evolve exactly as for the Hartree field are a little more complicated, and we end up with the result that each of the one-electrons functions φ_η should be an eigenfunction of the Hartree-Fock operator in the form:

$$\boldsymbol{F}_i = \boldsymbol{H}_i + \sum_\lambda \boldsymbol{J}_{\lambda\eta} - \sum_\lambda \boldsymbol{K}_{\lambda\eta}, \quad \text{with} \quad \lambda \neq \eta$$

where:

$$\boldsymbol{J}_{\lambda\eta}\varphi_\eta(i) = \frac{e^2}{4\pi\varepsilon_0} \int \frac{1}{r_{ij}} \mid \varphi_\lambda(j) \mid^2 \varphi_\eta(i) d\tau_j ;$$

$$\boldsymbol{K}_{\lambda\eta}\varphi_\eta(i) = \frac{e^2}{4\pi\varepsilon_0} \int \frac{1}{r_{ij}} \varphi_\lambda^*(j)\varphi_\lambda(i)\varphi_\eta(j) d\tau_j.$$

The investigation of the eigenfunctions of the Hartree-Fock operators is again carried out by successive approximations and the eigenvalues represent, except for the sign, the ionization energy corresponding to the removal of an electron from the level being treated. This is the general statement of the Koopmans theorem. For a more complete treatment see: R. Daudel, R. Lefebvre, and C. Moser, *Quantum Chemistry* (New York: Interscience Publishers, 1959), particularly Appendix II.

---

† The form (15) of the wave function used in the Hartree method assumes complete independence between the relative positions of the various electrons. This is an approximation, since two electrons *i* and *j* cannot be at the same point: (*a*) by virtue of the Pauli principle if they have the same spin; and (*b*) because of the electrostatic repulsion. Use of antisymmetric functions takes account of (*a*) but not of (*b*). As a result, Hartree-Fock energies are too low. The correction known as the correlation energy poses a difficult problem (O. Sinanoglu, "Many Electron Theory of Atoms, Molecules and their Interactions," *Advances in Chemical Physics* (New York: Interscience Publishers, 1964), Vol. VI).

12 CLASSIFICATION OF THE STATES OF MANY-ELECTRON SYSTEMS

The discussion in section 45 has shown that degeneracy of the electron levels frequently occurs in even the most simple states. This degeneracy arises in the first approximation where we ignore the electron interactions, but is at least partially lifted when we take the interactions into account, and we then get separation of the levels which are grouped together in the first approximation. We will study this problem from various points of view, depending on the nature of the system under examination.

The Many-Electron Atom 47

The problem here is to characterize the states of a many-electron system and to find a certain number of mutually compatible variables whose values can be fixed simultaneously. We already know two such variables: the energy, and the variable represented by the antisymmetrization operator $\mathscr{A}$, which has the value $n!$ in the states permitted under the Pauli principle. If there is no interaction between the electrons, the properties of the one-electron atom remain valid, with the result that the variables L_r^2 and $L_{z,r}$ relative to any electron with index r are well defined as is the spin component $S_{z,r}$.

Initially, it is necessary to note that even if we disregard the electron interactions which are the new terms of the hamiltonian, we must limit ourselves to the examination of variables remaining invariant under permutation of the electrons, and represented, as a result, by operators which commute with $\mathscr{A}$. We are thus led to introduce the variables defined for a set of N electrons:

$$(1) \qquad \mathscr{L}_x = \sum_{r=1}^{n} L_{x,r}; \qquad \mathscr{L}_y = \sum_{r=1}^{n} L_{y,r}; \qquad \mathscr{L}_z = \sum_{r=1}^{n} L_{z,r}$$

and also for the spin components:

$$(1') \qquad \mathcal{S}_x = \sum_{r=1}^{n} S_{x,r}; \qquad \mathcal{S}_y = \sum_{r=1}^{n} S_{y,r}; \qquad \mathcal{S}_z = \sum_{r=1}^{n} S_{z,r}.$$

The operators $\boldsymbol{L}_{x,r}$, $\boldsymbol{L}_{y,r}$, and $\boldsymbol{L}_{z,r}$ satisfy the same anticommutation relations for any r (Appendix VI-2), and we limit ourselves to one example:

$$\boldsymbol{L}_{x,r}\boldsymbol{L}_{y,r} - \boldsymbol{L}_{y,r}\boldsymbol{L}_{x,r} = i\hbar\boldsymbol{L}_{z,r}.$$

In addition, we have shown (§43) that the variables belonging to two distinct electrons are compatible and the corresponding operators must therefore commute. As a consequence, the same anticommutation relations can be written for the operators representing the variables defined by (1), so that:

$$\mathscr{L}_x \mathscr{L}_y - \mathscr{L}_y \mathscr{L}_x = i\hbar \mathscr{L}_z.$$

It follows that all of the results deduced (Appendix VI) for these relations are valid here, particularly that the operators:

$$\mathscr{L}^2 = \mathscr{L}_x^2 + \mathscr{L}_y^2 + \mathscr{L}_z^2 \qquad \text{and} \qquad \mathscr{L}_z$$

commute and have simultaneous eigenvalues of the form:

(2) $\mathscr{L}^2 = \mathrm{L}(\mathrm{L}+1)\hbar^2; \qquad \mathscr{L}_z = \mathrm{M}\hbar, \qquad \text{with} \qquad -\mathrm{L} \leqslant \mathrm{M} \leqslant \mathrm{L}.$

Similar considerations lead us to introduce the pair of operators $\mathscr{S}^2 = \mathscr{S}_x^2 + \mathscr{S}_y^2 + \mathscr{S}_z^2$ and $\mathscr{S}_z$ with the simultaneous eigenvalues:

(2′) $\mathscr{S}^2 = \mathrm{S}(\mathrm{S}+1)\hbar^2; \qquad \mathscr{S}_z = \mathrm{M}_s\hbar, \qquad \text{with} \qquad -\mathrm{S} \leqslant \mathrm{M}_s \leqslant \mathrm{S}.$

We will construct and study the following set of operators which commute with themselves and with $\mathscr{A}$:

$$\mathscr{L}^2, \mathscr{L}_z. \mathscr{S}^2, \mathscr{S}_z.$$

The second point of our study involves the relationship between these operators and the hamiltonian $\mathscr{H}$. In the approximation in which we ignore the spin-orbit coupling, $\mathscr{H}$ does not contain the spin variables, and, as a result, commutes with $\mathscr{S}^2$ and $\mathscr{S}_z$. We now look at the case of $\mathscr{L}^2$ and $\mathscr{L}_z$. From the classical point of view, the angular momentum of each of the electrons defined by the three components $\mathrm{L}_{x,r}$, $\mathrm{L}_{y,r}$, and $\mathrm{L}_{z,r}$ remains constant in the absence of a mutual electron interaction, but this is no longer the case when we take the interaction into account. In the latter case, it is the total angular momentum defined by $\mathscr{L}_x$, $\mathscr{L}_y$, and $\mathscr{L}_z$, that remains constant and provides the first integrals of the motion. Conversely, we assume that the operators associated with these variables commute with $\mathscr{H}$, which we can always verify by a direct calculation (see **Exercise 1).**

To summarize, in a state of given energy, the two variables:

$$\mathscr{L}^2 = \mathrm{L}(\mathrm{L}+1)\hbar^2; \qquad S^2 = \mathrm{S}(\mathrm{S}+1)\hbar^2$$

are known, with the result that the quantum numbers L and S characterize the state being examined. For a pair L and S we have 2L + 1 possible values of $\mathscr{L}_z$, with each of which can be associated one of the 2S + 1 possible values of S_z, hence a total degeneracy of the energy level equal to (2L + 1)(2S + 1). The nomenclature follows through the introduction of symbols known as spectroscopic terms.

(*a*) The name of the level is based on the value of the quantum number L, which, as we shall see, is always integral, and uses the capitals corresponding to the lower case letters used for the one-electron levels, namely, S(L = 0), P(L = 1), D(L = 2), and F(L = 3).

(*b*) We will see that the value of S is of the form $k/2$ with the integer k either even or odd. The value of S defines the multiplicity of the level equal to 2S + 1, and this value is indicated as a left-side superscript of the characteristic letter (^{3}S, for example). For S = 0 the multiplicity reduces to unity, and we say that we have a *singlet term*. We also get *doublets* (S = $\frac{1}{2}$), *triplets* (S = 1), quadruplets (S = $\frac{3}{2}$), etc.

Procedure for the study. We know that the values of $\mathscr{L}_z$ and S_z are quantized and, as a result, remain invariant when we make the interaction terms tend to zero. We thus get a reference state where the state of each of the electrons is known. It is possible to calculate M and M_s from L and S, as we will see in the different examples given by again using the case of the helium atom for which the many-electron wave functions have already been derived (§45).

Configuration $(1s)^2$. The Pauli principle allows only one system of values of m_r and σ_r, conforming to the following table which thus reduces to one line.

m_1	σ_1	m_2	σ_2	M	M_s
0	$\frac{1}{2}$	0	$-\frac{1}{2}$	0	0

We get a single value M = 0 which eliminates any other values of L, so that L = 0. This then gives an S term. For the same reason M_s = 0 is only compatible with S = 0, hence a singlet state. Finally, the state $(1s)^2$ is designated by the symbol ^{1}S.

Configuration $(1s)^1(2s)^1$. The Pauli principle no longer limits the combinations of quantum numbers and we get the table:

m_1	σ_1	m_2	σ_2	M	M_s
0	$\frac{1}{2}$	0	$\frac{1}{2}$	0	1
0	$\frac{1}{2}$	0	$-\frac{1}{2}$	0	0
0	$-\frac{1}{2}$	0	$\frac{1}{2}$	0	0
0	$-\frac{1}{2}$	0	$-\frac{1}{2}$	0	-1

The conclusions about M are the same as in the first case, and again we have S terms. We now look for the largest value of M_s, namely, $M_s = 1$, which necessarily requires a value $S = 1$, and thus a triplet. For $S = 1$ we have three possible values of M_s; 1, 0, and −1. Thus, there remains then in the table a state corresponding to $M_s = 0$ which can only be associated with $S = 0$, therefore to a singlet. In conclusion, this level can have the terms ^1S and ^3S.

This discussion allows us to understand the previous results (§45) in the calculation of the antisymmetric wave functions, which we rewrite adding the normalization factor $(1/\sqrt{n} = 1/\sqrt{2})$:

$$\begin{aligned} \frac{1}{\sqrt{2}}\mathcal{A}\psi_a &= \frac{1}{\sqrt{2}}[u_{1s}(1)\, u_{2s}(2) - u_{1s}(2)\, u_{2s}(1)]\, \alpha(1)\, \alpha(2) \\ \frac{1}{\sqrt{2}}\mathcal{A}\psi_b &= \frac{1}{\sqrt{2}}[u_{1s}(1)\, \alpha(1)\, u_{2s}(2)\, \beta(2) - u_{1s}(2)\, \alpha(2)\, u_{2s}(1)\, \beta(1)] \\ \frac{1}{\sqrt{2}}\mathcal{A}\psi_c &= \frac{1}{\sqrt{2}}[u_{1s}(1)\, \beta(1)\, u_{2s}(2)\, \alpha(2) - u_{1s}(2)\, \beta(2)\, u_{2s}(1)\alpha(1)] \\ \frac{1}{\sqrt{2}}\mathcal{A}\psi_d &= \frac{1}{\sqrt{2}}[u_{1s}(1)\, u_{2s}(2) - u_{1s}(2)\, u_{2s}(1)]\, \beta(1)\beta(2). \end{aligned} \tag{3}$$

The upper and lower functions evidently belong to the triplet with $M_s = 1$ and −1, respectively; the space function is the same, this being necessary since the two states must have the same energy. We should have a third function for which $M_s = 0$, which is immediately obtained by linear combination of the two remaining functions, so that:

$$\begin{aligned} &\frac{1}{2}(\mathcal{A}\psi_b + \mathcal{A}\psi_c) \\ &= \frac{1}{\sqrt{2}}[u_{1s}(1)\, u_{2s}(2) - u_{1s}(2)\, u_{2s}(1)] \cdot \frac{1}{\sqrt{2}}[\alpha(1)\beta(2) + \alpha(2)\beta(1)]. \end{aligned} \tag{4}$$

We again find the same spatial factor. It is easy to get another representation of the singlet state by using a different combination of the two functions corresponding to $M_s = 0$, namely:

$$\begin{aligned} &\frac{1}{2}(\mathcal{A}\psi_b - \mathcal{A}\psi_c) \\ &= \frac{1}{\sqrt{2}}[u_{1s}(1)\, u_{2s}(2) + u_{1s}(2)\, u_{2s}(1)] \cdot \frac{1}{\sqrt{2}}[\alpha(1)\beta(2) - \alpha(2)\beta(1)]. \end{aligned} \tag{5}$$

We reproduce the same spin functions as in (11-11), representing the singlet state $(1s)^2$.

Configuration $(1s)^1(2p)^1$. We have twelve states which we list in the following table:

m_1	σ_1	m_2	σ_2	M	M_s
0	$\frac{1}{2}$	1	$\frac{1}{2}$	1	1
0	$\frac{1}{2}$	1	$-\frac{1}{2}$	1	0
0	$-\frac{1}{2}$	1	$\frac{1}{2}$	1	0
0	$-\frac{1}{2}$	1	$-\frac{1}{2}$	1	-1
0	$\frac{1}{2}$	0	$\frac{1}{2}$	0	1
0	$\frac{1}{2}$	0	$-\frac{1}{2}$	0	0
0	$-\frac{1}{2}$	0	$\frac{1}{2}$	0	0
0	$-\frac{1}{2}$	0	$-\frac{1}{2}$	0	-1
0	$\frac{1}{2}$	-1	$\frac{1}{2}$	-1	1
0	$\frac{1}{2}$	-1	$-\frac{1}{2}$	-1	0
0	$-\frac{1}{2}$	-1	$\frac{1}{2}$	-1	0
0	$-\frac{1}{2}$	-1	$-\frac{1}{2}$	-1	-1

The maximum value $M = 1$ can only be associated with $L = 1$, hence a P term. The discussion of the preceding case can be repeated for M_s and leads to a triplet term and another singlet, giving two states with the symbols ^3P and ^1P, respectively.

The examples we have just treated are generalized without difficulty to an atom containing any number of electrons. The corresponding tables can in general be made smaller by taking note of the following:

(*a*) The *complete electron shells* give a zero contribution both to M and M_s, so that it is pointless to include them in the table. As a result, the study of the electron configuration of carbon $(1s)^2(2s)^2(2p)^2$ reduces to the examination of the pair of electrons in the $2p$ level.

(*b*) The nature of the terms can be predicted by application of the vector model already mentioned (§24). We assume the values of l_r of the different one-electron states are arranged in increasing order $l_1 \leqslant l_2 \leqslant l_3 \leqslant \ldots$. If we take into account only the first two electrons, we have for L either $l_2 + l_1$ or $l_2 - l_1$. The line of reasoning can be extended, adding the electrons one by one, but without any possibility of giving a general formulation of the result.

The case of the multiplicity is easier to treat, since all of the values σ_r are equal to $\pm\frac{1}{2}$. As a result, if we progressively build up the possible multiplicities by adding electron after electron, the application of the vector model leads to a simple representation (fig. 59). For the first electron the value $S=\frac{1}{2}$ is indicated on the ordinate. If we add a second electron, the vector diagram allows two values of S gotten by adding and subtracting $\frac{1}{2}$, this being

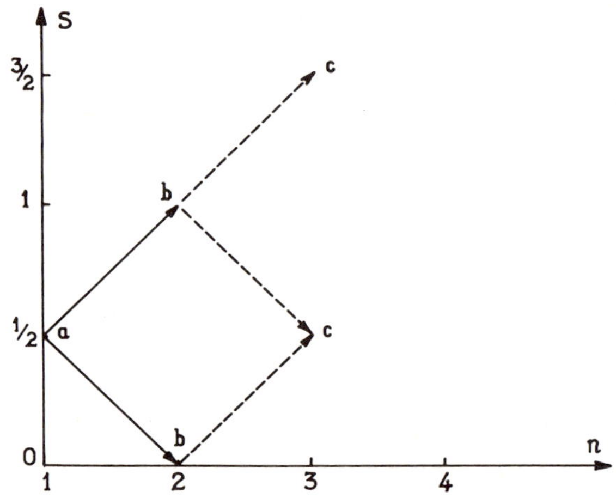

Figure 59. Application of the vector model in the prediction of spin states in a system of n electrons: $n=1$ (a); $n=2$ (b); $n=3$ (c).

shown by the points b in the diagram. This diagram is gradually extended, and we see that for three electrons one of the values, namely, $S=\frac{1}{2}$, is gotten twice (points c in the diagram). We include (fig. 60) the extended diagram, indicating at each node the number of times that the multiplicity occurs. In general, the Pauli principle prevents us from getting all of the possibilities shown in the diagram and, at the limit, when some number of electrons fills a shell exactly, only the singlet term is possible.

One other way of seeing the possible multiplet structures involves drawing up a table of different spin configurations, with each characterized by the known configuration for all electrons of index r with spin components $\sigma_r=\pm\frac{1}{2}$, giving the value M_s of the resultant. Under the assumption that all spin combinations are permitted, the value of M_s can go from $n/2$ to $-n/2$, where n is the number of electrons. An intermediate value:

$$M_s = \frac{1}{2}(a - b),$$

is gotten for the various combinations having a electrons with $\sigma = \frac{1}{2}$ and $b = n - a$ with $\sigma = -\frac{1}{2}$, the number of times this appears being given by the equation:

$$\binom{n}{a} = \frac{n!}{a!\,b!}.$$

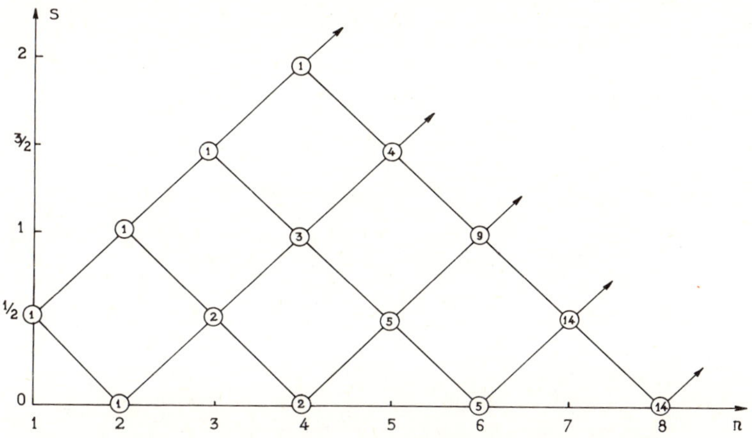

Figure 60. Diagram indicating for a given number of electrons, n, the maximal number of different multiplet states, each of these states being characterized by the value of S given on the ordinate.

The discussion presented earlier about this kind of table uses the fact that the number of times a given value of M_s occurs is equal to the number of multiplets of index higher or equal to $S = M_s$, assumed positive. To obtain the number of multiplets of order S it is necessary to determine the total number of combinations, the number of values of M_s arising from the multiplets of order $S \geqslant M_s - 1$ given by:

$$\binom{n}{a+1} = \frac{n!}{(a+1)!\,(b+1)!}.$$

Finally, the number of multiplets of order $S = \frac{1}{2}(a - b)$ is given by:

$$p(S) = \binom{n}{a} - \binom{n}{a+1} = \frac{n!}{a!\,b!}\left(1 - \frac{b}{a+1}\right).$$

In particular, the number of singlets obtained in the case of an even number of electrons ($n = 2k$) is calculated from the relation above by taking $a = b = k$, so that:

$$p(1) = p_k = \frac{(2k)!}{k!\,(k+1)!}.$$

It is easy to verify that this same equation allows us to evaluate the number of doublets in the case of an odd number of electrons $n = 2k - 1$, which is found to be equal to the number of singlets in the case above, in accord with the progressive building process of the diagram.

Exercises. 1. Verify that the hamiltonian of the helium atom commutes with the operator $\mathscr{L}_z$.
2. Verify, using the operators $\mathscr{L}_x \pm i\mathscr{L}_y$, and $\mathscr{S}_x \pm i\mathscr{S}_y$, that the wave function (11) defines a state corresponding to $L = S = 0$ (transpose the calculations from the spin-orbit coupling).
3. Study the action of the operators $\mathscr{S}_x \pm i\mathscr{S}_y$ and $\mathscr{S}_z$ on the spin functions introduced in the consideration of the level $(1s)^1(2s)^1$ namely:

$$\alpha(1)\alpha(2)\,;\ \frac{1}{\sqrt{2}}[\alpha(1)\beta(2) + \alpha(2)\beta(1)]\,;\ \beta(1)\beta(2).$$

Show that these three function belongs to the same eigenvalue of $\mathscr{S}^2$.

48 Ordering and Structure of the Levels

The preceding study has not treated the question of the ordering of the levels. We now return to the example of the $(1s)^1(2s)^1$ level in helium, which has two terms ^1S and ^3S. These levels have the same energy under the assumption that the electron interaction is negligible, but this is no longer true when we take this interaction into account, and the problem is to anticipate how separation of the levels is effected. Study of the electron interaction energy in the two cases is required so as to get the first order perturbations (§20). We get for the displacement of the ^3S level:

$$\Delta E(^3S) = \int \Psi^*_{^3S} \frac{e^2}{4\pi\varepsilon_0 r_{12}} \Psi_{^3S}\, d\tau_1 d\tau_2 \tag{6}$$

and $\Delta E(^1S)$ is similarly defined. The spin functions do not arise in the perturbation energy, with the result that we are required to consider only the space functions that we have written in the preceding paragraph. The functions u_{1s} and u_{2s} are real, and it is not necessary to use the complex conjugate symbol. As a result, we get:

$$\Delta E(^3S) = \frac{1}{2}\int [u_{1s}(1)\,u_{2s}(2) - u_{1s}(2)\,u_{2s}(1)]^2 \frac{e^2}{4\pi\varepsilon_0 r_{12}}\, d\tau_1 d\tau_2$$

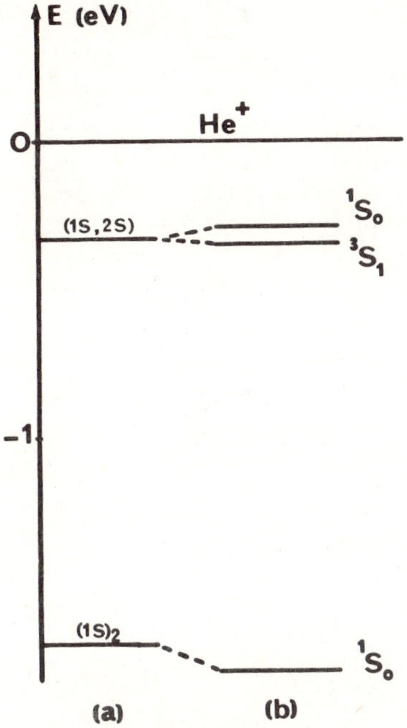

Fig. 61.

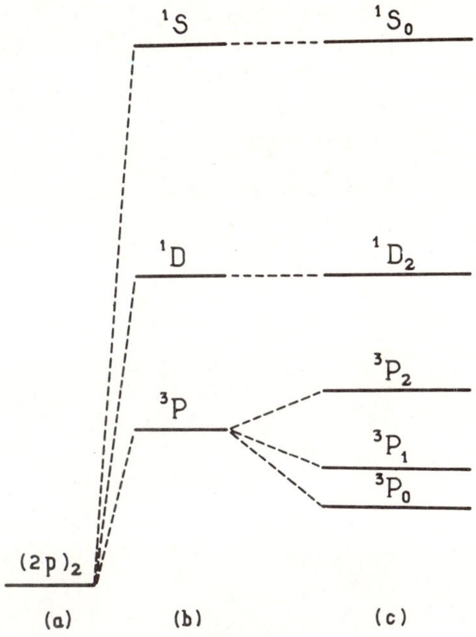

Fig. 62.

Figure 61. Energy levels for the helium atom in the states $(1s)^2$ and $(1s, 2s)$: (*a*) with the values calculated in the Slater approximation; (*b*) experimental values indicating the spectroscopic terms.

Figure 62. Distribution of the energy levels of carbon in the configuration $(2p)^2$: (*a*) in the Slater approximation; (*b*) with the mutual repulsion of the electrons being taken into account; (*c*) with spin-orbit coupling.

likewise:

$$\Delta E(^1S) = \frac{1}{2}\int [u_{1s}(1)\,u_{2s}(2) + u_{1s}(2)\,u_{2s}(1)]^2 \frac{e^2}{4\pi\varepsilon_0 r_{12}}\, d\tau_1 u\tau_2$$

where:

$$\Delta E(^1S) - \Delta E(^3S) = \frac{e^2}{2\pi\varepsilon_0}\int u_{1s}(1)\,u_{2s}(1)\,\frac{1}{r_{12}}\,u_{1s}(2)\,u_{2s}(2)d\tau_1 d\tau_2.$$

This difference is certainly positive, so that the ^3S level lies lower (fig. 61). As a general result, moreover, the levels are arranged in order of decreasing multiplicity. It should be noted that this rule in no way involves spin interaction, but is the interpretation of the electrostatic electron repulsion which differs from term to term.

An analogous rule allows us to find the ordering of the states corresponding to different values of L. For example, if we examine the carbon atom in its ground state $(1s)^2(2s)^2(2p)^2$, the analysis reduces to that of $(2p)^2$, as we saw in section 47, with the result that the 15 states that we can enumerate are distributed among three terms ^3P, ^1D, and ^1S. This is the order which the levels follow, the most stable being ^3P, which thus represents the carbon atom in the ground state (fig. 62).

All the preceding arguments exclude spin-orbit coupling. Such interaction puts the spin variables into the hamiltonian, and the operators $\mathscr{L}_r$ and $\mathscr{S}_r$ cease to commute separately with $\mathscr{H}$. We can no longer use them to exactly characterize the state, and must introduce the variable corresponding to the total angular momentum as in the case of spin-orbit coupling for one electron. The operators used here are defined by:

$$(7) \quad \mathscr{J}_x = \mathscr{L}_x + \mathscr{S}_x; \qquad \mathscr{J}_y = \mathscr{L}_y + \mathscr{S}_y; \qquad \mathscr{J}_z = \mathscr{L}_z + \mathscr{S}_z$$

and a state is characterized by the simultaneous eigenvalues of the operators $\mathscr{J}^2$ and $\mathscr{J}_z$, respectively, in the form:

$$\mathscr{J}^2 = J(J+1)\hbar^2: \qquad \mathscr{J}_z = M_J\hbar.$$

Taking the ^3P level as an example, the problem is to anticipate the shift in energy under the action of the spin-orbit coupling. The method is exactly the same as in the one-electron case, and the result can be simply forecast by use of the vector model. The possible values of J run from $L+S$ to $|L-S|$. In our example $(L=1, S=1)$, we have three terms corresponding respectively to $J=0, 1, 2$, indicated by giving the value of J as a subscript, so that we have $^3P_0$, $^3P_1$, and $^3P_2$. The nomenclature of the terms is essentially based on the values of L and S, and the shift due to spin-orbit coupling—expressed by the different values of J—represents a fine structure of the levels. As a standard example, we indicate the nomenclature and position of the levels in the case of a calcium atom (fig. 63), which can be thought of as an example of a two-electron system, since the filled shells K, L, and M, remain static during the transitions indicated on the diagram. As the discussion in section 47 indicated, the terms contain only singlets and triplets, each of the families being separated on the diagram, since optical transitions are only possible between terms of the same multiplicity. As a result, the observed spectra behave on the one hand like singlet states, and on the other, like triplets. One should also note the difference in order of magnitude between the separation of levels due to spin-orbit coupling, and that due to electron repulsion.

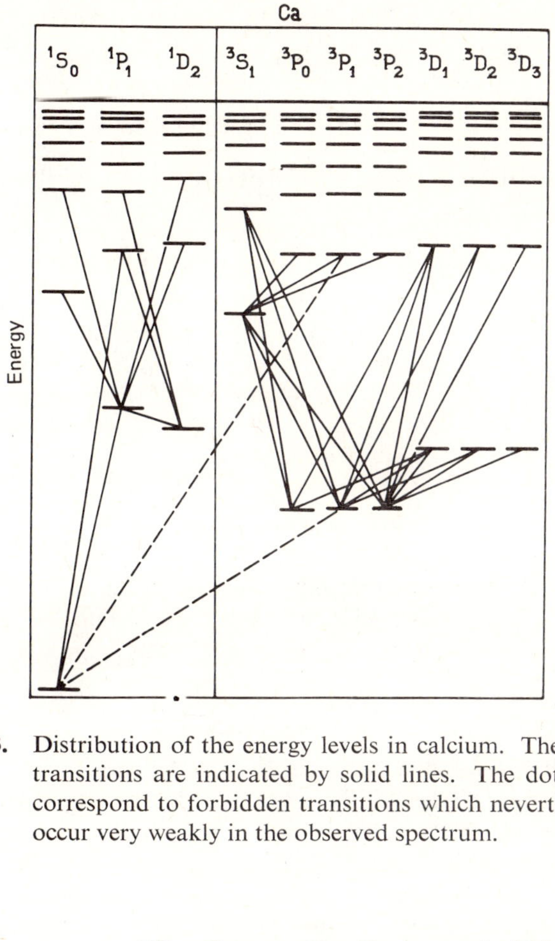

Figure 63. Distribution of the energy levels in calcium. The allowed transitions are indicated by solid lines. The dotted lines correspond to forbidden transitions which nevertheless do occur very weakly in the observed spectrum.

The States of a Symmetric Diatomic Molecule 49

The discussion parallels that for the many-electron atom (§47). The initial problem is the description of the states without the electron interactions. We have given the nomenclature (§31) as well as the mode of formation of the orbitals (MO) from those AO's of the bound atoms (§32). The products of such one-electron functions give a solution of the hamiltonian with interaction, and we will select antisymmetric orbitals as required by the Pauli principle using the operator $\mathscr{A}$. We take as an example the hydrogen molecule in its ground state. The one-electron bond orbital (§30) with symmetry g is written:

$$u_g = \frac{(\psi_{1s})_a + (\psi_{1s})_b}{\sqrt{2(1+S)}}. \tag{8}$$

We will not repeat the discussion in section 45, in which we demonstrated that we are only required to consider spin combinations consistent with the Pauli principle, namely:

$$u_g(1)\alpha(1)\, u_g(2)\beta(2) \quad \text{and:} \quad u_g(1)\beta(1)\, u_g(2)\alpha(2).$$

We get the same antisymmetric orbital from either of these two functions, namely:

$$\frac{1}{\sqrt{2}} \mathcal{A} u_g(1)\alpha(1)\, u_g(2)\beta(2) = \frac{1}{\sqrt{2}} u_g(1)\, u_g(2)[\alpha(1)\beta(2) - \alpha(2)\beta(1)].$$

We again find the spin function characteristic of a singlet state. The method is easily extended to other electron configurations, but we will first indicate the nomenclature of the states defined.

Nomenclature of the States of a Symmetric Diatomic Molecule

The principle is essentially the same as for atoms. As a result of the electron interaction, the components L_{z1}, L_{z2}, ..., L_{zn} of the angular momentum are no longer compatible with the energy and therefore cannot characterize a state of the system. We are then lead to examine the variable, symmetric with respect to electron exchange:

$$\mathcal{L}_z = L_{z1} + L_{z2} + \ldots + L_{zn}$$

and which corresponds to the total angular momentum neglecting the spin. The rules of quantization are the same as for each of the isolated components L_{zr}, so that:

$$\mathcal{L}_z = K\hbar, \qquad \text{with K an integer } (0, 1, 2, \ldots).$$

The nomenclature uses the same letters as in the description of the one-electron states, except that in this case capitals are used $\Sigma(K=0)$, $\Pi(K=1)$, $\Delta(K=2)$ .... Each of the Σ states is also characterized by the behavior of the wave function under reflection on the plane passing through the molecular axis. The alternatives pointed out in section 31 can both be found; the difference from the one-electron case is that we call the states in which the reflection leaves the state unchanged Σ^+, and use Σ^- where the sign changes.

In the case of symmetric molecules treated here, as in the one-electron case, the molecule can be classified by its behavior under inversion into even (g) or odd (u) states.

With regard to spin, the situation is the same as in the atomic case and S^2 and S_z again characterize the state.

From the practical point of view, the description of the molecular states is accomplished by starting with the approximation in which the interactions are negligible, where the different L_{zr} are well known. It then becomes possible to evaluate K, which remains invariant if we imagine the interaction as being progressively reestablished. This corresponds to a method of calculation, but clearly not to a real physical process. The g or u character is immediately determined, since the state function is a product of a number of functions, some of type g, others of type u. We give as examples of this method the various states of H_2 obtained by joining two hydrogen atoms in their $1s$ state.

Configuration $(\sigma_g 1s)^2$. We draw up the following table, which in accordance with the Pauli principle is limited to one line.

k_1	σ_1	k_2	σ_2	K	M_s
0	$\frac{1}{2}$	0	$-\frac{1}{2}$	0	0

We get only the value $K = 0$ implying a Σ state. Likewise $M_s = 0$, gives $S = 0$, a singlet state. The wave function is the product of two even functions and retains this character. We see also that this function is invariant under reflection on a plane through the molecular axis. Finally, the state thus defined is of the form Σ_g^+.

We now examine the other configurations, of which the energetically most favorable have only one electron in the bonding state, thus prohibiting the formation of a stable molecule.

Configuration $(\sigma_g 1s)^1(\sigma_u 1s)^1$. The table corresponding to this configuration is gotten without difficulty:†

k_1	σ_1	k_2	σ_2	K	M_s
0	$\frac{1}{2}$	0	$\frac{1}{2}$	0	1
0	$\frac{1}{2}$	0	$-\frac{1}{2}$	0	0
0	$-\frac{1}{2}$	0	$\frac{1}{2}$	0	0
0	$-\frac{1}{2}$	0	$-\frac{1}{2}$	0	-1

† Note that the two lines in the table $(0,\frac{1}{2};\ 0,-\frac{1}{2})$ and $(0,-\frac{1}{2};\ 0,\frac{1}{2})$ correspond to distinct states, since the two electrons occupy different orbitals $\sigma_g 1s$ and $\sigma_u 1s$. Because in the preceding case of $(\sigma_g 1s)^2$ these two configurations did not correspond to distinct states, only one should appear in the table.

Once again we have Σ states. The maximum value $M_s = 1$ implies the existence of a triplet state ($S = 1$) in addition to a singlet state. The product of two functions of the form g and u gives an odd function. The state function is invariant under reflection on any plane passing through the axis. We then get two states ${}^3\Sigma_u^+$ and ${}^1\Sigma_u^+$. It is easy to show (see **Exercise** p. 271) that the electron interaction is weakest in the triplet state, which therefore lies lowest.

Configuration $(\sigma_u 1s)^2$**.** We see immediately that we again have a ${}^1\Sigma_g^+$ term, as in the first case. The state function contains two electrons in an antibonding orbital, hence there is certainly a very strong repulsion between the two nuclei. As a result, this state is not taken into consideration if it is not of the same type as the fundamental state defined by $(\sigma_g 1s)^2$, which would then warrant a hybridization between the two states. Consequently, we can imagine *a priori* that the better two-electron wave function representing the fundamental state is not represented by $(\sigma_g 1s)^2$, but by a combination of this function with that representing $(\sigma_u 1s)^2$. We will give this function using the notation $(\psi_{1s})_a = a$ and $(\psi_{1s})_b = b$, on the assumption that the nuclear separation is sufficiently large that the overlap integral S—which occurs in (8) and in the analogous expression of the u type functions—is negligible; we then write:

$$u_g \simeq \frac{1}{\sqrt{2}}(a + b), \qquad u_u \simeq \frac{1}{\sqrt{2}}(a - b).$$

The space function representing the ground state level is of the form:

$$\frac{1}{2\sqrt{1+\lambda^2}}[(a_1 + b_1)(a_2 + b_2) + \lambda(a_1 - b_1)(a_2 - b_2)]$$

which we rewrite:

$$(9) \qquad \frac{1}{2\sqrt{1+\lambda^2}}[(a_1 b_2 + a_2 b_1)(1 - \lambda) + (a_1 a_2 + b_1 b_2)(1 + \lambda)].$$

For the value $\lambda = -1$ of the hybridization parameter, the wave function reduces to $(1/\sqrt{2})[a_1 b_2 + a_2 b_1]$, and represents the hybridization of the two states defined respectively by $a_1 b_2$ and $a_2 b_1$. In the first of these, electron 1 is on the H_a atom, and electron 2 on H_b, with the inverse situation for the second, thus giving the usual representation of the two-electron bond. If we now look at the wave function for $\lambda = 1$, we get $(1/\sqrt{2})[a_1 a_2 + b_1 b_2]$. The corresponding state is described as a hybridization of the two states, defined respectively by $a_1 a_2$ and $b_1 b_2$, which represent the ionic forms H_a, H_b^+ and H_a^+, H_b. We then must expect that the best orbital representing the ground state is gotten for a value near $\lambda = -1$, showing relatively small participation of the ionic form.

We complete this study with a second example, which suggests the difficulties that arise in the analysis of the electron states of molecules. We will consider the normal state of the C_2 molecule whose spectrum is well known (Swann bands). The one-electron description of this molecule is obtained by distributing the twelve electrons on the MO's, and progressively filling them in the order in which they are arranged in the correlation table (p. 170). We then get the structure:

$$|(\sigma_g 1s)^2 (\sigma_u 1s)^2|, \quad |(\sigma_g 2s)^2 (\sigma_u 2s)^2|, \quad (\pi_u 2p)^4,$$

by separating the terms corresponding to the associated bonding and antibonding orbitals which do not contribute to the bond between the two carbon atoms. This configuration is suggested by the Mulliken correlation diagram (Table II), which, by taking the limiting case of the unitary atom, places the $(\pi_u 2p)$ level below the $(\sigma_g 2p)$. We can nevertheless note that these MO's are different in type, and the non-crossing rule does not hold; nor do we have any evidence that the ordering determined by the unitary atom holds for all internuclear distances. Since the two orbitals are bonding, we can imagine two other possible configurations of the bond electrons, namely:

$$(\pi_u 2p)^3 (\sigma_g 2p)^1 \quad \text{and} \quad (\pi_u 2p)^2 (\sigma_g 2p)^2.$$

The reason already invoked in the case of O_2 can be retained here to show that the distribution of the electrons on the several orbitals corresponds to a spatial distribution having the least electron repulsion energy and, as a result, to a lowering of the level.

In the first configuration $(\pi_u 2p)^4$, filling the two π_u levels by electrons of opposite spin is possible only in one way, hence a $^1\Sigma$ term. The corresponding wave function is gotten by applying the antisymmetrization operator to one of the functions permitted by the Pauli principle and gives two electrons on the level $k = 1(\pi_u^+)$ and two others on the level $k = -1(\pi_u^-)$, with opposite spin, so that:†

$$(\pi_u^+)_1 (\pi_u^+)_2 (\pi_u^-)_3 (\pi_u^-)_4 \alpha_1 \beta_2 \alpha_3 \beta_4. \tag{10}$$

† This notation, where the number of the electron in the wave function expression is a subscript: $(\pi_u^+)_2$, should not be confused with the notation for an electronic structure, where the number of electrons on a level is given by a superscript: $(\pi_u 2p)^4$.

The spatial function is the product of four odd functions, and thus remains invariant under inversion, so that the level has g character. We now examine the effect of a reflection.

The behavior under a reflection follows from the AO's $2p_1$ and $2p_{-1}$, respectively, in the form:

$$(2p_1) = (x + iy)f(r)\,; \qquad (2p_{-1}) = (x - iy)f(r).$$

The reflection $\boldsymbol{P}$ on the xz-plane leads to:

$$\boldsymbol{P}(2p_1) = (x - iy)f(r) = (2p_{-1})\,; \quad \boldsymbol{P}\,(2p_{-1}) = (2p_1).$$

The effect is then to exchange the two functions:

$$(\pi_u^+) = (2p_1)_a + (2p_1)_b \quad \text{and:} \quad (\pi_u^-) = (2p_{-1})_a + (2p_{-1})_b.$$

The function gotten from (10) by the reflection is written:

$$(\pi_u^-)_1(\pi_u^-)_2(\pi_u^+)_3(\pi_u^+)_4\alpha_1\beta_2\alpha_3\beta_4.$$

This last function can be derived from (10) by permutation of electrons 1 and 3 on the one hand, and 2 and 4 on the other, and leads to the same antisymmetric function. We see that the state function remains invariant under this reflection, so that the term is Σ_g^+.

For the second configuration, $(\pi_u 2p)^3(\sigma_g 2p)^1$, we prepare the table of values of K and M_s by noting that the one of two π_u levels is filled by two electrons with opposite spins, and contributes nothing to K and M_s, so that we have a two-electron case corresponding to the following table:

k_1	σ_1	k_2	σ_2	K	M_S
1	$\frac{1}{2}$	0	$\frac{1}{2}$	1	1
1	$\frac{1}{2}$	0	$-\frac{1}{2}$	1	0
1	$-\frac{1}{2}$	0	$\frac{1}{2}$	1	0
1	$-\frac{1}{2}$	0	$-\frac{1}{2}$	1	-1
-1	$\frac{1}{2}$	0	$\frac{1}{2}$	-1	1
-1	$\frac{1}{2}$	0	$-\frac{1}{2}$	-1	0
-1	$-\frac{1}{2}$	0	$\frac{1}{2}$	-1	0
-1	$-\frac{1}{2}$	0	$-\frac{1}{2}$	-1	-1

The only values of K are ± 1, implying that this is a Π state. The values of M_s correspond to a singlet and a triplet state. The wave function formed from the product of three odd functions retains the odd character, so we have two states $^3\Pi_u$ and $^1\Pi_u$.

We end with the analysis of the $(\pi_u 2p)^2(\sigma_g 2p)^2$ configuration. The determination of the term does not use $(\sigma_g 2p)$, which is filled, so that we again have a two-electron system with the following table.

k_1	σ_1	k_2	σ_2	K	M_s
1	$\frac{1}{2}$	1	$-\frac{1}{2}$	2	0
1	$\frac{1}{2}$	-1	$\frac{1}{2}$	0	1
1	$\frac{1}{2}$	-1	$-\frac{1}{2}$	0	0
1	$-\frac{1}{2}$	-1	$\frac{1}{2}$	0	0
1	$-\frac{1}{2}$	-1	$-\frac{1}{2}$	0	-1
-1	$\frac{1}{2}$	-1	$-\frac{1}{2}$	-2	0

Examination of the table immediately allows us to see a $^1\Delta$ state and two Σ states—a triplet and a singlet. All of these states are even. The most difficult point is to determine the action of a reflection on a plane passing through the molecular axis and requires the formation of the corresponding wave functions. The calculation parallels that performed (§47) for the helium atom, and simply requires transposition of equations (3), (4), and (5). We get the following spatial functions for the two states:

$$^3\Sigma_g : \frac{1}{\sqrt{2}}[(\pi_u^+)_1(\pi_u^-)_2 - (\pi_u^+)_2(\pi_u^-)_1]$$

$$^1\Sigma_g : \frac{1}{\sqrt{2}}[(\pi_u^+)_1(\pi_u^-)_2 + (\pi_u^+)_2(\pi_u^-)_1].$$

We see that the reflection has the effect of permutating the wave functions (π_u^+) and (π_u^-), with the result that the wave function is invariant in the singlet state and changes sign in the triplet. Finally, the analysis of the configuration leads to three terms $^1\Delta_g$, $^3\Sigma^-{}_g$, and $^1\Sigma_g^+$.

It thus appears relatively simple to find the terms corresponding to a given electron configuration. On the other hand, we have set aside two tricky problems: the determination of the ordering of the levels, and the dissociation products obtained from each state.

These problems are particularly important in the interpretation of the observed band spectrum. We give Herzberg's diagram† of C_2 (fig. 64). We note in particular that bringing together two

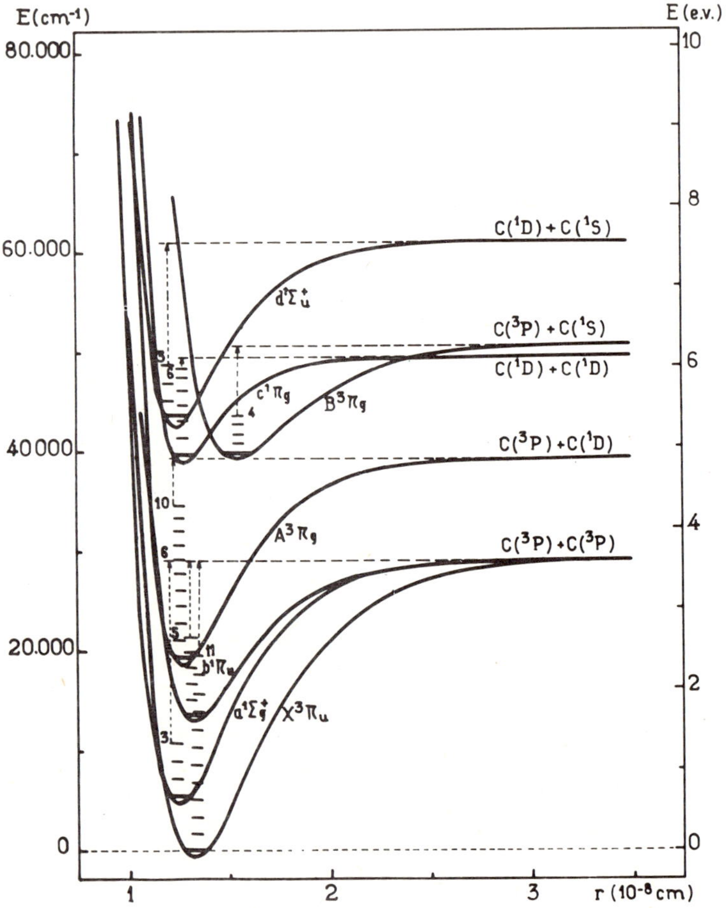

Figure 64. The potential energy curve as a function of the internuclear distance r for the molecule C_2 (after Herzberg).

carbon atoms in the ^{3}P ground state gives rise to a molecule in one of the states, $^3\Pi_u$, $^1\Sigma_g^+$ and $^1\Pi_u$, discussed in the first two configurations studied. We note, in addition, the crossing of the potential energy curves relative to terms of a different kind.

† Herzberg, *Spectra of Diatomic Molecules* (New York: D. Van Nostrand, 1951), p. 454.

Exercise. Compare the interaction of the π electrons of the ethylene molecule in the excited triplet state and in the singlet. The treatment follows that for helium (§47) and requires expression of the π_g and π_u MO's as a function of the atomic orbitals $(2p_y)_a$ and $(2p_y)_b$ of the bound atoms (§34).

Polyatomic Molecules 50

In a given system, the classification of the states is tied to the symmetry operations that leave the system unchanged, as was shown in the different examples we presented for atoms and diatomic molecules. In the atomic case it is the three-dimensional rotation group which gives the nomenclature. The group of rotations about an axis which leaves a diatomic molecule unchanged is a subgroup of the preceding group, which leads to the correspondence between the nomenclature in the two cases. In addition, both for atoms and diatomic molecules, the hamiltonian does not contain the spin variables in the first approximation, so that the spin states are formed in the same way as for atoms by use of the operators S^2 and S_z (§47).

In general, a polyatomic molecule has less symmetry than that which governs the preceding cases, and this usually implies little degeneracy in the energy levels. As a consequence, the analysis of the ground state shows that most of the time it corresponds to a saturated level which gives only a singlet term. The excited state due to an electron transition from the highest occupied level to the nearest free level involves use of the line of reasoning often followed in this chapter, and we get a singlet term and a triplet term which generally lies lower because of the electron interactions. It is only when the molecule has symmetry operations of order greater than two that we apparently see degenerate levels, and, consequently, a more complex term structure corresponding to a given electron configuration. But the nomenclature of the one-electron levels, as well as the analysis of the terms, require the resources of group theory. Without going into the details of this theory we cite the case of the distribution of the π electrons in the benzene molecule. We have calculated (§35) the MO's on which the six electrons are distributed in the ground state and the symmetry types of these orbitals a_{2u} and e_{1g}, the latter being degenerate of order two, as well as the level nearest free. As a result, there are two electron configurations of the excited state (fig. 65), each having a singlet term and a triplet term. This analysis is possible in an extremely simple way not requiring group theory and using the free electron model (§38). The one-electron levels are given by:

$$\mathrm{E} = \frac{k^2\hbar^2}{8m\mathrm{L}^2}, \quad \text{with :} \quad k = 0, 1, 2.,.$$

All of these levels are twofold degenerate, except the first, which is non-degenerate. This result is intuitive if we assume that we are dealing with a planar rotor with angular momentum along the z-axis of $\mathrm{L}_z = k\hbar$. The distinct values k and $-k$, if $k \neq 0$, correspond to two possible senses of the movement of the electron on its trajectory. We can then imagine that the molecule has in its set states defined the total component $\mathscr{L}_z = \sum_r \mathrm{L}_{z,r}$, the sum being extended to all π electrons. But as in the preceding treatments,

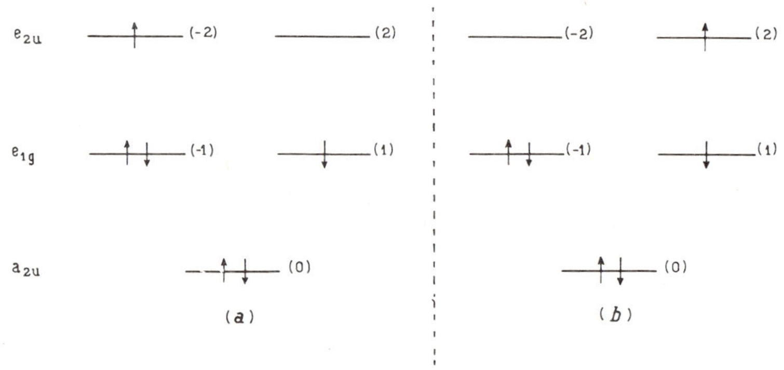

Figure 65. Diagram showing the two possibilities corresponding to the $a_{2u} \rightarrow e_{1g}$ transition in benzene: (*a*) K = –3; (*b*) K = 1.

it is useless to include the electrons filling the level $k = 0$—whose contribution to L_z is zero—in the calculations. We prepare the table of different electron states characterized by the values of k for each of the remaining electrons, as well as those of K which govern $\mathscr{L}_z$. We omit here the values of the spin components, whose effect is simply to multiply by four the number of lines of the table, leading for each of the terms to a singlet and a triplet state.

k_1	k_2	k_3	k_4	K
1	1	− 1	2	3
1	− 1	− 1	2	1
1	1	− 1	− 2	− 1
1	− 1	− 1	− 2	− 3

Here it is a question of an electron transition from $k = 1$ to the $k = 2$ level. In the general case, the transition occurs between the levels n and $n + 1$, and the corresponding table gives $K = \pm 1$ and $K = \pm(2n + 1)$. Platt† has introduced the following nomenclature based on the values of K: A($K = 0$), B($K = 1$), C($K = 2$), ..., K($K = 2n$), L($K = 2n + 1$), ...

With this nomenclature, the fundamental state of benzene is of the ^{1}A type, and two optical transitions starting from this state are $^1A \rightarrow {}^1B$ and $^1A \rightarrow {}^1L$, limiting our consideration to singlet levels, since the triplets generally do not have observable absorptive transitions. The Platt classification extends to all those condensed aromatic hydrocarbons characterized by the possibility, as in benzene, of having a closed trajectory for the π electrons.

---

† J. R. Platt, *J. Chem. Phys.* (1949), **17**, 484.

13 THE COVALENT BOND

General Principles: Bond Functions 51

In Chapter 7 we treated the problem of the covalent bond for the case of diatomic molecules, extending the discussion to the π electrons of polyatomic molecules in Chapter 8. In both cases we used one-electron wave functions.

This approach is not the most general and for the case of H_2 we were lead (§48) to assume that the description of the ground state using the AO's of the bound atoms requires hybridization of the $(\sigma_g 1s)^2$ and $(\sigma_u 1s)^2$ configurations, both of which correspond to the ${}^1\Sigma_g^+$ term. The description of this state using only the $(\sigma_g 1s)$ MO is not altogether satisfactory, and the better wave function for two electrons came from the hybridization of the state $(1/\sqrt{2})[a_1 b_2 + a_2 b_1]$, with the functions representing the ionic forms $H_a H_b^+$ and $H_a^+ H_b$. We can predict *a priori* that these ionic forms participate only slightly in the real structure, so that we have two ways of describing the ground state of H_2:
(*a*) Starting with the MO which leads to:

$$\text{(1)} \qquad \Theta = \frac{1}{2}(a_1 + b_1)(a_2 + b_2)\frac{1}{\sqrt{2}}(\alpha_1\beta_2 - \alpha_2\beta_1).$$

(*b*) Starting with the two-electron function:

$$\text{(2)} \qquad \Theta = \frac{1}{2}(a_1 b_2 + a_2 b_1)\frac{1}{\sqrt{2}}(\alpha_1\beta_2 - \alpha_2\beta_1).$$

These two approximations can be improved, the first by hybridization with the function representing $(\sigma_u 1s)^2$, the second by hybridization with the ionic form; in both cases we get the same result so that the two methods are equivalent. It is nevertheless predictable that the function (2) will represent a better approximation to the electron distribution than (1), and this is confirmed by comparison with the experimental results† (see Table IV below). The function (1) remains interesting, however, since it corresponds to a one-electron—and therefore the most simple—approach. In addition, the integrals introduced in the calculation are chosen so as to predict certain experimental results (see Chapter 8), so that we can get a fairly satisfactory representation of a large number of

† A. A. Frost and J. Braunstein, *J. Chem. Phys.* (1951), **19**, 1133.

molecules simply. The use of the wave function (2) leads to much more difficult calculations but corresponds to a representation somewhat closer to reality, additionally it forms the basis of the mesomeric method. This is why we limit the discussion, as in the case of the molecular orbitals, to the study of π electrons in unsaturated hydrocarbons.

TABLE IV

	R(Å)	D_e (eV)	Z
MO Method—simple wave function (1)	0.84	2.65	1
Same method with adjustment of Z	0.73	3.47	1.19
Two-electron function (2)	0.80	3.14	1
Same method adjusting Z	0.76	3.76	1.17
Hybridization $(\sigma_g 1s)^2$ and $(\sigma_u 1s)^2$ with adjustment of Z	0.77	4.00	1.19
Experimental results	0.74	4.47	

However, to clarify the method, we will begin with the most simple case previously treated in another way, the H_2 molecule. The state function (2), written for the case of no overlap between a and b orbitals,† can be written:

$$\Theta = \frac{1}{2}\left[(a_1\alpha_1 b_2\beta_2 - a_2\alpha_2 b_1\beta_1) + (a_2\beta_2 b_1\alpha_1 - a_1\beta_1 b_2\alpha_2)\right]$$

$$= \frac{1}{\sqrt{2}}(\Psi_{\mathrm{I}} + \Psi_{\mathrm{II}})$$

with:

$$\Psi_{\mathrm{I}} = \frac{1}{\sqrt{2}}(a_1\alpha_1 b_2\beta_2 - a_2\alpha_2 b_1\beta_1)$$

$$\Psi_{\mathrm{II}} = \frac{1}{\sqrt{2}}(a_2\beta_2 b_1\alpha_1 - a_1\beta_1 b_2\alpha_2).$$

The state described by Θ can thus be thought of as a hybrid of the two states defined by Ψ_{I} and Ψ_{II}. The antisymmetric function Ψ_{I} defines a state such that one of the electrons is localized on atom a, with spin component $S_z = \frac{1}{2}\hbar$, and the other on b, with

† This approximation is more legitimate than in the MO case, since the exact calculation of the normalization coefficient replaces $1/\sqrt{2}$ by $1/\sqrt{2(1 + S^2)}$, and these differ by very little since:

$$S = \int ab\,d\tau \sim 0.25.$$

spin component $S_z = -\frac{1}{2}\hbar$, and which we can characterize by the notation $a\alpha b\beta$. Likewise Ψ'_{II} defines the state $a\beta b\alpha$. We say that the states represented by Ψ'_I and Ψ'_{II} are *spin states.* The wave function Θ can then be written as a hybrid of two spin states, $a\alpha b\beta$ and $a\beta b\alpha$, which are the only ones that give a singlet state, and we can set aside the other two combinations, $a\alpha b\alpha$ and $a\beta b\beta$, which only occur in the description of the triplet state.

This method of formation of the state function by representing the ground state of the molecule through hybridization of the spin state functions can be extended to the case of a general molecule. The formalism is the same, that is, either it is a question of the bond between a certain number of hydrogen atoms, or of the behavior of the π electrons in a substance with the same number of carbon atoms having conjugated double bonds; the former does not correspond to any real situation, since such molecules are not found for $n > 2$. It is from the latter domain that we will draw our examples.

Spin State Method: Butadiene 52

We propose to describe the behavior of the electrons of butadiene by means of the wave functions constructed from the $2p_z$ AO's of each of the four carbon atoms, designated respectively by a, b, c, and d. The immediate generalization of the method developed for H_2 involves considering a basis state where each electron is localized on one of the atoms, which can lead to the 16 following possibilities. We write these using the notation of the preceding paragraph.

$a\alpha\ b\alpha\ c\alpha\ d\alpha$	$a\beta\ b\alpha\ c\alpha\ d\alpha$	$a\beta\ b\alpha\ c\alpha\ d\beta$	$a\beta\ b\alpha\ c\beta\ d\beta$
$a\alpha\ b\alpha\ c\alpha\ d\beta$	$a\alpha\ b\alpha\ c\beta\ d\beta$	$a\beta\ b\alpha\ c\beta\ d\alpha$	$a\beta\ b\beta\ c\alpha\ d\beta$
$a\alpha\ b\alpha\ c\beta\ d\alpha$	$a\alpha\ b\beta\ c\alpha\ d\beta$	$a\beta\ b\beta\ c\alpha\ d\alpha$	$a\beta\ b\beta\ c\beta\ d\alpha$
$a\alpha\ b\beta\ c\alpha\ d\alpha$	$a\alpha\ b\beta\ c\beta\ d\alpha$	$a\alpha\ b\beta\ c\beta\ d\beta$	$a\beta\ b\beta\ c\beta\ d\beta$

Each of these 16 configurations corresponds to a many-electron wave function, the first of which we write indexing the electron numbers as $a_1\alpha_1 b_2\alpha_2 c_3\alpha_3 d_4\alpha_4$. The antisymmetrization operator $\mathscr{A}$ allows us to derive the antisymmetric function required by the Pauli principle. The description of a state of the molecular π electrons of butadiene must be obtained by hybridization of the 16 functions we have constructed and, in the case of an n-atom molecule, of 2^n functions. This number can be reduced, as in the case of H_2, by considering only the configurations corresponding to a total spin zero. This result can be stated in general in the description of the ground state of the system with the same justification as in the H_2 case. We need only imagine the system described by the MO method, the construction of the ground state being accomplished by progressive filling of the orbitals—starting from the lowest—by electrons with opposite spins, so that in general we get a resultant with spin zero ($M_s = 0$). The example of C_2 shows that there are exceptions to this law, but these do not arise in stable molecules with an even number of π electrons.

Finally we will keep only 6 of the 16 possible configurations, namely, only those which have an equal number of α and β spins.

$$\text{(3)} \qquad \begin{array}{lll} \mathrm{I} : a\alpha\, b\beta\, c\beta\, d\alpha\,; & \mathrm{II} : a\beta\, b\alpha\, c\alpha\, d\beta\,; & \mathrm{III} : a\alpha\, b\beta\, c\alpha\, d\beta \\ \mathrm{IV} : a\beta\, b\alpha\, c\beta\, d\alpha : & \mathrm{V} : a\alpha\, b\alpha\, c\beta\, d\beta\,; & \mathrm{VI} : a\beta\, b\beta\, c\alpha\, d\alpha. \end{array}$$

We designate by $\Psi_{\mathrm{I}}, \ldots, \Psi_{\mathrm{VI}}$ the antisymmetric functions corresponding to these various spin states, so that, for example:

$$\text{(4)} \qquad \Psi_{\mathrm{II}} = \frac{1}{\sqrt{4\,!}} \mathcal{A} a_1\beta_1 b_2\alpha_2 c_3\alpha_3 d_4\beta_4.$$

The hybrid orbital representing the ground state is determined in principle by the variational method, by looking for the energy extremum and taking the normalization into account. Even in this simple case of four electrons the calculations are very long, since this is a variational problem with six variables. We can reduce the number of basis functions by noting that the six functions $\Psi_{\mathrm{I}}, \ldots, \Psi_{\mathrm{VI}}$ are all of the same kind, such that in each $S_z = 0$, but this condition alone is insufficient to define a singlet state, since the singlet also requires $S^2 = 0$. We will describe a method allowing the *a priori* introduction of combinations of the preceding functions corresponding to the singlet term, known as the *valence bond method* (VB), still using the example of butadiene.

One approach involves starting with the result gotten in the case of the diatomic molecule in the preceding paragraph, which gives the function representing the ground singlet state, and which we rewrite without the normalization factor:

$$\text{(5)} \qquad (a_1 b_2 + a_2 b_1)(\alpha_1\beta_2 - \beta_1\alpha_2).$$

This function can be thought of as generated by:

$$a_1 b_2(\alpha_1\beta_2 - \beta_1\alpha_2),$$

by introducing the antisymmetrization condition realized by application of the operator $\mathcal{A}$.

Given this, we consider among the six configurations (3) those having opposite spins on *a* and *b*, for example, I. In this configuration, the spins on *c* and *d* are also necessarily opposed, since the functions α and β are equal in number. We can assume that this configuration occurs in the description of a bond between c and d and we can introduce the generating function of the bond:

$$c_3 d_4(\alpha_3\beta_4 - \beta_3\alpha_4).$$

We thus form a function representing two independent bonds in the molecules a—b and c—d by the product of the preceding two functions, with electrons 1 and 2 assuring the first bond, and electrons 3 and 4 the second.

$$a_1 b_2 c_3 d_4(\alpha_1\beta_2 - \alpha_2\beta_1)(\alpha_3\beta_4 - \alpha_4\beta_3).$$

Finally, the wave function associated with configuration I is gotten by application of the antisymmetrization operator, so that:

$$\Theta = k\mathcal{A} a_1 b_2 c_3 d_4(\alpha_1\beta_2 - \alpha_2\beta_1)(\alpha_3\beta_4 - \alpha_4\beta_3). \tag{6}$$

The process of uniting the functions representing the singlet state guarantees that function (6), whose structure we will now study, is necessarily a singlet function (see p. 294).

Note in the first place that the function (6) does not characterize configuration I, but is gotten to within the sign from all other configurations (3) having opposite spins for a and b on the one hand, and c and d on the other, namely, I, II, III, and IV. As we will show, this gives rise to the possibility of getting (6) from the functions Ψ_{I}, Ψ_{II}, Ψ_{III}, and Ψ_{IV}. Thus we note that in (6) $\beta_1\alpha_2$ derives from $\alpha_1\beta_2$ by spin permutations on atoms a and b, which we denote by:

$$\beta_1\alpha_2 = \boldsymbol{T}_{ab}\alpha_1\beta_2,$$

introducing the operator $\boldsymbol{T}_{ab}$ which acts exclusively on the spins of a and b. The second parenthesis of (6) likewise uses $\boldsymbol{T}_{cd}$, independent of $\boldsymbol{T}_{ab}$ and commuting with it. We can then write (6) in the equivalent form:

$$\Theta = (1 - \boldsymbol{T}_{ab})(1 - \boldsymbol{T}_{cd})k\mathcal{A} a_1\alpha_1 b_2\beta_2 c_3\alpha_3 d_4\beta_4$$

so that by referring to table (3):

$$\Theta = k(1 - \boldsymbol{T}_{ab})(1 - \boldsymbol{T}_{cd})\Psi_{\mathrm{III}}.$$

This calculation uses the permutation properties of $\mathcal{A}$, which act on the electron numbers, and of $\boldsymbol{T}_{ab}$ and $\boldsymbol{T}_{cd}$, which act on the spin functions and are therefore independent. The expression for Θ can also be written:

$$\Theta = k(\Psi_{\mathrm{III}} - \boldsymbol{T}_{ab}\Psi_{\mathrm{III}} - \boldsymbol{T}_{cd}\Psi_{\mathrm{III}} + \boldsymbol{T}_{ab}\boldsymbol{T}_{cd}\Psi_{\mathrm{III}})$$

$T_{ab}\Psi_{\text{III}}$ is the function associated with the configuration derived from III by permutation of the spins a and b, namely Ψ_{II}. Similarly, action of the other terms in the parentheses leads to different Ψ_r corresponding to various spin configurations, such that spins a and b are opposed, as are c and d. The parity of the spin permutations gives the sign of the corresponding function. The function gotten should be thought of as representing the bonding scheme a—b and c—d, and is written:

(7) $$\Theta_1 = k(\Psi_{\text{III}} - \Psi_{\text{II}} - \Psi_{\text{I}} + \Psi_{\text{IV}}).$$

Other bonding schemes can be established by associating all possible pairs of atoms. To reach a general formulation it is not necessary to concern ourselves with determining whether or not these diagrams correspond to real bonds, which can only occur if the bound atoms are adjacent. In the case of the four atoms of butadiene, we have the following three diagrams (fig. 66):

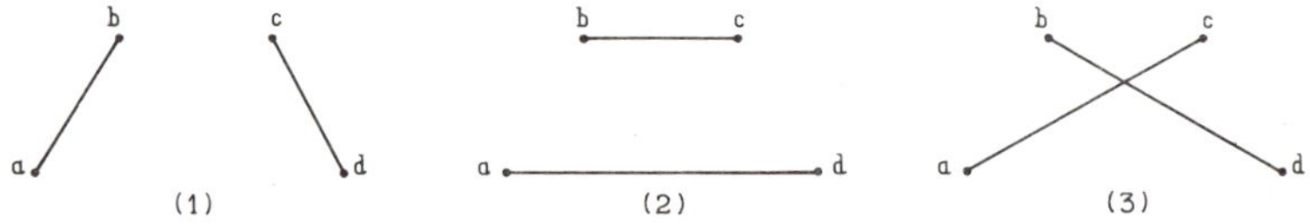

Figure 66. Bond diagrams for the π electrons in butadiene.

The method we have given allows us to derive immediately the state function with diagram (2), namely:†

(8) $$\Theta_2 = k(1 - T_{ad})(1 - T_{bc})\Psi_{\text{III}} = k(\Psi_{\text{III}} - \Psi_{\text{VI}} - \Psi_{\text{V}} + \Psi_{\text{IV}}).$$

Likewise for diagram (3):

(9) $$\Theta_3 = k(1 - T_{ac})(1 - T_{bd})\Psi_{\text{I}} = k(\Psi_{\text{I}} - \Psi_{\text{VI}} - \Psi_{\text{V}} + \Psi_{\text{II}}).$$

The three functions Θ_1, Θ_2, and Θ_3 are not independent, and we can easily verify that:

$$\Theta_3 = \Theta_2 - \Theta_1.$$

Finally, the function representing the ground state can be represented as the hybrid of two functions Θ_1 and Θ_2, defined by (7) and (8) and dependent upon only one parameter and the normalization. The problem is thus considerably simplified with respect to the initial state where that function was described as a superposition of six functions $\Psi_{\text{I}}, \ldots, \Psi_{\text{VI}}$. This is a general method which applies to the case of a molecule with any number of atoms, and whose rules of application we will now establish.

† It is necessary to start from one of the configurations (3) having opposite spins on b and c and on a and d, which is the case for III.

General Rules for the Formation of the Bond Functions 53

As we have already seen, the method applies both to the study of a general molecule with $n = 2k$ hydrogen atoms, and to the behavior of the electrons of a hydrocarbon having an equal number of mobile electrons. We will ultimately show the extension to the case of a molecule with an odd number of electrons in which the ground state is a doublet.

The first part of the discussion involves the formation of tables of different bonding schemes, and it is possible to show† that the results gotten for butadiene are general, namely, that to get a system of independent singlets we can limit ourselves to diagrams which do not have intersections between the lines joining the atoms.

Figure 67. Construction of the bond diagrams in the case of a general molecule with $2k$ atoms which we distribute on a circle in arbitrary order.

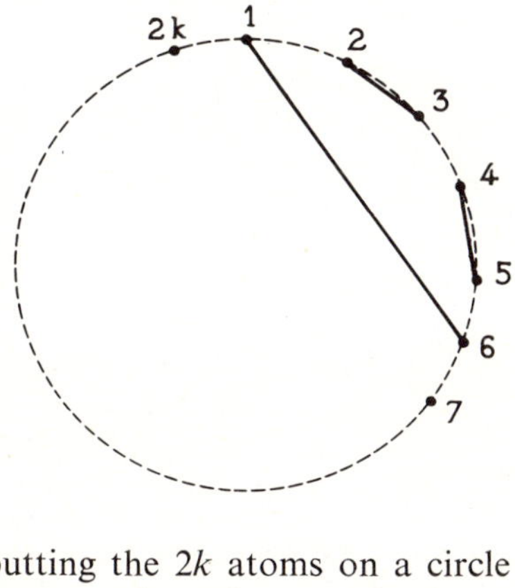

A convenient method involves putting the $2k$ atoms on a circle in any *a priori* order, although interpretation of the results in the language of mesomerism is facilitated by having ordering as close as possible to the order of the atoms in the molecule. It is then necessary to determine all of the configurations in which the atoms are joined in pairs by non-crossing lines. The number $f(k)$ of such schemes gotten for $2k$ atoms obeys a simple recurrence relation. For this we join one of the atoms, 1 for example, to the atom $2r$ randomly chosen from among the k atoms of even index in the diagram (fig. 67) (here $2r = 6$). The different possible diagrams involving the $1 - r$ bond without any presumption as to the relative position of the corresponding atoms in the molecule are gotten by joining the two possible configurations of the set of atoms situated on either side of the $1 - r$ bond. In the example in figure 67, the number of these configurations on one side is $f(2)$,

† Rumer, *Göttinger Nachr.* (1932), 377.

and on the other $f[(2k-6)/2] = f(k-3)$, hence a number of combinations equal to $f(k-3)$. By hypothesis, $f(2)$ cannot cross 1—6. By varying r from 0 to $2k-1$, we get all the possible configurations the number of which is given by:

$$f(k) = \sum_{r=0}^{k-1} f(k-r-1) \cdot f(r).$$

The function $f(k+1)$ is expressed from the functions of lower index by the recurrence relation:

$$f(k+1) = \sum_{r=0}^{k} f(k-r) \cdot f(r). \tag{10}$$

By taking $f(0) = 1$ and $f(1) = 1$, equation (10) allows us to calculate $f(k)$ term by term for any value of k. We then get:

$$f(2) = 2\,; \quad f(3) = 5\,; \quad f(4) = 14...$$

We get exactly the number of possible singlet terms† given by the analysis in section 47, namely:

$$p_k = \frac{(2k)\,!}{k\,!\,(k+1)\,!},$$

enabling us to anticipate that the process we have just described will allow us to get all of the singlet states. For a rigorous analysis we refer to Rumer's paper, cited previously.

† The recurrence relation can be used in an indirect way by introducing the function of an auxiliary variable ξ:

$$F(\xi) = \sum_{0}^{\infty} f(k)\xi^k.$$

The recurrence relation allows us to verify the identity:

$$\xi F^2 = F - f(0).$$

Taking $f(0) = 1$, the equation of the second order defining the roots of F is regular for $\xi = 0$, so that:

$$F = \frac{1}{2\xi}\left[1 - \sqrt{1 - 4\xi}\right],$$

whose expansion by the binomial theorem leads to:

$$F = \sum_{0}^{\infty} \frac{(2k)\,!}{k\,!\,(k+1)\,!}\,\xi^k$$

hence, by identity we have:

$$f(k) = p_k.$$

As an example we use the five diagrams gotten for benzene ($k = 3$). These turn out to be the two forms of Kékulé, and the three forms of Dewar, since the distribution of the atoms on the circle corresponds to their natural order in the molecule (fig. 68).

The bond function representing one of the diagrams is obtained exactly as for butadiene, namely, by taking one configuration in which two atoms joined by a *bond* (this term being used here for convenience) have opposing spins. The various possible spin combinations associated with the diagram under consideration are derived from the diagram by taking the spin permutation of the

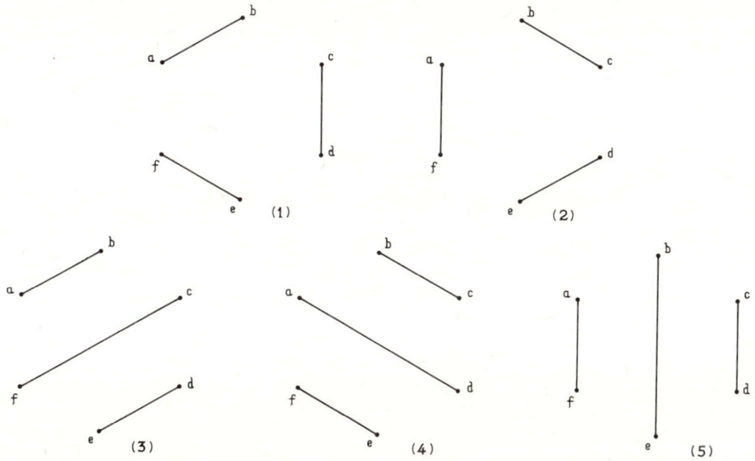

Figure 68. Bond diagrams for the benzene π electrons: (1) and (2) Kékulé structures; (3), (4), and (5), Dewar structures.

different bonds of the diagram with the factor ± 1, depending on the parity of the permutations. We use the shorthand notation:

$$\frac{1}{\sqrt{6!}}\mathcal{A}a_1\alpha_1 b_2\beta_2 c_3\alpha_3 d_4\beta_4 e_5\alpha_5 f_6\beta_6 = \begin{pmatrix} a & b & c & d & e & f \\ \alpha & \beta & \alpha & \beta & \alpha & \beta \end{pmatrix}.$$

With this notation, the bonding function associated with diagram 1, for example, is written:

$$\Theta_1 = k\left[\begin{pmatrix} a & b & c & d & e & f \\ \alpha & \beta & \alpha & \beta & \alpha & \beta \end{pmatrix} - \begin{pmatrix} a & b & c & d & e & f \\ \beta & \alpha & \alpha & \beta & \alpha & \beta \end{pmatrix} - \begin{pmatrix} a & b & c & d & e & f \\ \alpha & \beta & \beta & \alpha & \alpha & \beta \end{pmatrix} - \begin{pmatrix} a & b & c & d & e & f \\ \alpha & \beta & \alpha & \beta & \beta & \alpha \end{pmatrix}\right.$$
$$\left. + \begin{pmatrix} a & b & c & d & e & f \\ \alpha & \beta & \beta & \alpha & \beta & \alpha \end{pmatrix} + \begin{pmatrix} a & b & c & d & e & f \\ \beta & \alpha & \alpha & \beta & \beta & \alpha \end{pmatrix} + \begin{pmatrix} a & b & c & d & e & f \\ \beta & \alpha & \beta & \alpha & \alpha & \beta \end{pmatrix} - \begin{pmatrix} a & b & c & d & e & f \\ \beta & \alpha & \beta & \alpha & \beta & \alpha \end{pmatrix}\right].$$

The first part of the problem, to construct the bonding functions corresponding to the different possible cases, is then resolved, and by hybridization we are able to get the function corresponding to the ground state of the molecule being studied.

Exercise 1. Form the 14 bonding diagrams for an eight-electron molecule using the process which leads to (10). Note the number of forms with four bonds between consecutive atoms; those with

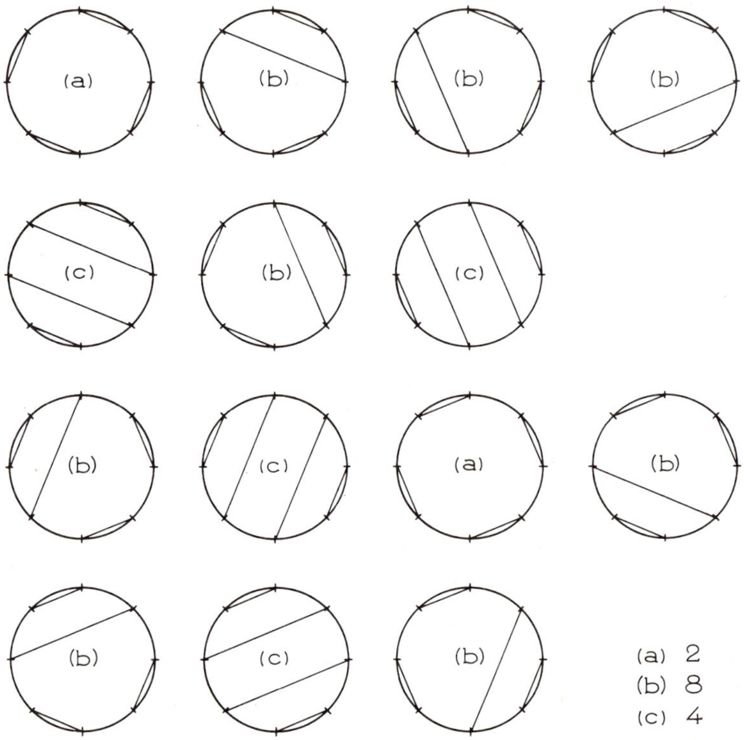

Figure 69. Bond diagrams for a molecule with eight electrons obtained by means of the construction indicated in figure 67.

three such bonds (occasionally called "excited"); and the forms with two such bonds (double excited) (fig. 69).

Exercise 2. Form the five possible diagrams for the benzene molecule by adopting a numbering system different from the natural order in the chain. Note the difference between the distribution of the bonds in Rumer's circular scheme, which obeys the non-crossing rule, and in this molecule, where this is not necessarily the case.

Determination of the Ground State: Calculation of the Energy 54

We now turn to the second part of the problem, the determination of the function representing the ground state as a hybrid of the functions obtained above, and we get this by taking the extremum of the energy, following the method whose principle we now review.

We take butadiene as an example. The function representing the ground state is a linear combination of the two bonding functions Θ_1 and Θ_2 defined by (7) and (8), so that:

$$\Phi = c_1\Theta_1 + c_2\Theta_2.$$

The mean energy in this state is calculated by means of (9-16), which we write in the shortened form:

$$\text{(11)} \qquad E = \sum_{\sigma_1\sigma_2\sigma_3\sigma_4} \int \Phi^* \mathcal{H} \Phi d\tau_1 d\tau_2 d\tau_3 d\tau_4 = (\Phi, \mathcal{H}\Phi)$$

with a sum over the different possible combinations of the spin variable ($\sigma_r = \pm\frac{1}{2}$) and an integral over the spatial coordinates of the four electrons. In this case, as in most of the examples, the functions used are all real, and we can limit ourselves to the case of real coefficients c_r. The energy E is then in a quadratic homogeneous form of the coefficients c_r, which we write:

$$\text{(12)} \qquad E = c_1^2 H_{11} + 2c_1c_2H_{12} + c_2^2 H_{22}$$

with:

$$H_{11} = (\Theta_1, \mathcal{H}\Theta_1);$$

$$H_{12} = (\Theta_1, \mathcal{H}\Theta_2) = (\Theta_2, \mathcal{H}\Theta_1);$$

$$H_{22} = (\Theta_2, \mathcal{H}\Theta_2).$$

The function Φ must be normalized, hence the condition:

$$\text{(13)} \qquad (\Phi, \Phi) = c_{11}^2\Delta_{11} + 2c_1c_2\Delta_{12} + c_2^2\Delta_{22} = 1$$

with:

$$\Delta_{11} = (\Theta_1, \Theta_1); \quad \Delta_{12} = (\Theta_1, \Theta_2); \quad \Delta_{22} = (\Theta_2, \Theta_2).$$

The extremum of E, along with the normalization condition, is treated by introducing the Lagrange multiplier (Appendix III):

$$\text{(14)} \qquad \begin{aligned} c_1H_{11} + c_2H_{12} &= \lambda(c_1\Delta_{11} + c_2\Delta_{12}) \\ c_1H_{12} + c_2H_{22} &= \lambda(c_1\Delta_{12} + c_2\Delta_{22}). \end{aligned}$$

The term by term addition of these two equations after multiplication by c_1 and c_2, respectively, gives the quadratic forms (12) and (13) and, as a consequence shows the significance of the Lagrange multiplier: $\lambda = \mathrm{E}$. The energy E is fixed by using the compatibility condition on the system of linear homogeneous equations (14), namely:

$$(15) \qquad \begin{vmatrix} \mathrm{H}_{11} - \mathrm{E}\Delta_{11} & \mathrm{H}_{12} - \mathrm{E}\Delta_{12} \\ \mathrm{H}_{12} - \mathrm{E}\Delta_{12} & \mathrm{H}_{22} - \mathrm{E}\Delta_{22} \end{vmatrix} = 0.$$

We get two roots $\mathrm{E}' < \mathrm{E}''$, and the lower value, E', defines the ground state level. With this we can determine the coefficients c_1 and c_2 using (14) and the normalization condition (13).

This variational calculation requires the six integrals H_{rs} and Δ_{rs}, defined by (12) and (13), and which are found by using the functions Θ_1 and Θ_2 expressed in terms of $\Psi_{\mathrm{I}}, \ldots, \Psi_{\mathrm{VI}}$ by equations (7) and (8). We are thus required to evaluate 42 integrals of the type $(\Psi_r, \mathscr{H}\Psi_s)$, or (Ψ_r, Ψ_s), using notation (11), and with $r, s = \mathrm{I}, \mathrm{II}, \ldots, \mathrm{VI}$. We will consider the method of calculation of such integrals using the example of $(\Psi_{\mathrm{I}}, \mathscr{H}\Psi_{\mathrm{II}})$. By referring to the method of formation of the functions Ψ_r we have:†

$$(\Psi_{\mathrm{I}}, \mathcal{H}\Psi_{\mathrm{II}}) = \frac{1}{4!}\sum_{\sigma}\int \mathcal{A}a_1\alpha_1 b_2\beta_2 c_3\beta_3 d_4\alpha_4 \mathcal{H}\mathcal{A}a_1\beta_1 b_2\alpha_2 c_3\alpha_3 d_4\beta_4 d\tau.$$

We will show that it is possible to get rid of one of the two antisymmetrization operators. To do this we explicitly take the second of these and write two of the terms of the expansion which has 4! terms:

$$\frac{1}{4!}\Big[\sum_{\sigma}\int \mathcal{A}a_1\alpha_1 b_2\beta_2 c_3\beta_3 d_4\alpha_4 \mathcal{H} a_1\beta_1 b_2\alpha_2 c_3\alpha_3 d_4\beta_4 d\tau$$

$$- \sum_{\sigma}\int \mathcal{A}a_1\alpha_1 b_2\beta_2 c_3\beta_3 d_4\alpha_4 \mathcal{H} a_2\beta_2 b_1\alpha_1 c_3\alpha_3 d_4\beta_4 d\tau + \ldots\Big].$$

We consider the second integral and execute the change in variables corresponding to a permutation of the numbers of the electrons, which we write in shortened form: $1' = 2$; $2' = 1$; $3' = 3$; $4' = 4$. This odd permutation, applied to the operator $\mathscr{A}$, leads to a change in sign, so that the second integral is written:

$$\sum_{\sigma}\int \mathcal{A}a_{1'}\alpha_{1'} b_{2'}\beta_{2'} c_{3'}\beta_{3'} d_{4'}\alpha_{4'} \mathcal{H} a_{1'}\beta_{1'} b_{2'}\alpha_{2'} c_{3'}\alpha_{3'} d_{4'}\beta_{4'} d\tau$$

† σ and $d\tau$ stand for $\sigma_1, \sigma_2, \sigma_3, \sigma_4$ and $d\tau_1\, d\tau_2\, d\tau_3\, d\tau_4$, respectively.

and thus is identical to the first integral, since the naming of the integration variables is without significance. The same reasoning applies to each of the 4! terms of the expansion, which then allows us to write:

$$(\Psi_I, \mathscr{H}\Psi_{II}) = \sum_\sigma \int \mathrm{A} a_1 \alpha_1 b_2 \beta_2 c_3 \beta_3 d_4 \alpha_4 \mathscr{H} a_1 \beta_1 b_2 \alpha_2 c_3 \alpha_3 d_4 \beta_4 \, d\tau$$

using only one antisymmetrization operator.

A second reduction can be made by noting that the operator $\mathscr{H}$ does not contain the spin variables, the only terms of the expansion of $\mathscr{A} a_1 \alpha_1 b_2 \beta_2 c_3 \beta_3 d_4 \alpha_4$ that have a non-zero contribution to the integral being those involving the same distribution of spins as in the second factor $a_1 \beta_1 b_2 \alpha_2 c_3 \alpha_3 d_4 \beta_4$, and which are 4 in number in place of the $4! = 24$ expansion terms in the expansion of the antisymmetric function. We then get:

$$\begin{aligned}(\Psi_I, \mathscr{H}\Psi_{II}) = \sum_\sigma \int [& a_2\alpha_2 b_1\beta_1 c_4\beta_4 d_3\alpha_3 \mathscr{H} a_1\beta_1 b_2\alpha_2 c_3\alpha_3 d_4\beta_4 \\ & - a_3\alpha_3 b_1\beta_1 c_4\beta_4 d_2\alpha_2 \mathscr{H} a_1\beta_1 b_2\alpha_2 c_3\alpha_3 d_4\beta_4 \\ & - a_2\alpha_2 b_4\beta_4 c_1\beta_1 d_3\alpha_3 \mathscr{H} a_1\beta_1 b_2\alpha_2 c_3\alpha_3 d_4\beta_4 \\ & + a_3\alpha_3 b_4\beta_4 c_1\beta_1 d_2\alpha_2 \mathscr{H} a_1\beta_1 b_2\alpha_2 c_3\alpha_3 d_4\beta_4] d\tau.\end{aligned}$$

The spin functions are normalized, with the result that after doing the sum over the spin variables we get:

$$\begin{aligned}(16) \quad (\Psi_I, \mathscr{H}\Psi_{II}) = \int [& a_2 b_1 c_4 d_3 \mathscr{H} a_1 b_2 c_3 d_4 - a_3 b_1 c_4 d_2 \mathscr{H} a_1 b_2 c_3 d_4 \\ & - a_2 b_4 c_1 d_3 \mathscr{H} a_1 b_2 c_3 d_4 + a_3 b_4 c_1 d_2 \mathscr{H} a_1 b_2 c_3 d_4] d\tau.\end{aligned}$$

Equation (16) is a sum of four integrals of the form:

$$\int a_r b_s c_t d_u \mathscr{H} a_1 b_2 c_3 d_4 \, d\tau.$$

The usual approximation retains only those of the integrals for which the order r, s, t, u, differs from the natural order 1, 2, 3, 4, at best by a permutation of two numbers. We must then consider first of all the integrals called *coulomb integrals*, given by:

$$\mathrm{C} = \int a_1 b_2 c_3 d_4 \mathscr{H} a_1 b_2 c_3 d_4 \, d\tau$$

and the *exchange integrals* of the form:

$$\mathrm{A}_{ab} = \int a_2 b_1 c_3 d_4 \mathscr{H} a_1 b_2 c_3 d_4 \, d\tau.$$

In this approximation we get $(\Psi_I, \mathscr{H}\Psi_{II}) = 0$.

The same calculation can be repeated for all of the integrals $(\Psi_r, \mathcal{H}\Psi_s)$, as well as for the overlap integrals (Ψ_r, Ψ_s). In this latter case we are required to evaluate integrals for the form:

$$\int a_r b_s c_t d_u a_1 b_2 c_3 d_4 d\tau$$

and we retain only those in which the order r, s, t, u, is in the natural order, and which are then equal to unity.

Although not difficult in principle, such calculations remain somewhat long and tedious, and various authors have sought rules for evaluating the integrals $(\Theta_r, \mathcal{H}\Theta_s)$ and (Θ_r, Θ_s) relative to two bond diagrams of r and s. The most used (Pauling) method† introduces the superposition diagram of the r and s bonding diagram (fig. 66). In the case of butadiene we get (fig. 70):

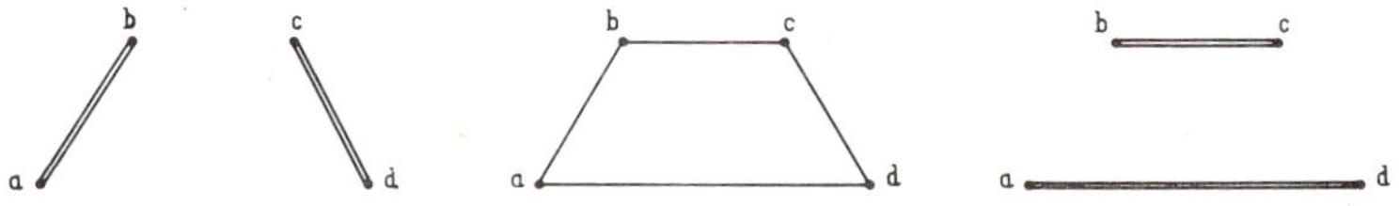

Figure 70. Superposition diagram of the bond schemes (figure 66) of butadiene.

In the general case, these superposition diagrams are composed of a certain number of closed polygons or *islands*, none having a common vertex. These islands have an even number of sides and are eventually reducible to a double line, as in the case of butadiene (1 + 1) and (2 + 2). Once these diagrams are traced, we see that for a system of $2n$ electrons and a superposition diagram of r and s containing a number of islands equal to i, we have:

$$\mathrm{H}_{rs} = (\Theta_r, \mathcal{H}\Theta_s) = \frac{2^i}{2^n}\left[\mathrm{C} + \sum_{p>q} a_{pq}\mathrm{A}_{pq}\right]$$

with:

$$\mathrm{C} = \int a_1 b_2 c_3 \ldots \mathcal{H} a_1 b_2 c_3 \ldots d\tau$$

$$\mathrm{A}_{ab} = \int a_2 b_1 c_3 \ldots \mathcal{H} a_1 b_2 c_3 \ldots d\tau, \ \ldots$$

† L. Pauling, *J. Chem. Phys.* (1933), **1**, 280.

This sum is extended to all the pairs of atoms p and q with the following rules:

If p and q belong to distinct islands $a_{pq} = -\frac{1}{2}$.
If p and q belong to the same island and:

The number of sides between p and q is odd $a_{pq} = 1$.
The number of sides between p and q is even $a_{pq} = -2$.

Counting of the number of sides in one sense or the other is unimportant, since the number of sides of the island is always even.

Application to butadiene ($n = 2$) gives the following results:

$$H_{11} = C + A_{ab} + A_{cd} - \frac{1}{2}(A_{ac} + A_{ad} + A_{bc} + A_{bd})$$

$$H_{12} = \frac{1}{2}[C + A_{ab} + A_{bc} + A_{cd} + A_{da} - 2(A_{ac} + A_{bd})]$$

$$H_{22} = C + A_{bc} + A_{ad} - \frac{1}{2}(A_{ab} + A_{ac} + A_{bd} + A_{cd}).$$

The value of the exchange integrals A_{pq} evidently depends on the relative position of atoms p and q in the molecule. The most simple approximation involves retaining only those between adjacent atoms, and taking all others equal (aromaticity hypothesis). In the butadiene example all of these integrals are zero, with the exception of $A_{ab} = A_{bc} = A_{cd} = \alpha$.

The rules to be elucidated for the calculation of the matrix elements H_{rs} also allow calculation of $\Delta_{r,s} = (\Theta_r, \Theta_s)$. In the most simple approximation possible the same relations are used, by taking $C = 1$ and $A_{pq} = 0$, whatever p and q may be.

We can now treat the problem of the π electrons of butadiene. The six quantities H_{rs} and Δ_{rs} occurring in equation (15) are, through application of the rules above:

$$H_{11} = C + \frac{3}{2}\alpha; \quad H_{12} = \frac{1}{2}(C + 3\alpha); \quad H_{22} = C$$

$$\Delta_{11} = 1; \quad \Delta_{12} = \frac{1}{2}; \quad \Delta_{22} = 1.$$

Equation (15) is then written, taking $E - C = x$:

$$x^2 - 3\alpha^2 = 0, \text{ so that: } x = \pm\alpha\sqrt{3}.$$

The parameter α is negative, as is C, with the result that the ground state should correspond to the positive root, namely:

$$E = C + 1.732\alpha.$$

This value can be compared with that of the form (1) alone:

$$E = C + 1.50\alpha$$

hence the increase in energy due to the hybridization is equal to $R = 0.232\alpha$, which corresponds to the definition of the resonance energy of the two mesomeric forms 1 and 2.

The participation of the state function (2) can be calculated by determining the coefficients c_1 and c_2 from (14) and the normalization condition (13). We get then:

$$\Phi = 0.8165\Theta_1 + 0.2988\Theta_2.$$

This indicates a preponderance of bonding scheme (1) (fig. 66) in the description of the ground state of the molecule, but it is difficult to give a definite weight to each of these two forms since Θ_1 and Θ_2 are not orthogonal.

We note, finally, that the calculation uses only one of the roots of equation (15), and one might ask about the physical significance of the second, which in principle should also define a singlet state of the molecule. We could suppose that it is an excited level, but we must note that all of the calculations done here exclude the ionic form in which two electrons are localized on the same atom. We examined this approximation in the case of the ground state of the diatomic molecule (§49), where it appears satisfactory, but it is far from being so for the first excited singlet state and, as a consequence, we must expect this second root to lead to a very much elevated energy for this level, so that the deviations are, in a sense, the reverse of those gotten by the molecular orbital method. In effect, that method (see **Exercise 1**) leads to an excited ionic state, but implies an increased participation by the ionic forms in the ground state.

Exercise 1. Repeat the problem of the ethylene molecule and form the antisymmetric two-electron function representing the ground state and the excited singlet, taking the MO (§34). Note that the excited state is given as a hybrid of the ionic forms and can only be described using the same spin states as in the ground state.

55 Various Applications of the Method

290 The study of the behavior of the π electrons in benzene is a problem of six electrons and we have shown that the five basis functions can be related to the two forms of Kékulé and the three Dewar forms (fig. 68), so that the state function is given by:

$$\Phi = c_1\Theta_1 + c_2\Theta_2 + c_3\Theta_3 + c_4\Theta_4 + c_5\Theta_5.$$

The five coefficients c_r are, in principle, fixed by the variational calculation, but the symmetry of the molecule leads to simplification of the calculation. In effect, the discussion (§35) using the MO method demonstrated that the ground state corresponds to the filling of the first two one-electron levels by the π electrons, and we find that this corresponds to a simple ^1A term using Platt's notation (§50). As a consequence, a rotation of $\pi/3$ which leaves the molecule uncharged should preserve the equation of state, which implies $c_1 = c_2 = c$, and $c_3 = c_4 = c_5 = d$, with the result that the ground state is represented by a function of two parameters:

$$\Phi = c(\Theta_1 + \Theta_2) + d(\Theta_3 + \Theta_4 + \Theta_5).$$

The calculation therefore evolves as in the case of butadiene by giving the energy expression, so that:

$$\begin{aligned} E = (\Phi, \mathcal{H}\Phi) &= c^2(H_{11} + 2H_{12} + H_{22}) \\ &+ 2cd(H_{13} + H_{14} + H_{15} + H_{23} + H_{24} + H_{25}) \\ &+ d^2(H_{33} + H_{44} + H_{55} + 2H_{34} + 2H_{35} + 2H_{45}). \end{aligned}$$

The elements H_{rs} are calculated from the superposition diagrams, which are of five different types (fig. 71). Using the Pauling rules, we get:

$$\begin{aligned} &H_{11} = H_{22} = C + \frac{3}{2}\alpha \\ &H_{12} = \frac{1}{4}C + \frac{3}{2}\alpha \\ &H_{13} = H_{14} = H_{15} = H_{23} = H_{24} = H_{25} = \frac{1}{2}C + \frac{3}{2}\alpha \\ &H_{33} = H_{44} = H_{55} = C \\ &H_{34} = H_{35} = H_{45} = \frac{1}{4}C + \frac{3}{2}\alpha. \end{aligned}$$

It is not necessary to generate the table of Δ_{rs}, as we need only take the coefficient of C in the corresponding expression for H_{rs}. Finally, we get the following energy expressions and the normalization condition:

$$E = C^2\left(\frac{5}{2}C + 6\alpha\right) + 2cd(3C + 9\alpha) + d^2\left(\frac{9}{2}C + 9\alpha\right)$$

$$1 = \frac{5}{2}c^2 + 6cd + \frac{9}{2}d^2.$$

The extremum of E is determined as for butadiene and the condition to compatibility is written taking $E - C = x$:

$$x^2 + 2\alpha x - 12\alpha^2 = 0$$

with an acceptable root here of:

$$x = (-1 + \sqrt{13})\alpha = 2.61\alpha.$$

As we indicated (§35), the resonance energy is defined by reference to one of the forms of Kékulé ($H_{11} = C + 1.5\alpha$), from which we get its value:

$$R = 1.11\alpha.$$

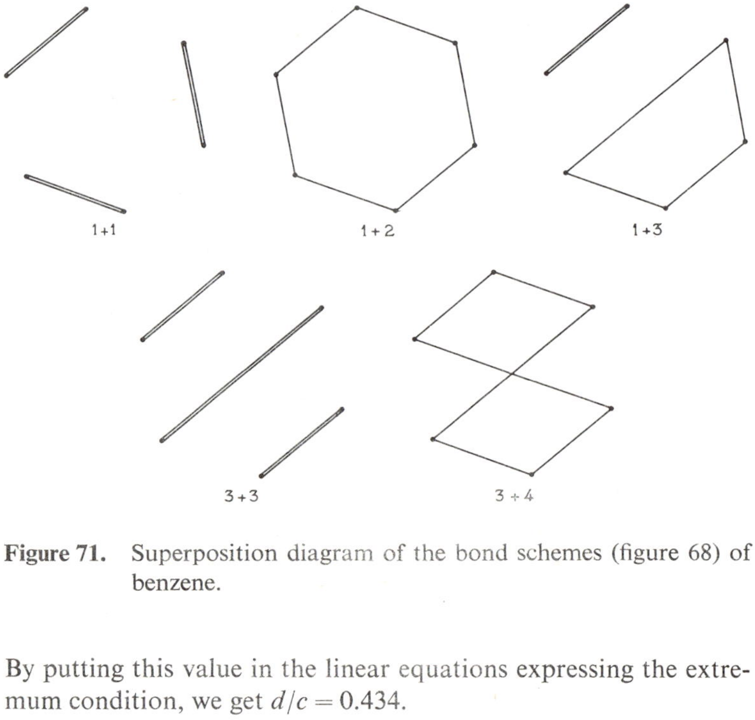

Figure 71. Superposition diagram of the bond schemes (figure 68) of benzene.

By putting this value in the linear equations expressing the extremum condition, we get $d/c = 0.434$.

It is possible to determine the contribution of the Dewar forms by repeating the calculation using only the Kékulé forms. The variational calculation is not needed here, since we have $c_1 = c_2 = c$, and the normalization gives $\frac{5}{2}c^2 = 1$. We then get:

$$E = (H_{11} + 2H_{12} + H_{22})c^2 = C + 2.4\alpha$$

so that we have a value of 0.9α for the resonance energy due essentially to the hybridization of the two Kékulé forms.

The Three-Electron Problem

This is the problem posed by the non-observable molecule H_3, which we take as a model for the case where the number of electrons is odd. It is nevertheless possible to reduce it to the preceding case, since the diagram of the multiplet states (fig. 60) shows that a system of $2k$ electrons has a number of singlet terms equal to the number of doublets of a system of $2k-1$ electrons, hence the procedure consists of adding an electron and an artificial atom to the latter system. In the case of a system of three electrons, we use the preceding calculations relative to butadiene and disregard all of the exchange integrals involving the artificial atom d. We then get:

$$H_{11} = C + A_{ab} - \frac{1}{2}(A_{ac} + A_{bc})$$

$$H_{12} = \frac{1}{2}(C + A_{ab} + A_{bc} - 2A_{ac})$$

$$H_{22} = C + A_{bc} - \frac{1}{2}(A_{ab} + A_{ac})$$

with the same values of Δ_{rs}, so that $\Delta_{11} = \Delta_{22} = 1$, and $\Delta_{12} = \frac{1}{2}$.

To extend the calculation it is convenient to make an assumption about the value of the exchange integrals. If we take a system of three nuclei aligned with the order abc, we can ignore A_{ac}, and take $A_{ab} = A_{bc} = \alpha$. We then get:

$$H_{11} = C + \frac{1}{2}\alpha; \qquad H_{12} = \frac{1}{2}C + \alpha; \qquad H_{22} = C + \frac{1}{2}\alpha.$$

Using the same notation, so that $E = C + x$, the determinant giving the energy values is written:

$$\begin{vmatrix} \frac{\alpha}{2} - x & \alpha - \frac{1}{2}x \\ \alpha - \frac{1}{2}x & \frac{1}{2}\alpha - x \end{vmatrix} = 0$$

where the acceptable root $x = \alpha$ indicates that in our approximation the energy of H_3 is no different than H_2, which demonstrates the impossibility of its existence as a stable molecule.

Note on the Formation of the Singlet Functions

(See Section 52)

The proof (p. 279) that equation (6) leads to a singlet state can be justified by showing that it involves an eigenfunction of

$\mathscr{S}^2 = \mathscr{S}_1^2 + \mathscr{S}_2^2 + \mathscr{S}_3^2 + \mathscr{S}_4^2$ having eigenvalue zero. The commutability of $\mathscr{S}^2$ and $\mathscr{A}$ (§47) allows the expansion of $\mathscr{S}^2\Theta$ to give the quantity:

$$\begin{aligned}
&\mathscr{S}^2(\alpha_1\beta_2 - \alpha_2\beta_1)(\alpha_3\beta_4 - \alpha_4\beta_3) \\
&\quad = [(\mathscr{S}_1^2 + \mathscr{S}_2^2)(\alpha_1\beta_2 - \alpha_2\beta_1)](\alpha_3\beta_4 - \alpha_4\beta_3) \\
&\qquad + (\alpha_1\beta_2 - \alpha_2\beta_1)[(\mathscr{S}_3^2 + \mathscr{S}_4^2)(\alpha_3\beta_4 - \alpha_4\beta_3)].
\end{aligned}$$

Each of the two quantities between the brackets is zero, since the corresponding spin functions define singlet states, which certainly implies $\mathscr{S}^2\Theta = 0$.

14 SPECIAL PROBLEMS

A APPLICATIONS OF THE VARIATIONAL METHOD

The operator identity (7) of section 37, applying to the case of a one-electron state function, is easily generalized to the case of a many-electron system. The hamiltonian operator for a system of ν electrons is:

$$\mathcal{H} = -\frac{\hbar^2}{2m_0}(\Delta_1 + \Delta_2 + \ldots + \Delta_\nu) + V.$$

If x_r is the coordinate of electron r, we have:

$$\mathcal{H}x_r\Psi - x_r\mathcal{H}\Psi = -\frac{\hbar^2}{2m_0}(\Delta_r x_r - x_r\Delta_r)\Psi$$

identical to that of the one-electron case. Effectively, we get the same operational equation:

$$\mathcal{H}x_r - x_r\mathcal{H} = -\frac{\hbar^2}{m_0}\frac{\partial}{\partial x_r}. \tag{1}$$

We have already seen that calculations on many-electron systems require symmetric functions of the electron coordinates; adding the equations (1) for the ν electrons of the system leads to:

$$\mathcal{H}X - X\mathcal{H} = -\frac{\hbar^2}{m_0}\sum_{r=1}^{\nu}\frac{\partial}{\partial x_r}, \quad \text{with} \quad X = \sum_{r=1}^{\nu} x_r. \tag{2}$$

We apply this equation to the solution of the Schrödinger equation where Ψ satisfies $\mathcal{H}\Psi = E\Psi$; the procedure for calculation is indicated in section 37, so that integration following multiplication by $X\Psi^\star$ gives:

$$(X\Psi, \mathcal{H}X\Psi) - E\langle X^2\rangle = \frac{\nu\hbar^2}{2m_0} \tag{3}$$

with

$$\langle X^2\rangle = (X\Psi, X\Psi) = \sum_\sigma \int X^2\Psi^\star\Psi d\tau.$$

The only complication with respect to the calculations in section 37 concerns the use of antisymmetric orbitals, and we have shown (§54) how to reduce the corresponding integrals. We now treat a few examples.†

† J. Barriol and J. Regnier, *Advances in Chemical Physics* (New York: Interscience Publishers, 1965), Vol. VIII.

56 Calculation of the Transition Energy: Benzene

In the ground state, the six π electrons are distributed on the three one-electron orbitals, whose expression we rewrite as a function of the $2p_z$ AO's of the carbon atoms (§35):

$$a = \varphi_0 = \frac{1}{\sqrt{6(1+2S)}}(\psi_1 + \psi_2 + \psi_3 + \psi_4 + \psi_5 + \psi_6)$$

$$(4) \qquad b = \varphi_1' = \frac{1}{\sqrt{12(1+S)}}(\psi_1 - \psi_2 - 2\psi_3 - \psi_4 + \psi_5 + 2\psi_6)$$

$$c = \varphi_1'' = \frac{1}{\sqrt{4(1+S)}}(\psi_1 + \psi_2 - \psi_4 - \psi_5).$$

The antisymmetric function giving the ground state is then written:

$$(5) \qquad \Phi = \frac{1}{\sqrt{6\,!}}\mathcal{A}a_1\alpha_1a_2\beta_2b_3\alpha_3b_4\beta_4c_5\alpha_5c_6\beta_6,$$

as a function of the MO's a, b, and c, which are real and orthonormal.

As in the one-electron case, the calculation of the transition energy is based upon the assumption of the existence of a unique transition from the ground state, which allows us to assume that the excited state is represented to within the normalization factor by $X\Phi$. Equation (3) then gives the transition energy:

$$(6) \qquad E' - E = \frac{\nu\hbar^2}{2m_0\langle X^2\rangle} = \frac{ch}{\lambda}$$

with the result that we are required to calculate $\langle X^2\rangle$. The rules indicated in section 54 for the calculations using antisymmetric orbitals, also apply here. First of all, it is possible to drop one of the antisymmetrization operators, which leads to:

$$\sum_\sigma \int X^2\Phi^2 d\tau$$

$$(7) \qquad = \sum_\sigma \int \mathcal{A}a_1\alpha_1a_2\beta_2b_3\alpha_3b_4\beta_4c_5\alpha_5c_6\beta_6 X^2 a_1\alpha_1a_2\beta_2b_3\alpha_3b_4\beta_4c_5\alpha_5c_6\beta_6 d\tau$$

with:

$$X^2 = \sum_r x_r^2 + \sum_{r\neq s} x_r x_s = \sum_r x_r^2 + 2\sum_{r>s} x_r x_s.$$

In the calculation of the contribution of the square terms, the orthogonal character of the spin functions on the one hand, and of the a, b, and c orbitals on the other, shows that the only term of the antisymmetric expansion giving a non-zero result is for the factor x_1^2:

$$\sum_{\sigma} \int a_1\alpha_1 a_2\beta_2 b_3\alpha_3 b_4\beta_4 c_5\alpha_5 c_6\beta_6 \, x_1^2 a_1\alpha_1 a_2\beta_2 b_3\alpha_3 b_4\beta_4 c_5\alpha_5 c_6\beta_6 d\tau$$

$$= \int a_1^2 x_1^2 d\tau_1 = \int a^2 x^2 d\tau,$$

without any necessity of giving the number of the electron being considered. The same calculation can be made for each of the

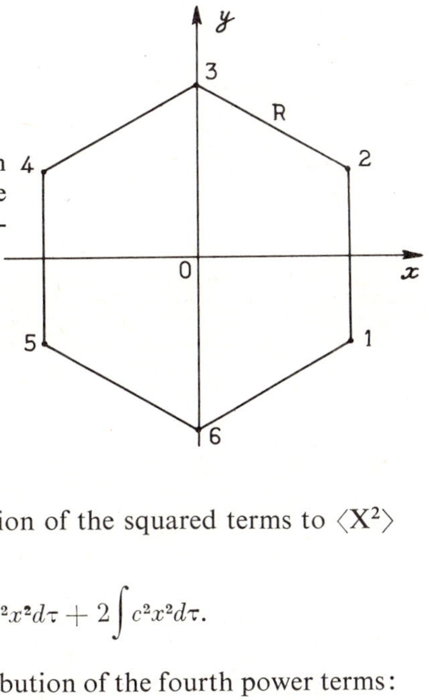

Figure 72. System of axes used in the calculation of the wave functions of benzene.

factors x_r^2, so that the contribution of the squared terms to $\langle X^2 \rangle$ is written:

$$2\int a^2x^2 d\tau + 2\int b^2x^2 d\tau + 2\int c^2x^2 d\tau.$$

We now examine only the contribution of the fourth power terms: for $x_1 x_2$ the preceding calculation gives:

$$\int a_1 a_2 x_1 x_2 a_1 a_2 d\tau_1 d\tau_2 = \left(\int a^2 x d\tau\right)^2 = 0$$

this result being implied by the symmetry of the orbitals a, b, and c, which is such that a^2, b^2, and c^2 remain unchanged under a reflection on the plane normal to the x-axis (fig. 72). For the factor $x_1 x_3$ the only term of the expansion to be considered is that for which permutation of electrons 1 and 3 preserves the spin configuration and gives:

$$-\int a_1 b_3 x_1 x_3 a_3 b_1 d\tau_1 d\tau_3 = -\left(\int xabd\tau\right)^2$$

which can eventually be non-zero.

Finally, we get for the result of integral (7):

$$(8) \qquad \langle X^2 \rangle = 2\int x^2a^2d\tau + 2\int x^2b^2d\tau + 2\int x^2c^2d\tau$$
$$- 4\left[\left(\int xabd\tau\right)^2 + \left(\int xacd\tau\right)^2 + \left(\int xbcd\tau\right)^2\right].$$

To evaluate these various integrals it is necessary to express the MO's by equations (4), which leads to integrals involving the $2p_z$ AO's, $\psi_1, \ldots, \psi_6$. As in section 37, for the integrals containing ψ_r^2 we take the corresponding nucleus as the origin, which involves:

$$\xi^2 = \int x_r^2 \psi_r^2 d\tau.$$

For the terms in $\psi_r\psi_s$ we retain only those corresponding to the neighboring atoms $|r-s| = 1$, and we take the middle of the corresponding bond as a new origin. We indicate the results of this calculation using the same approximation as in section 37, that is, by ignoring the following two integrals:†

$$\eta_x^2 = \int x^2\psi_r\psi_s d\tau; \qquad \eta_y^2 = \int y^2\psi_r\psi_s d\tau.$$

If R is the length of the benzene bond, and S the overlap integral of the two adjacent AO's ψ_r and ψ_s, we have:

$$\int x^2a^2d\tau = \frac{1}{1+2S}\left[\xi^2 + \frac{R^2}{2}\left(1+\frac{3S}{2}\right)\right]$$

$$\int x^2b^2d\tau = \frac{1}{1+S}\left[\xi^2 + \frac{R^2}{4}\right]$$

$$\int x^2c^2d\tau = \frac{1}{1+S}\left[\xi^2 + \frac{3R^2}{4}(1+S)\right]$$

$$\int xabd\tau = \int xbcd\tau = 0; \qquad \left(\int xacd\tau\right)^2 = \frac{(2+3S)^2R^2}{8(1+S)(1+2S)}.$$

The numerical calculations have been carried out using the Slater form for the ψ_r AO, with $Z^\star = 3.43$. We get $S = 0.22$ and

† These integrals are written with the origin at the middle of the r—s bond.

$\xi^2 = 0.143 \times 10^{-16}$ cm². Substituting in equation (6), we get $\lambda = 2{,}600$ Å, in quite good agreement with the experimental value generally taken to be 2,550 Å. We need not have excessive illusions about the significant character of this agreement, which results from the compensation for various sources of error, namely: suppression of the integrals η_x^2 and η_y^2, which are of comparable magnitude to ξ^2; too simple a representation of the state equation by means of the MO's *a*, *b*, and *c*; and finally, too rudimentary an expression for the function representing the excited state by $X\Phi$. It is possible to improve on these various points at the obvious expense of requiring more complicated calculations.

Variational Calculation of the Electric Polarizability 57

We consider a system of ν electrons whose ground state is governed by the Schrödinger equation:

$$\mathcal{H}_0\Phi_0 = E_0\Phi_0.$$

Putting this system into a uniform electric field f directed along the x-axis gives the additional potential energy:

$$V = -\sum_r q_r x_r f,$$

where x_r gives the abscissa of the particle (either the nucleus or electron) with charge q_r. The term relative to the nuclei disappears by taking the origin at the center of mass of the positive charges, and there remains:

$$V = -ef\sum_r x_r = -efX,$$

e being the electron charge ($e = 4.80 \times 10^{-10}$ esu).

The hamiltonian is then written:

$$\mathcal{H} = \mathcal{H}_0 - efX.$$

The energy of the perturbed system is quickly found by the variational method, taking the following state function dependent on the parameter λ:

$$\Phi = N(1 + \lambda X)\Phi_0$$

where N is found by normalization. The systems that we usually consider do not have a permanent electric moment, which implies:

$$\int X\Phi_0^2 d\tau = 0.$$

As a consequence, the normalization takes the form:

$$N^2(1 + \lambda^2 \langle X^2 \rangle) = 1.$$

The variational calculation involves, as a first step, the formation of the energy in the perturbed state, namely:

$$E = (\Phi, \mathcal{H}\Phi) = N^2((1 + \lambda X)\Phi_0, \mathcal{H}(1 + \lambda X)\Phi_0)$$

so that, by expanding:

$$E = N^2[(\Phi_0, \mathcal{H}\Phi_0) + \lambda(X\Phi_0, \mathcal{H}\Phi_0)$$
$$+ \lambda(\Phi_0, \mathcal{H}X\Phi_0) + \lambda^2(X\Phi_0, \mathcal{H}X\Phi_0)].$$

We express the different terms in this expansion in the second order limit, which requires us to ignore the terms of order higher than λf and λ^2. We then get:

$$(\Phi_0, \mathcal{H}\Phi_0) = (\Phi_0, \mathcal{H}_0\Phi_0) - ef(\Phi_0, X\Phi_0) = E_0.$$

The two coefficients of λ are equal as a result of the hermiticity of H, so that:

$$(X\Phi_0, \mathcal{H}\Phi_0) = (\Phi_0, \mathcal{H}X\Phi_0)$$
$$= (X\Phi_0, \mathcal{H}_0\Phi_0) - ef(X\Phi_0, X\Phi_0) = -ef \langle X^2 \rangle.$$

In our approximation the coefficient of λ^2 is written:

$$(X\Phi_0, \mathcal{H}X\Phi_0) \simeq (X\Phi_0, \mathcal{H}_0X\Phi_0) = E_0 \langle X^2 \rangle + \frac{\nu\hbar^2}{2m_0},$$

by using equation (3).

Finally we get:

$$E = N^2\left[E_0 - 2\lambda ef \langle X^2 \rangle + \lambda^2\left(E_0 \langle X^2 \rangle + \frac{\nu\hbar^2}{2m_0}\right)\right]$$

so that:

$$E = E_0 + N^2\left[\frac{\nu\hbar^2}{2m_0}\lambda^2 - 2ef \langle X^2 \rangle \lambda\right].$$

In the second order limit in λ we can take $N^2 = 1$. The extremum condition on E is then written:

$$\frac{\nu\hbar^2}{m_0}\lambda = 2ef \langle X^2 \rangle$$

where:

$$E = E_0 - \frac{2m_0e^2}{\nu\hbar^2} \langle X^2 \rangle^2 f^2$$

which is of the form:†

$$E = E_0 - \frac{1}{2}\alpha_x f^2,$$

hence an electric polarizability:

$$\alpha_x = \frac{4m_0e^2}{\nu\hbar^2}\langle X^2\rangle^2. \tag{9}$$

We get the same expressions for α_y and α_z, assuming that the axes of the reference system are taken along the principal axis of the electric polarizability tensor of the system.

Equation (9) is applicable to any non-polar system. The calculations are particularly simple in the case of ions having the electron structure of a rare gas, and for which the $^1S_0$ ground state is represented by a unique many-electron function. The calculation of $\langle X^2\rangle$ is made using the same methods as in the preceding paragraph. We will indicate the results by limiting ourselves to considering, for each of the types of ions, the electrons in the outer shell:

ions with the He structure:

$$\alpha = \frac{8a_0^3}{Z^{\star 4}}$$

ions with the Ne structure:

$$\alpha = \frac{9{,}800}{9}\cdot\frac{a_0^3}{Z^{\star 4}}$$

ions with the A structure:

$$\alpha = 17{,}860.5\,\frac{a_0}{Z^{\star 4}}.$$

In these equations, a_0 designates the radius of the first hydrogen orbital. The calculations are made using Slater orbitals. We note the inverse dependence of α on $Z^{\star 4}$, and thus, for ions of the same structure, on r^4 (Wasastjerna's law). In numerical problems the choice of $Z^\star$ poses a tricky problem. The equation applied here to the case of He, Ne, and A, shows the necessity of taking relatively low values of $Z^\star$, so that, for example, $Z^\star = 1.56$ for He in place of 1.69, required in the energy calculation. Likewise, for Ne we take $Z^\star = 4.55$ instead of 5.85. These variations are natural and are explained by the decisive role of the exterior regions of the orbitals in the mechanism of electric polarization.

† See the end of section 57 for the definition of the polarizability α in the case of an isotropic system.

Definition of the Polarizability

We have used the equation $E = E_0 - \frac{1}{2}\alpha f^2$ to introduce the polarizability of the system. This classical electrostatic equation is immediately obtained by taking a simple model of the polarizable molecule formed of two opposite charges $+q$ and $-q$ elastically bound at a distance of ξ with a force constant k. In the absence of electrical fields, the potential energy of the system is:

$$U = \frac{1}{2} k\xi^2,$$

with a minimum corresponding to $\xi = 0$, and thus a zero moment. The action of an electrical field $\vec{f}$ introduces a supplementary energy $-\vec{\mu}\cdot\vec{f} = -q\vec{\xi}\cdot\vec{f}$, hence:

$$U = \frac{1}{2} k\xi^2 - q\vec{\xi}\cdot\vec{f}.$$

The equilibrium defined by the minimum of the energy implies:

$$k\vec{\xi} = q\vec{f}, \quad \text{where} \quad \vec{\mu} = q\vec{\xi} = q^2\vec{f}/k = \alpha\vec{f};$$

$$U = -\, q^2f^2/2k = -\, \alpha f^2/2.$$

58 London Intermolecular Forces

We will apply the preceding method to the calculation of the interaction energy of two non-polar molecules. Although each of these molecules has no permanent electrical moment, the displacement of the interior electrons of each gives them an electrical moment:

$$\mu_x = e\sum_r x_r = eX\,; \qquad \mu_y = e\sum_r y_r = eY\,;$$

$$\mu_z = e\sum_r z_r = eZ$$

whose magnitude and direction follow the fluctuations of their electron motion which varies rapidly in time. It is to the interactions of these fluctuating moments of the two molecules that we can attribute the attractive forces in the absence of permanent moments. We will carry out the calculation of the interaction energy between molecules 1 and 2 with their positive charges separated by a distance R. The potential energy in the presence of two electric dipoles† is written, assuming point dipoles:

(10) $$V = \frac{e^2}{4\pi\varepsilon_0 R^3}[X_1X_2 + Y_1Y_2 - 2Z_1Z_2].$$

The mean value of this potential energy is written with Ψ_1 and Ψ_2 representing the ground state of each of these two molecules when they are isolated:

(11) $$\int V\Psi_1^2\Psi_2^2 d\tau_1 d\tau_2.$$

This quantity is zero if:

$$\int X_1\Psi_1^2 d\tau_2 = \int X_2\Psi_2^2 d\tau_2 = 0,$$

as are the other integrals involving Y and Z, which is consistent with the hypothesis of a zero permanent moment. There is no interaction between molecules 1 and 2 when the fluctuations of these two dipoles are independent. We can therefore say that the first order perturbation energy (§20), given by (11), is zero. To treat the next approximation it is necessary to introduce the modification of the state functions due to the interaction. This can be done by means of classical second order perturbation theory. We will use the variational method, taking the state function formed in the absence of the perturbation as the product of the functions Ψ_1 and Ψ_2 of the isolated systems, and introducing a linear factor relative to the components of the two dipoles, so that:

$$\Theta = N\Psi_1\Psi_2(1 + \lambda X_1X_2 + \mu Y_1Y_2 + \nu Z_1Z_2)$$

is dependent on the three parameters λ, μ, and ν. The coefficient N is found by normalization:

$$(\Theta, \Theta) = 1.$$

In the expansion of this expression, the coefficient of λ is written:

$$2N^2\int X_1X_2\Psi_1^2\Psi_2^2 d\tau_1 d\tau_2 = 2N^2\int X_1\Psi_1^2 d\tau_1 \int X_2\Psi_2^2 d\tau_2 = 0,$$

given the separation of variables between the two molecules. This is also true for the coefficients of μ and ν and, consequently, for the fourth-order terms. Finally, normalization uses only the square terms and is written:

(11) $$N^2[1 + \lambda^2\langle X_1^2\rangle\langle X_2^2\rangle + \mu^2\langle Y_1^2\rangle\langle Y_2^2\rangle + \nu^2\langle Z_1^2\rangle\langle Z_2^2\rangle] = 1.$$

† See the end of the section.

The state function Θ must satisfy the extremum condition on the energy defined by:

$$(12) \quad E = (\Theta, \mathcal{H}\Theta), \qquad \text{with} \qquad \mathcal{H} = \mathcal{H}_1 + \mathcal{H}_2 + V = \mathcal{H}^0 + V.$$

We will expand equation (12), limiting ourselves to second-order terms, the perturbation energy V and the parameters λ, μ, and ν being infinitesimally small in the first order. The expansion involves the following terms:

Zero order in λ:

$$N^2(\Psi_1\Psi_2, \mathcal{H}^0\Psi_1\Psi_2) = N^2[(\Psi_1, \mathcal{H}_1\Psi_1) + (\Psi_2, \mathcal{H}_2\Psi_2)]$$
$$= N^2(E_1 + E_2)$$

by taking:

$$\mathcal{H}_1\Psi_1 = E_1\Psi_1; \qquad \mathcal{H}_2\Psi_2 = E_2\Psi_2.$$

First order in λ:

$$2N^2(X_1X_2\Psi_1\Psi_2, V\Psi_1\Psi_2)$$

Only the terms $X_1^2X_2^2$ give a non-zero contribution, which is written:

$$2N^2\frac{e^2}{4\pi\varepsilon_0 R^3}(X_1X_2\Psi_1\Psi_2, X_1X_2\Psi_1\Psi_2) = 2N^2\frac{e^2}{4\pi\varepsilon_0 R^3}\langle X_1^2\rangle\langle X_2^2\rangle.$$

Second order in λ:

$$N^2(X_1X_2\Psi_1\Psi_2, \mathcal{H}^0X_1X_2\Psi_1\Psi_2)$$
$$= N^2[(X_1\Psi_1, \mathcal{H}_1X_1\Psi_1)\langle X_2^2\rangle + (X_2\Psi_2, \mathcal{H}_2X_2\Psi_2)\langle X_1^2\rangle].$$

These two expressions are evaluated by means of equation 3. All other integrals in the expansion are zero, and by regrouping terms we get:†

$$E = N^2(E_1 + E_2)(1 + \lambda^2\langle X_1^2\rangle\langle X_2^2\rangle$$
$$+ \mu^2\langle Y_1^2\rangle\langle Y_2^2\rangle + \nu^2\langle Z_1^2\rangle\langle Z_2^2\rangle)$$
$$+ \frac{2e^2N^2}{4\pi\varepsilon_0 R^3}(\lambda\langle X_1^2\rangle\langle X_2^2\rangle + \mu\langle Y_1^2\rangle\langle Y_2^2\rangle - 2\nu\langle Z_1^2\rangle\langle Z_2^2\rangle)$$
$$+ \frac{\hbar^2}{2m_0}N^2[\lambda^2(\nu_2\langle X_1^2\rangle + \nu_1\langle X_2^2\rangle) + \mu^2(\nu_2\langle Y_1^2\rangle + \nu_1\langle Y_2^2\rangle)$$
$$+ \nu^2(\nu_2\langle Z_1^2\rangle + \nu_1\langle Z_2^2\rangle)].$$

† ν_1 and ν_2 represent the number of electrons in the interacting molecules as introduced in equation 3.

Taking account of the normalization (11), the first term reduces to $E_1 + E_2$ as long as we can take $N^2 = 1$ in the second order. We then get three trinomials of the second degree in λ, μ, and ν, whose extremum is found immediately. We then get:

$$E = E_1 + E_2 - \frac{2m_0}{\hbar^2}\left(\frac{e^2}{4\pi\varepsilon_0 R^3}\right)^2 \left[\frac{(\langle X_1^2\rangle\langle X_2^2\rangle)^2}{\nu_2\langle X_1^2\rangle + \nu_1\langle X_2^2\rangle} + \frac{(\langle Y_1^2\rangle\langle Y_2^2\rangle)^2}{\nu_2\langle Y_1^2\rangle + \nu_1\langle Y_2^2\rangle} + \frac{4(\langle Z_1^2\rangle\langle Z_2^2\rangle)^2}{\nu_2\langle Z_1^2\rangle + \nu_1\langle Z_2^2\rangle}\right].$$

In the case where the two molecules can be considered as isotropic, the means values of X^2, Y^2, and Z^2 are the same, hence we get the following form for the interaction energy:

$$E - (E_1 + E_2) = \Delta E = -\frac{12m_0}{\hbar^2}\left(\frac{e^2}{4\pi\varepsilon_0 R^3}\right)^2 \frac{(\langle X_1^2\rangle\langle X_2^2\rangle)^2}{\nu_2\langle X_1^2\rangle + \nu_1\langle X_2^2\rangle}.$$

The variation as $-1/R^6$ shows that we have an attraction between the molecules under consideration. The formula we get can be expressed in many forms by giving the quantity $\langle X_1^2\rangle\langle X_2^2\rangle$ as a function of experimentally measurable variables. We can, for example, use the previous results (6) and (9) related to the transition energy and polarizability. For molecule 1 we then get:

$$\Delta E_1 = \frac{\nu_1\hbar^2}{2m_0\langle X_1^2\rangle}; \qquad \alpha_1 = \frac{4m_0e^2}{\nu_1\hbar^2}\langle X_1^2\rangle^2.$$

By dividing the numerator and the denominator of the interaction energy equation by $\langle X_1^2\rangle\langle X_2^2\rangle$, we get:

(13) $$\Delta E = -\frac{3}{2}\left(\frac{1}{4\pi\varepsilon_0 R^3}\right)^2 \alpha_1\alpha_2 \frac{\Delta E_1 \cdot \Delta E_2}{\Delta E_1 + \Delta E_2}.$$

In applications we generally replace the transition energies ΔE_1 and ΔE_2 by the corresponding ionization energies, and we then get one of the typical forms for the interaction energy:

(14) $$\Delta E = -\frac{3}{2}\alpha_1\alpha_2\frac{I_1 I_2}{I_1 + I_2}\cdot\frac{1}{R^6}, \text{ in esu.}$$

For molecules of the same species we get:

(15) $$\Delta E = -\frac{3}{4}\frac{\alpha^2 I}{R^6}.$$

The quantity I is obviously the least well defined, but equation (6) with $E' - E = I$ allows us to eliminate it, giving:

$$\alpha I^2 = \frac{\nu\hbar^2 e^2}{m_0} = \nu a_0 e^4,$$

by introducing the first Bohr radius. The interaction energy is then written:

$$\Delta E = -\frac{3}{4}\frac{e^2}{R^6}\alpha\sqrt{\nu a_0 \alpha} \tag{16}$$

which we generally use by considering only the valence electrons of the molecule.

Form of the Perturbation Function

This is gotten immediately by expanding the electrostatic energy expression for a pair of dipoles (fig. 73). We then get:

$$V = \frac{q^2}{4\pi\varepsilon_0}\left(\frac{1}{R} - \frac{1}{|\vec{R} - \vec{\xi}_1|} - \frac{1}{|\vec{R} + \vec{\xi}_2|} + \frac{1}{|\vec{R} - \vec{\xi}_1 + \vec{\xi}_2|}\right).$$

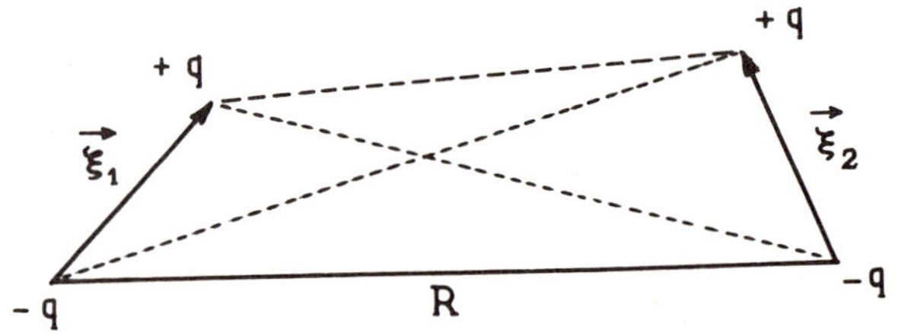

Figure 73. A system of two electric dipoles. The interaction energy is calculated by taking the charges in pairs along the lines indicated.

For each of these terms we have the expansion:

$$\frac{1}{|\vec{R} + \vec{u}|} = [(\vec{R} + \vec{u})^2]^{-\frac{1}{2}} = \frac{1}{R}\left(1 + 2\frac{\vec{R}\cdot\vec{u}}{R^2} + \frac{u^2}{R^2}\right)^{-1/2},$$

$$\simeq \frac{1}{R}\left(1 - \frac{\vec{R}\cdot\vec{u}}{R^2} - \frac{u^2}{2R^2} + \frac{3}{2}\frac{(\vec{R}\cdot\vec{u})^2}{R^4}\right).$$

In the regrouping, the first-order terms disappear and we get the classical relation:

$$V = \frac{1}{4\pi\varepsilon_0}\left(\frac{\vec{\mu}_1\cdot\vec{\mu}_2}{R^3} - \frac{3(\vec{R}\cdot\vec{\mu}_1)(\vec{R}\cdot\vec{\mu}_2)}{R^5}\right)$$

which leads to the form (10) used when the z-axis is taken along the direction of R.

B THE VIRIAL THEOREM

The Classical Concept of the Virial 59

This theorem was first introduced in classical mechanics and used in the kinetic theory of gases, where this name is often used to designate the various terms in the series expansion of pv as a function of $1/v$. We take a system of point masses for which the dynamical equations are:

$$m_r\ddot{x}_r = X_r\,; \qquad m_r\ddot{y}_r = Y_r\,; \qquad m_r\ddot{z}_r = Z_r.$$

We combine these equations in the following way:

$$m_r(x_r\ddot{x}_r + y_r\ddot{y}_r + z_r\ddot{z}_r) = x_rX_r + y_rY_r + z_rZ_r.$$

We take the mean value of two terms in this expression over an arbitrarily large time interval τ. Each of the terms of the first quantity gives:

$$m_r\frac{1}{\tau}\left[\int_0^\tau d(x_r\dot{x}_r) - \int_0^\tau \dot{x}_r^2 dt\right] = -\,m_r\overline{\dot{x}_r^2},$$

taking into account the fact that the integrals vanish for a confined system of particles whose speed is bounded,† which is the case for an enclosed gas at a given temperature. By taking the sum of the expressions corresponding to all of the particles of the system, we get the mean kinetic energy of the system, and this becomes:

$$-\,2\overline{T} = \sum_r \overline{(x_rX_r + y_rY_r + z_rZ_r)}$$

with the expression on the right being termed the *virial of the applied forces*. A particularly important form for the generalization into quantum mechanics occurs when we have only coulomb forces between the particles. The potential energy of the system is given by:

$$V = \sum_{i>j}\frac{q_iq_j}{r_{ij}}, \quad \text{with} \quad r_{ij} = [(x_i - x_j)^2 + (y_i - y_j)^2 + (z_i - z_j)^2]^{\frac{1}{2}}.$$

† This quantity has as its value $(m_r/\tau)\,[(x_r\dot{x}_r)_\tau - (x_r\dot{x}_r)_0]$. The coordinates of the particles in an enclosure are limited in absolute value $|x_r| < a$. If the magnitude of the velocity is also limited, $|\dot{x}_r| < b$, we have as an upper limit to the modulus of our expression $2m_rab/\tau$, which tends to zero for an infinite time τ.

The components of the forces exerted on particles *i* and *j* are then in the form:

$$X_i = -\frac{\partial V}{\partial x_i} = \sum_j \frac{q_i q_j}{r_{ij}^2} \frac{(x_i - x_j)}{r_{ij}}$$

$$X_j = -\frac{\partial V}{\partial x_j} = \sum_i \frac{q_i q_j}{r_{ij}^2} \frac{(x_j - x_i)}{r_{ij}}.$$

The contribution of the pair *i* and *j* to the expression whose mean constitutes the virial is written:

$$(x_i X_i + x_j X_j + \ldots) = \frac{q_i q_j}{r_{ij}^3} [(x_i - x_j)^2 + (y_i - y_j)^2 + (z_i - z_j)^2]$$

$$= \frac{q_i q_j}{r_{ij}}.$$

By adding the contributions of the different pairs of particles *i* and *j*, we get the expression for the potential energy of the system, hence the value of the virial equal to the mean value $\overline{V}$ of the potential energy. Equation (17) can then be written:

(18) $$\overline{V} = -2\overline{T}.$$

This equation constitutes the *virial theorem* for a system of particles with coulomb interactions. The equation requires only mean values, which *a priori* renders plausible its extension into quantum mechanics.

60 The Virial Theorem in Quantum Mechanics

We will treat the problem of a diatomic molecule, although the following arguments are useful for any system of nuclei and electrons in the approximation in which only coulomb forces occur, that is, assuming that we can neglect spin coupling. The equation of state of the system under study is written:

$$-\frac{\hbar^2}{2m_0}\sum_i \Delta_i \Phi + \left(\sum_{i>j} \frac{e^2}{r_{ij}} - \sum_i \frac{Z_a e^2}{r_{ai}} - \sum_i \frac{Z_b e^2}{r_{bi}} + \frac{Z_a Z_b e^2}{R}\right)\Phi = E\Phi$$

the molecule being made up of two nuclei *a* and *b* at a distance R with respective charges $-Z_a e$ and $-Z_b e$. In principle, the equation can be solved for any value of R which plays the role of a parameter in the energy eigenvalue E(R) as in the equation of state $\Phi(q_i, R)$, q_i being used to represent the set of electron coordinates. These two functions of R are assumed continuous and the equilibrium position of the molecule is defined by $dE/dR = 0$.

Given this, we will introduce a family of parametric functions involving the transformation of the function Φ by a one-to-one mapping in a ratio λ or, as we have previously said, by a scale change. If the transformation maps M onto M′ ($\vec{r}_{M'} = \lambda\vec{r}_M$), the transformed function Ψ should take in M′ the value which Φ has in M, which by definition corresponds to the following equation:

(19) $$\Psi(\lambda q_i ; \lambda R) = \Phi(q_i ; R)$$

or inversely:

$$\Psi(q_i ; R) = \Phi(\lambda^{-1} q_i ; \lambda^{-1} R).$$

The function Ψ so defined is not normalized, since:

$$\int |\Psi(q_i ; R)|^2 d\tau_1 \ldots d\tau_n = \int |\Phi(\lambda^{-1} q_i ; \lambda^{-1} R)|^2 d\tau_1 \ldots d\tau_n.$$

The change in variables $x_i = \lambda x_i', \ldots$ leads to:

$$\lambda^{3n} \int |\Phi(q_i' ; \lambda^{-1} R)|^2 d\tau_1' \ldots d\tau_n' = \lambda^{3n},$$

since Φ is assumed normalized for all values of R, and therefore for $\lambda^{-1}R$.

We have thus defined a family of normalized functions depending on two parameters λ and R, or $s = \lambda^{-1}R$, so that:

(20) $$\lambda^{-\frac{3n}{2}} \Psi(q_i ; R) = s^{\frac{3n}{2}} \Phi(s q_i ; sR).$$

These functions become the solution of the equation of state for $s = 1$, where we again get $\Phi(q_i, R)$.

We will calculate the mean value of the kinetic and potential energies in the state defined by (20). For the kinetic energy we get:†

$$\overline{T}_s(R) = s^{3n} \int \Phi^*(s q_i ; sR)\left(-\frac{\hbar^2}{2m_0} \sum_i \Delta_i\right) \Phi(s q_i ; sR) d\tau_1 \ldots d\tau_n.$$

The same change of variables as used above gives:

$$\overline{T}_s(R) = s^2 \int \Phi^*(q_i' ; sR)\left(-\frac{\hbar^2}{2m_0} \sum_i \Delta_i'\right) \Phi(q_i' ; sR) d\tau_1' \ldots d\tau_n'$$

† It is immaterial whether we write $\overline{T}_s$ or $\langle T_s \rangle$. The first notation is preferred here for ease in writing.

so that:

(21) $$\overline{T}_s(R) = s^2\overline{T}(sR).$$

This mean kinetic energy is thus tied in a simple way to the mean kinetic energy in a real state of the system.

The calculation of the mean potential energy is made in an analogous way. We get by definition:

$$\overline{V}_s = s^{3n}\int \Phi^*(sq_i\,;\,sR)\Big(\sum_{i>j}\frac{e^2}{r_{ij}} - \sum_i \frac{Z_a e^2}{r_{ai}} - \sum_i \frac{Z_b e^2}{r_{bi}} + \frac{Z_a Z_b e^2}{R}\Big)\Phi(sq_i\,;\,sR)d\tau_1\ldots d\tau_n.$$

We will again make the same change of variables, but the transformation of the potential energy only reduces to a multiplication if we take at the same time $R' = sR$. We then get:

$$\overline{V}_s(R) = s\int \Phi^*(q_i'\,;\,R')\Big(\sum_{i>j}\frac{e^2}{r_{ij}} - \sum_i \frac{Z_a e^2}{r_{ai}} - \sum_i \frac{Z_b e^2}{r_{bi}} + \frac{Z_a Z_b e^2}{R}\Big)\Phi(q_i'\,;\,R')d\tau_1' \ldots d\tau_n'$$

so that:

(22) $$\overline{V}_s(R) = s\overline{V}(R') = s\overline{V}(sR).$$

Finally, we get the following expression for the mean energy in the state defined by (20):

(23) $$\overline{E}_s(R) = s^2\overline{T}(sR) + s\overline{V}(sR).$$

By hypothesis, equation (20) gives an exact solution of the equation of state for $s = 1$. This value should then correspond to an extremum of $\overline{E}_s(R)$ whatever R may be. We then have:

$$\frac{\partial \overline{E}_s}{\partial s} = 2s\overline{T}(sR) + \overline{V}(sR) + s^2R\overline{T}'(sR) + sR\overline{V}'(sR)$$

from which the extremum condition:

(24) $$2\overline{T}(R) + \overline{V}(R) + R[\overline{T}'(R) + \overline{V}'(R)] = 0$$

hence, by introducing the energy $E = \overline{T} + \overline{V}$:

(25) $$2\overline{T}(R) + \overline{V}(R) + R\frac{dE}{dR} = 0.$$

Combined with the energy value we then get the two Slater equations:

$$\overline{T} = -E - R\frac{dE}{dR}; \qquad \overline{V} = 2E + R\frac{dE}{dR}. \tag{26}$$

For the equilibrium position of the molecule, these relations lead to:

$$-2\overline{T} = \overline{V} = 2E. \tag{27}$$

We again get the equation (18) of classical mechanics. The virial theorem has numerous applications. Equations (26) and (27) allow calculation of the mean kinetic and potential energy when we know

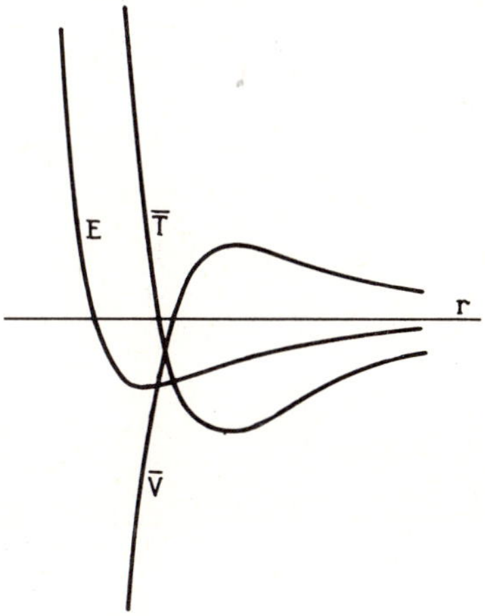

Figure 74. Curve representing the energy E of a diatomic molecule ($n = 9$, $m = 5$) as a function of the internuclear distance. The potential energy $\overline{V}$ and the kinetic energy $\overline{T}$, which are obtained from equation (26), are also given.

a form of the function E(R), which can be an empirical relation (fig. 74) relative, for example, to a form of the energy:

$$E = E_\infty - \frac{a}{r^m} + \frac{b}{r^n}, \quad \text{with } n = 9, \ m = 5.$$

Equation (27) is also useful as a criterion in the approximation of the states of an atom or molecule, but we must note that this equation is necessary but not sufficient. A classic example involves the hydrogen molecule, for which an approximate solution is written neglecting the spin factor:

$$\Phi = a(1)b(2) + a(2)b(1) = N\left[e^{-\alpha(r_{a_1}+r_{b_2})} + e^{-\alpha(r_{a_2}+r_{b_1})}\right],$$

with $\alpha = 1/a_0$.

The energy minimum corresponding to this state function occurs for $R = 1.52a_0$.

The calculation leads to the following values:

$$E(R_e) = -30.20 \text{ eV} ; \qquad -\overline{T}(R_e) = -22.79 \text{ eV} ;$$

$$\frac{1}{2}\overline{V}(R)_e = -26.49 \text{ eV}$$

whereas these three quantities should be equal. If we consider the state function:

$$\Phi(sq_i ; R) = N'[e^{-s\alpha(r_{a_1}+r_{b_2})} + e^{-s\alpha(r_{a_2}+r_{b_1})}]$$

the minimal energy is gotten for $s = 1.166$ and $R = 1.41a_0$, and we get $E = \overline{T} = \frac{1}{2}\overline{V} = -30.83$ eV for the minimum. The extremum condition relative to the variation of the mapping parameters insures that the virial criterion is automatically satisfied.

We will give one final example relative to the treatment of many-electron atoms in the Slater approximation. The many-electron function Φ is formed as the product of one-electron hydrogen functions so that:

$$\Phi = \Pi \psi_{n_i l_i}(q_i)$$

the functions $\psi_{n_i l_i}$ satisfy by hypothesis the one-electron state equation:

$$-\frac{\hbar^2}{2m_0}\Delta\psi_{n_i l_i} - \frac{Z_i^\star e^2}{r}\psi_{n_i l_i} = E_i\psi_{n_i l_i} \tag{28}$$

whose solution is:

$$E_i = \frac{Z_i^{\star 2}}{n_i^2} E_0.$$

The virial theorem then allows us to write:

$$\overline{T}_i = -E_i = -\frac{Z_i^{\star 2}}{n_i^2} E_0.$$

The mean kinetic energy in the state defined by Φ is expressed:

$$\overline{T} = \left(\Phi, -\frac{\hbar^2}{2m_0}\sum_i \Delta_i\Phi\right) = \sum_i \overline{T}_i.$$

A second application of the virial theorem permits evaluation of the total electron energy:

$$E = -\sum_i \overline{T}_i = \sum_i \frac{Z_i^{\star 2}}{n_i^2} E_0$$

which shows that the Z*'s governing the best Slater orbitals determine the contribution of the corresponding electrons to the total electron energy. Since the $\psi_{n_i l_i}$ are the best hydrogenic functions for construction of a Slater many-electron orbital Φ, the functions $\psi_{n_i l_i}$ can also be thought of as the approximate one-electron solutions of the self-consistent field equations, which have a form different from (28) for the potential energy (see the calculations of §27 relative to He). Such an equation, involving the interaction with the other particles, does not fulfil the conditions of the virial theorem, with the result that it is natural that the corresponding values ε_i are different from the values E_i relative to (28), the significant physical difference being that they represent the ionization energy of the electrons in the states being considered.

The Hellmann-Feynman Theorem 61

We treat this theorem following our discussion of the virial, because it too involves an extension of a classical mechanical property, which we will now review, to quantum mechanics.

Classical Point of View

We consider a mechanical system whose potential energy contains a variable parameter ξ. We cite for example the motion of a particle in a potential well, where we vary the diameter L. An example of a macroscopic system is given by a simple pendulum whose pendular mass is a magnet which oscillates in the field B of a solenoid (fig. 75*a*). By contrast, we exclude the case of the simple pendulum whose length we can vary continuously, a ring serving as the point of suspension (fig. 75*b*). In this latter case not only the potential energy but also the kinetic energy $\frac{1}{2}l\dot{\theta}^2$ depend explicitly on the variable parameter l.

The proof of this theorem involves the Hamiltonian function H introduced (Appendix I-10′) and defined by:

$$H = \sum_k p_k \dot{q}_k - L \quad \text{with} \quad p_k = \frac{\partial T}{\partial \dot{q}_k}, \qquad L = T - V.$$

By hypothesis, the kinetic energy T does not explicitly contain the parameter ξ and, as a result, takes the form of a homogeneous quadratic function for which Euler's theorem allows us to write:

$$\sum_k p_k \dot{q}_k = \sum_k \frac{\partial T}{\partial \dot{q}_k} \dot{q}_k = 2T$$

hence:

$$H = 2T - L = T + V = E.$$

The most general variation of Hamilton's function is written:

$$dH = \sum_k p_k d\dot{q}_k + \sum_k \dot{q}_k dp_k - \sum_k \frac{\partial L}{\partial q_k} dq_k - \sum_k \frac{\partial L}{\partial \dot{q}_k} d\dot{q}_k - \frac{\partial L}{\partial \xi} d\xi.$$

Taking account of:

$$p_k = \frac{\partial T}{\partial \dot{q}_k} = \frac{\partial L}{\partial \dot{q}_k},$$

the preceding equation reduces to:

$$dH = \sum_k \dot{q}_k dp_k - \sum_k \frac{\partial L}{\partial q_k} dq_k + \frac{\partial V}{\partial \xi} d\xi.$$

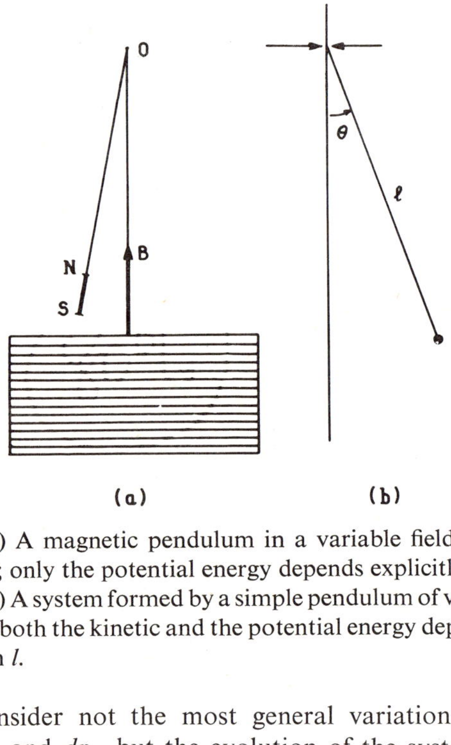

Figure 75. (*a*) A magnetic pendulum in a variable field of induction B; only the potential energy depends explicitly on B.
(*b*) A system formed by a simple pendulum of variable length *l*; both the kinetic and the potential energy depend explicitly on *l*.

We now consider not the most general variation of H, with arbitrary dq_k and dp_k, but the evolution of the system with time governed by the Lagrange equations (Appendix I), which implies:

$$\frac{d}{dt}\left(\frac{\partial L}{\partial \dot{q}_k}\right) - \frac{\partial L}{\partial q_k} = 0, \text{ so that } \dot{p}_k = \frac{\partial L}{\partial q_k}.$$

We then get:

$$dH = \sum_k \dot{q}_k dp_k - \sum_k \dot{p}_k dq_k + \frac{\partial V}{\partial \xi} d\xi$$

and during the motion with $d\dot{p}_k = p_k dt$, and $d\dot{q}_k = q_k dt$, we get the final equation:

$$dH = \frac{\partial V}{\partial \xi} d\xi,$$

where V is a function of q_k and of ξ.

This equation will allow us to get the Hellmann-Feynman theorem if we consider an extremely slow variation in time between t_0 and t_1 of the parameter ξ. The variation in energy in this period is then given by:

$$\Delta E = \int_{t_0}^{t_1} \frac{\partial V}{\partial \xi} \dot{\xi} dt.$$

The difficulty in the evaluation of this integral rests on the fact that V depends on q_k, which varies in time. One conclusion can be arrived at in the case where the domain of variation of q_k is limited, and where the motion of $E = \text{const.}$ satisfies the ergodic, or the quasi-ergodic, hypothesis, which implies that the hypersurface $H(p_k, q_k) = E$ is completely swept by the point in the finite time interval which we take to equal τ. If the speed of variation $\dot{\xi}$ of ξ is sufficiently slow so that it can be regarded as constant during this time interval, we can write:

$$\int_t^{t+\tau} \frac{\partial V(q_k, \xi)}{\partial \xi} \dot{\xi} dt = \dot{\xi}\tau \cdot \frac{1}{\tau} \int_t^{t+\tau} \frac{\partial V}{\partial \xi} (q_k, \xi) dt = \frac{\partial \overline{V}}{\partial \xi} \dot{\xi}\tau$$

by introducing the mean value of $\partial V/\partial \xi$ on the hypersurface $H = E$. This quantity no longer depends on ξ, and we get finally:

$$\Delta E = \sum_\tau \frac{\partial \overline{V}}{\partial \xi} \dot{\xi}\tau = \int_{t_0}^{t_1} \frac{\partial \overline{V}}{\partial \xi} \dot{\xi} dt = \int_{\xi_0}^{\xi_1} \frac{\partial \overline{V}}{\partial \xi} d\xi$$

and its differential form:

$$\frac{dE}{d\xi} = \frac{\partial \overline{V}}{\partial \xi}$$

which can be regarded as the classical form of the Hellmann-Feynman theorem.

Quantum Point of View

We will now consider a many-electron system governed by the state equation:

$$\mathcal{H}\Phi = E\Phi.$$

By hypothesis, the hamiltonian depends on a parameter ξ through the potential energy, with the result that the energy E and the wave function Φ are functions of ξ, which we will assume differentiable. We can write for each value of ξ:

$$\mathrm{E} = \int \Phi^* \mathcal{H} \Phi d\tau$$

the integration can be combined with a finite sum on the spin variables. We get by differentiation relative to ξ:

$$\frac{d\mathrm{E}}{d\xi} = \int \frac{\partial \Phi^*}{\partial \xi} \mathcal{H} \Phi d\tau + \int \Phi^* \frac{\partial \mathcal{H}}{\partial \xi} \Phi d\tau + \int \Phi^* \mathcal{H} \frac{\partial \Phi}{\partial \xi} d\tau.$$

The hermitian character of $\mathscr{H}$ allows us to write:

$$\int \Phi^* \mathcal{H} \frac{\partial \Phi}{\partial \xi} d\tau = \int \frac{\partial \Phi}{\partial \xi} \mathcal{H} \Phi^* d\tau.$$

By hypothesis, Φ is an eigenfunction of $\mathscr{H}$ corresponding to the real eigenvalue E, which allows us to write:

$$\mathcal{H}\Phi = \mathrm{E}\Phi; \qquad \mathcal{H}\Phi^* = \mathrm{E}\Phi^*$$

with the result that the preceding equation can be written:

$$\frac{d\mathrm{E}}{d\xi} = \int \Phi^* \frac{\partial \mathcal{H}}{\partial \xi} \Phi d\tau + \mathrm{E} \int \left(\Phi^* \frac{\partial \Phi}{\partial \xi} + \Phi \frac{\partial \Phi^*}{\partial \xi} \right) d\tau.$$

The coefficient of E is none other than $(\partial/\partial\xi) \int \Phi^* \Phi \, d\tau$ which is zero, since the state function Φ is normalized to a constant. Since, by hypothesis, the operator $\mathscr{H}$ depends on ξ only through the potential energy, we get the final expression:

$$\frac{d\mathrm{E}}{d\xi} = \int \frac{\partial \mathrm{V}}{\partial \xi} \Phi^* \Phi d\tau = \left\langle \frac{\partial \mathrm{V}}{\partial \xi} \right\rangle,$$

which involves the mean value of $\partial \mathrm{V}/\partial \xi$ in the configuration considered. This is the Hellmann-Feynman equation which is identical in form to the preceding classical equation.

Exercise. The Hellmann-Feynman equation is useful in the case of the one-dimensional potential well where we limit ourselves to the infinitely deep well. The variable parameter will be the width L of the well. The force exerted on one of the walls by the particle of mass m will be found equal to $\mathrm{F} = 2\mathrm{E}/\mathrm{L} = mv^2/\mathrm{L}$.

The same equation can again be gotten by purely classical considerations by repeating the usual calculation of the kinetic theory of gases (**Exercise 2**, §1) giving the mean force exerted by the particle on the walls.

C THE BORN-OPPENHEIMER APPROXIMATION

The Born-Oppenheimer Theorem 62

All of the studies of atoms and molecules made in the second and third parts of this book are looked at as uniquely electronic problems, the position of the nuclei being assumed fixed over the course of time. This generally adopted approximation is based on the Born-Oppenheimer theorem, which allows us to separate the motion of the electrons from that of the nuclei, the validity of this approximation being essentially tied to the fact that the mass of the nucleus is much larger than that of the electron. In the limit, the nucleus of infinite mass remains immobile, and the purely electronic treatment becomes rigorous. This fact is taken as the initial point in the Born-Oppenheimer proof, which involves the expansion of the complete equation of state on the increasing powers of m_0/M, m_0 being the electron mass and M the nuclear mass. The zero order of this expansion corresponds to a fixed nucleus [M. Born and J. R. Oppenheimer, *Ann. der Physik* (1929), **84**, 457]. We will not attempt a complete treatment here and we will follow the treatment of J. C. Slater, *Quantum Theory of Molecules* and *Solids*, Appendix II, p. 252.

In the first place the equation of state of a system of nuclei and electrons is written using the same principles as in the case of a purely electronic system (§43), starting with the classical expression for the energy of the system, and introducing the operators representing the components of the momentum of the different particles.

$$(31) \qquad \mathcal{H} = \sum_i -\frac{\hbar^2}{2M_i}\Delta_i + \sum_j -\frac{\hbar^2}{2m_0}\Delta_j + V(X_i, x_j)$$

the index i referring to the nucleus with coordinates X_i and the index j to the electrons with coordinates x_j.

The wave function solutions of $\mathcal{H}\Phi = \mathcal{E}\Phi$ are the functions $\Phi(X_i, x_j)$, for which we will examine the approximated form, derived from the eigenfunctions of the hamiltonian $\mathcal{H}^0$ gotten from $\mathcal{H}$ by suppressing the terms involving the nuclear kinetic energy, and which is then written:

$$\mathcal{H}^0 = \sum_j -\frac{\hbar^2}{2m_0}\Delta_j + V(X_i, x_j).$$

In this equation, the coordinates X_i of the nuclei remain in the potential energy expression, with the result that the eigenvalues

E and the wave functions Ψ depend on these coordinates, and can then be written $E(X_i)$ and $\Psi(X_i, x_j)$. Since the X_i play the role of parameters, the Hellmann-Feynman theorem is applicable, which involves the equation:

$$\frac{\partial E}{\partial X_i} = \left\langle \frac{\partial V}{\partial X_i} \right\rangle.$$

Now $\partial V/\partial X_i = -F_i$ represents one of the components of the force applied to nucleus i, with the result that $\langle \partial V/\partial X_i \rangle$ leads to the mean force F_i applied to this nucleus. The preceding equation can be written:

$$\overline{F}_i = -\frac{\partial E}{\partial X_i}$$

so that $E(X_i)$ represents the mean potential energy in the study of the dynamics of the nuclei. The corresponding wave functions $\Theta(X_i)$ are solutions of the equation:

$$\sum_i -\frac{\hbar^2}{2M_i} \Delta_i \Theta + E(X_i)\Theta = \mathcal{E}_0 \Theta. \tag{33}$$

With this, the Born-Oppenheimer theorem establishes that the product:

$$\Phi_0 = \Psi(X_i, x_j)\Theta(X_i)$$

in general forms a good approximation to the solution Φ of the complete equation, and $\mathscr{E}_0$ a good approximation to $\mathscr{E}$. An approximate justification of this can be gotten by studying the action of the operator $\mathscr{H}$ defined by (31) on the function $\Psi\Theta$. Taking into account the fact that $\Psi(X_i, x_j)$ is an eigenfunction of $\mathscr{H}^0$ corresponding to the eigenvalue $E(X_i)$, when Θ is a solution of (33), we get:

$$\mathscr{H}\Psi\Theta = \mathcal{E}_0 \Psi\Theta - \sum_i \frac{\hbar^2}{2M_i}\left(2\frac{\partial\Psi}{\partial X_i}\frac{\partial\Theta}{\partial X_i} + \Theta\frac{\partial^2\Psi}{\partial X_i^2}\right). \tag{34}$$

The validity of the Born-Oppenheimer theorem then depends on the importance of the term:

$$-\sum_i \frac{\hbar^2}{2M_i}\left(2\frac{\partial\Psi}{\partial X_i}\frac{\partial\Theta}{\partial X_i} + \Theta\frac{\partial^2\Psi}{\partial X_i^2}\right)$$

and the conditions under which it can be considered negligible. This discussion is tricky, and we limit ourselves to consideration of the mean value of this term defined by:

$$\int \Psi^* \Theta^* (\mathscr{H} - \mathcal{E}_0)\Psi\Theta \, d\tau_n d\tau_e = -\sum_i \frac{\hbar^2}{2M_i} \int \Psi^* \Theta^* \left(2\frac{\partial\Psi}{\partial X_i}\frac{\partial\Theta}{\partial X_i} + \Theta\frac{\partial^2\Psi}{\partial X_i^2}\right) d\tau_n d\tau_e \tag{35}$$

denoting by $d\tau_n d\tau_e$ the integration elements related respectively to the nuclear and electron coordinates. We limit ourselves to the case of a non-degenerate level defined by real wave functions. The first term of the sum to be studied is then written:

$$-\sum_i \frac{\hbar^2}{M_i} \int \Theta \frac{\partial\Theta}{\partial X_i} \left[\int \Psi \frac{\partial\Psi}{\partial X_i} d\tau_e \right] d\tau_n.$$

The integral between the brackets is written:

$$\int \Psi \frac{\partial\Psi}{\partial X_i} d\tau_e = \frac{1}{2} \frac{\partial}{\partial X_i} \int \Psi^2 d\tau_e = 0$$

since $\Psi(X_i, x_j)$ is normalized. There is no need to take this term into account in the evaluation of the mean energy. The second term is written:

$$-\sum_i \frac{\hbar^2}{2M_i} \int \Theta^2 \left[\int \Psi \frac{\partial^2\Psi}{\partial X_i^2} d\tau_e \right] d\tau_n.$$

An integration by parts used in the evaluation of the quantity between the brackets involves a definite integral, which is zero, and there remains:

$$-\int \left(\frac{\partial\Psi}{\partial X_i} \right)^2 d\tau_e.$$

The order of magnitude of the contribution of this term can be found by assuming that the function $\Psi(X_i, x_j)$ must remain invariant under a translation of the set, which implies:†

$$\sum_i \frac{\partial\Psi}{\partial X_i} + \sum_j \frac{\partial\Psi}{\partial x_j} = 0. \tag{36}$$

This equation leads us to assume without proof that the quantities $\partial\Psi/\partial X_i$ and $\partial\Psi/\partial X_j$ should be comparable to within an order of magnitude, which allows us to write:

$$\int \left(\frac{\partial\Psi}{\partial X_i} \right)^2 d\tau_e \sim \int \left(\frac{\partial\Psi}{\partial x_j} \right)^2 d\tau_e. \tag{37}$$

This last equation appears in the evaluation of the kinetic energy of the electrons:

$$\langle T_e \rangle = -\frac{\hbar^2}{2m_0} \sum \int \left(\frac{\partial\Psi}{\partial x_j} \right)^2 d\tau_e,$$

† The equation above involves the coordinates X_i and x_j of the various particles of the system along the same axis. The invariance of $\mathscr{H}$ under a translation a along this axis, implies $\Psi(X_i + a, x_j + a) = \Psi(X_i, x_j)$, whatever a may be. The expansion to within the second order leads to equation (36).

which according to the virial theorem, approximately represents the total energy $\mathscr{E}$ of the system.† Finally, we can assume that to within an order of magnitude each of the terms of the sum (35) is given by:

$$\mathcal{E}\frac{m_0}{M_i}\int \Theta^2 d\tau_n = \frac{m_0}{M_i}\mathcal{E}$$

which permits us to declare that the Born-Oppenheimer approximation is justified when the ratio m_0/M_i can be considered as negligible. Such assumptions are obviously very abstract and only involve non-degenerate levels. We set aside completely the question of the matrix elements of higher order and, as a consequence, everything about the influence of these terms on the transitions between different electron states.

† The virial equation has $\langle T \rangle = -\mathscr{E}$. In the total kinetic energy $\langle T \rangle$, the contribution of the nuclei is generally negligible relative to the total electron energy $\mathscr{E}$ which justifies $\langle T_e \rangle \sim -\mathscr{E}$.

APPENDIX I

THE THEORETICAL BASIS OF WAVE MECHANICS

The transition from the classical mechanical view of a point particle to the wave model follows from the variational form of the laws of mechanics.

Hamilton's Principle of Least Action

The motion of a point mass, m, moving in a field of force derived from a time-independent potential $V(x, y, z)$, is usually studied by integration of Newton's equation, which is written:

(1) $$m\frac{d^2x_r}{dt^2} = m\ddot{x}_r = X_r = -\frac{\partial V}{\partial x_r}, \quad \text{with} \quad r = 1,\ 2,\ 3.$$

The integration of this system of second order linear differential equations requires the evaluation of six arbitrary constants. Usually this is done by specifying the initial position and velocity of the particle. We may, however, make a second choice of the conditions by fixing the position $M_0(x_r^0)$ at time $t = 0$ and a final position $M_1(x_r^1)$ at time t_1. We can, in principle, find the solutions to (1) using these boundary conditions, written:

(2) $$x_r(t_0) = x_r^0\ ; \qquad x_r(t_1) = x_r^1.$$

to evaluate the constants.

With these boundary conditions the determination of the motion of the system becomes a problem in variational calculus. We define the *Lagrangian* as:

(3) $$L(\dot{x}_r, x_r) = T - V = \frac{1}{2}m(\dot{x}_1^2 + \dot{x}_2^2 + \dot{x}_3^2) - V(x_1, x_2, x_3)$$

where T is the kinetic energy of the particle. L should not be confused with the total energy E defined as $E = T + V$.

Knowing the Lagrangian allows us to calculate the definite integral:

(4) $$W = \int_{t_0}^{t_1} L\,dt.$$

known as the *Jacobi action integral.*

To find W we need to know the three functions $x_r(t)$. The Lagrangian $L(\dot{x}_r, x_r)$ thus becomes a function of time and W represents a definite integral taken along the trajectory of the particle between M_0 and M_1.

The calculation of the action W does not imply that the functions $x_r(t)$ correspond to a real motion of the particle, and may be accomplished by a random choice of these functions, constrained only by their being differentiable so as to provide a definite T. In particular, we can introduce a family of functions $x_r(t, \lambda)$ dependent on a parameter λ. By hypothesis, these curves constitute a family passing between M_0 and M_1 (fig. 76) and with the same values of t, t_0, and t_1, respectively. Thus we have:

(5) $$x_r(t_0, \lambda) = x_r^0; \qquad x_r(t_1, \lambda) = x_r^1, \quad \text{for any } \lambda.$$

The action integral can be calculated for any value of λ, and as a result we get W(λ) as a function of this parameter.

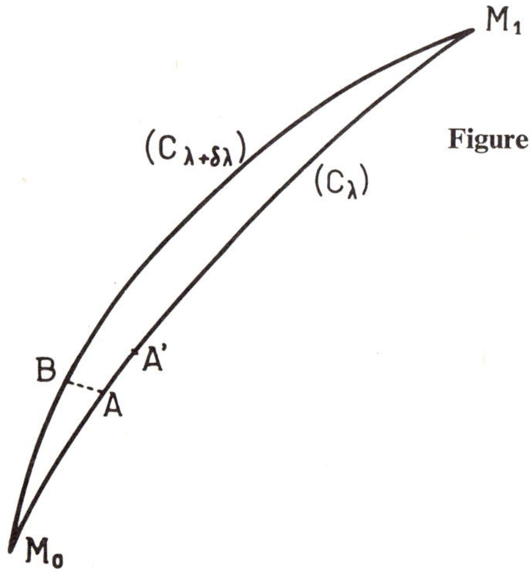

Figure 76. (C_λ) effective trajectory for the particle of mass m in going from M_0 at t_0 to M_1 at t_1; $(C_{\lambda+\delta\lambda})$ the varied trajectory with the same limits. $\overline{AA'}$ represents the displacement in time dt, $\overline{AB}$ the variation at constant t defined by δλ.

We can now state *Hamilton's principle*: *If for a certain value* $\lambda = \lambda_0$ *of the parameter, the functions* $x_r(t, \lambda_0)$ *define the real motion of the particle, that is to say, are solutions of* (1), *this value of* λ_0 *corresponds to an extremum of* $W(\lambda)$ *and*:

(6) $$\left(\frac{dW}{d\lambda}\right)_{\lambda_0} = 0.$$

This statement is equivalent to saying that, if for λ_0 we get the real trajectory of a particle going from M_0 at t_0 to M_1 at t_1, then for a value of λ close to λ_0 we get an infinitely close curve defining an artificial (non-real) motion *taking the same time interval.* We can state: *The real trajectory followed by a particle of mass m corresponds to an extremum of the action integral* (4), *that is, is invariant to within the second order when we cross to the real trajectory from an infinitely close curve requiring the same time interval.*

The equivalence of this principle with the dynamic equations (1) is shown through the use of the calculus of variations. We demonstrate this employing the following usage: the letter d in this case defines a displacement along the curve of integration; we use δ to denote a variation when we pass at fixed time to an infinitely close curve. If A is a point on the curve C_λ (fig. 76) corresponding to t, A′ the point on the same curve corresponding to $t + dt$, and B the point of $C_{\lambda+\delta 2}$ corresponding to the same value of t as A, we can write:

$$\text{(A')} \qquad x_r(t + dt, \lambda) = x_r(t, \lambda) + \frac{\partial x_r}{\partial t} dt = x_r(t, \lambda) + dx_r$$

$$\text{(B)} \qquad x_r(t, \lambda + \delta\lambda) = x_r(t, \lambda) + \frac{\partial x_r}{\partial \lambda} \delta\lambda = x_r(t, \lambda) + \delta x_r.$$

When C_λ is a real trajectory, $\partial x_r/\partial t$ is simply $\dot{x}_r$.

Using this notation we develop an explicit expression for W from (4), namely:

$$W(\lambda) = \int_{t_0}^{t_1} \Big[\sum_r \frac{1}{2} m\dot{x}_r^2 - V(x_r) \Big] dt.$$

Differentiation under the integral sign with respect to λ yields:

$$\text{(7)} \qquad \delta W = \frac{dW}{d\lambda} \delta\lambda = \int_{t_0}^{t_1} \Big[\sum_r m\dot{x}_r \frac{\partial \dot{x}_r}{\partial \lambda} - \frac{\partial V}{\partial \lambda} \Big] \delta\lambda \, dt.$$

The function V depends on λ only through the functions $x_r(t, \lambda)$ where,

$$\frac{\partial V}{\partial \lambda} \delta\lambda = \sum_r \frac{\partial V}{\partial x_r} \frac{\partial x_r}{\partial \lambda} \delta\lambda = \sum_r \frac{\partial V}{\partial x_r} \delta x_r.$$

The first part of the sum (7) can be transformed through use of the chain rule:

$$\frac{\partial \dot{x}_r}{\partial \lambda} \delta\lambda = \frac{\partial}{\partial \lambda} \left(\frac{\partial x_r}{\partial t} \right) \delta\lambda = \frac{\partial}{\partial t} \left(\frac{\partial x_r}{\partial \lambda} \right) \delta\lambda = \frac{\partial}{\partial t} (\delta x_r).$$

Each term of this first sum can be written:

$$\int_{t_0}^{t_1} m\dot{x}_r \frac{\partial \dot{x}_r}{\partial \lambda} \delta\lambda dt = \int_{t_0}^{t_1} m\dot{x}_r \frac{\partial}{\partial t} (\delta x_r) dt = \int_{t_0}^{t_1} m\dot{x}_r d(\delta x_r)$$

and, integrating by parts:

$$\int_{t_0}^{t_1} d(m\dot{x}_r \delta x_r) - \int_{t_0}^{t_1} \delta x_r m\ddot{x}_r dt.$$

The first of these integrals has an exact differential and (7) can be written in the form:

$$\text{(8)} \quad \delta W = \sum_r \big[m\dot{x}_r \delta x_r \big]_{t_0}^{t_1} - \sum_r \int_{t_0}^{t_1} \left(m\ddot{x}_r + \frac{\partial V}{\partial x_r} \right) \delta x_r dt, \quad \text{with } r = 1, 2, 3.$$

We have retained the assumption that the trajectory varies yet passes M_0 at $t = t_0$ and M_1 at $t = t_1$, which implies $\delta x_r = 0$ at the two limits. The value (8) of δW thus reduces to:

(9) $$\delta W = -\sum_r \int_{t_0}^{t_1} \left(m\ddot{x}_r + \frac{\partial V}{\partial x_r} \right) \delta x_r dt.$$

It is evident that if the dynamical equations (1) are satisfied by $x_r(t)$ at all points of the trajectory, δW is zero, whatever the system of variations δx_r may be. The inverse is more difficult to establish. The

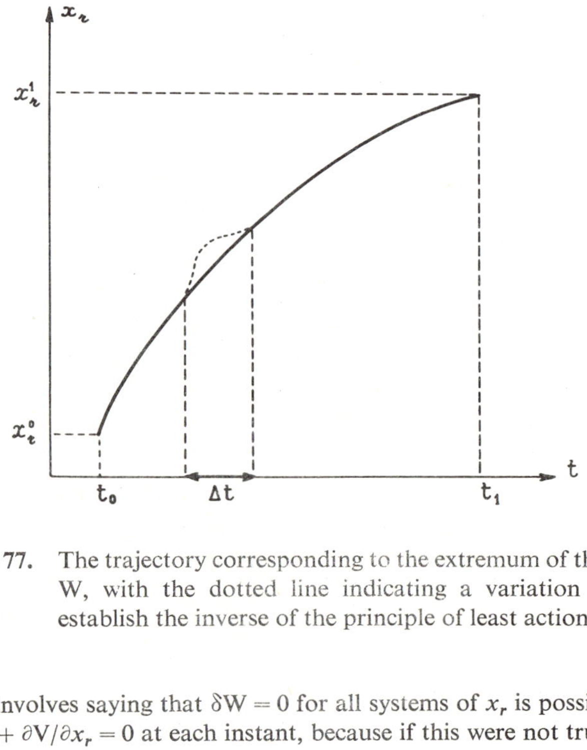

Figure 77. The trajectory corresponding to the extremum of the action W, with the dotted line indicating a variation used to establish the inverse of the principle of least action.

proof involves saying that $\delta W = 0$ for all systems of x_r is possible only if $m\ddot{x}_r + \partial V/\partial x_r = 0$ at each instant, because if this were not true under the continuity assumption, it would be possible to find an interval of time during which this sum would have a constant sign, and a judicious choice of δx_r taken equal to zero on the exterior of this interval and with fixed sign on the interior (fig. 77), which would allow δW to be zero, which is contrary to the hypothesis.

The preceding calculation was carried out in cartesian coordinates, but we are free to change to any coordinate system, and the Lagrangian function $L = T - V$ becomes a function of the generalized coordinates q_k and their derivatives with respect to time $\dot{q}_k$, and eventually time t. We will write the dynamical laws in a form to be established by the extremum of the Jacobi action W, but using the new variables. The variational calculation is similar to the preceding one. We begin with:

$$\mathrm{W} = \int_{t_0}^{t_1} \mathrm{L}(q_k, \dot{q}_k, t)dt.$$

The variation $\delta\mathrm{W}$ is calculated as before and written using the same notation:

$$\delta\mathrm{W} = \sum_k \int_{t_0}^{t_1} \left(\frac{\partial \mathrm{L}}{\partial q_k} \delta q_k + \frac{\partial \mathrm{L}}{\partial \dot{q}_k} \delta \dot{q}_k\right) dt.$$

The evaluation of $\delta\dot{q}_r$ is similar to that for $\delta\dot{x}_r$ and we get:

$$\delta\dot{q}_k = \frac{\partial}{\partial\lambda}\left(\frac{\partial q_k}{\partial t}\right)\delta\lambda = \frac{\partial}{\partial t}\left(\frac{\partial q_k}{\partial\lambda}\right)\delta\lambda = \frac{\partial}{\partial t}(\delta q_k)$$

so that:

$$\int_{t_0}^{t_1} \frac{\partial \mathrm{L}}{\partial \dot{q}_k} \frac{\partial}{\partial t}(\delta q_k) dt = \int_{t_0}^{t_1} \frac{\partial \mathrm{L}}{\partial \dot{q}_k} d(\delta q_k)$$

and integrating by parts:

$$\int_{t_0}^{t_1} \frac{\partial \mathrm{L}}{\partial \dot{q}_k} d(\delta q_k) = \int_{t_0}^{t_1} d\left(\frac{\partial \mathrm{L}}{\partial \dot{q}_k} \delta q_k\right) - \int_{t_0}^{t_1} \frac{d}{dt}\left(\frac{\partial \mathrm{L}}{\partial \dot{q}_k}\right) \delta q_k dt$$

and finally:

$$(8') \qquad \delta\mathrm{W} = \sum_k \left[\frac{\partial \mathrm{L}}{\partial \dot{q}_k} \delta q_k\right]_{t_0}^{t_1} + \sum_k \int_{t_0}^{t_1} \left[\frac{\partial \mathrm{L}}{\partial q_k} - \frac{d}{dt}\left(\frac{\partial \mathrm{L}}{\partial \dot{q}_k}\right)\right] \delta q_k dt$$

and the extremum condition on W with the same limits as before ($\delta q_k = 0$, for $t = t_0$ and $t = t_1$) implies:

$$\frac{d}{dt}\left(\frac{\partial \mathrm{L}}{\partial \dot{q}_k}\right) - \frac{\partial \mathrm{L}}{\partial q_k} = 0$$

for which $\delta\mathrm{W} = 0$ with an arbitrary choice of variations δq_k, which are by hypothesis the independent variables. The equations we get are called Lagrange's equations, and immediately give (1) when used with cartesian coordinates.

Properties of the Action Function

The action W is defined by the integral (4) relative to a motion of a particle of mass m from point $\mathrm{M}_0(x_r^0)$ at t_0 to $\mathrm{M}_1(x_r^1)$ at t_1. If we take M_0 and t_0 as fixed, the action can be thought of as a function of the initial conditions of the trajectory, and of the time at which some other point is reached. We write $\mathrm{W}(x_r, t)$ and use x_r and t to determine its value relative to the extremes of the trajectory, which are now considered as variables.

The differential of W as defined above is obtained by examining the two neighboring points M and M′ (fig. 78) reached respectively at t and $t' = t + dt$ on the real trajectories (C) and (C′). We put:

$$dW = \int_{t_0(C')}^{t'} L dt - \int_{t_0(C)}^{t} L dt.$$

Equation (8) can be used to calculate the variation of W when we alter the trajectory, but under the condition that the ends of the path are reached simultaneously on both curves. We introduce the point M_1' of

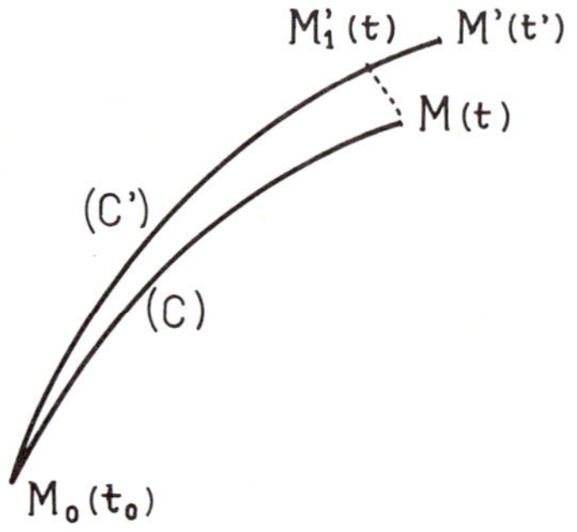

Figure 78. Trajectories C and C′ corresponding to the same initial conditions used to define the action function W(M, t) and its differential.

(C′) reached at the same time t as M is reached. We can then write dW in the form:

$$dW = \left(\int_{t_0(C')}^{t} L dt - \int_{t_0(C)}^{t} L dt\right) + \int_{t(C')}^{t+dt} L dt.$$

The difference term (between parentheses) can be calculated by use of (8) and gives:

$$\int_{t_0(C')}^{t} L dt - \int_{t_0(C)}^{t} L dt = \sum_r m\dot{x}_r \delta x_r \Big|_{t_0}^{t} - \sum_r \int_{t_0(C)}^{t} \left(m\ddot{x}_r + \frac{\partial V}{\partial x_r}\right) \delta x_r dt.$$

The integral on the right-hand side is zero since (C) is by hypothesis a real trajectory. In addition, (C) and (C′) start from the same point M_0 at t_0, with the result that the equation reduces to:

$$\int_{t_0(C')}^{t} L dt - \int_{t_0(C)}^{t} L dt = \sum_r m\dot{x}_r \delta x_r.$$

The components δx_r of $\overrightarrow{MM_1'}$ are inferred from the dx_r of $\overrightarrow{MM'}$ by noting that the passage from M_1' to M′ corresponds to motion of the particle on a real trajectory during the time $t' - t = dt$, where:

$$dx_r = \delta x_r + \dot{x}_r dt.$$

The expression for dW yields, in addition, an integral over the interval dt, and we now get:

$$dW = \sum_r m\dot{x}_r(dx_r - \dot{x}_r dt) + L dt.$$

This gives explicitly:

$$\sum_r m\dot{x}_r^2 = 2\mathrm{T},$$

and we have:†

(10) $$dW = \sum_r m\dot{x}_r dx_r - (2\mathrm{T} - \mathrm{L})dt = \sum_r m\dot{x}_r dx_r - \mathrm{E}dt,$$

$\mathrm{E} = 2\mathrm{T} - \mathrm{L} = \mathrm{T} + \mathrm{V}$ gives the total energy of the particle.

Equation (10) shows that knowledge of the Jacobi action function:

$$\mathrm{W}(x_1, x_2, x_3, t),$$

allows us to find the speed and energy of the particle which characterizes the motion of the particle to $\mathrm{M}(x_r)$ at t under the initial conditions of starting from M_0 at t_0:

(11) $$p_r = m\dot{x}_r = \frac{\partial \mathrm{W}}{\partial x_r}; \qquad \mathrm{E} = -\frac{\partial \mathrm{W}}{\partial t}.$$

In other words, this action function can be thought of as a characteristic function of the mechanical system with variables x_r and t in the same sense as a thermodynamic state function.

But, as in the thermodynamic case, other variables can be used, and instead of determining the time of the motion from M_0 to M_1 we can determine the trajectory from M_0 to M_1 by means of the energy E. Use of E takes advantage of the energy invariance at all points on the trajectory and is the more common way of treating the problem.‡

The construction of the characteristic function corresponding to this new choice of variables is carried out by means of a typical thermodynamic method, the introduction of a new function:

(12) $$\mathrm{S} = \mathrm{W} + \mathrm{E}t,$$

whose differential is written, using the result (10):

(13) $$d\mathrm{S} = d\mathrm{W} + \mathrm{E}dt + td\mathrm{E} = \sum_r m\dot{x}_r dx_r + td\mathrm{E}.$$

---

† The calculation in generalized coordinates q_k evolves in the same way from (8′) by taking $p_k = \partial \mathrm{L}/\partial \dot{q}_k$ we get:

(10′) $$d\mathrm{W} = \sum_k p_k dq_k - \mathrm{H}dt, \quad \text{with} \quad \mathrm{H} = \sum_k p_k \dot{q}_k - \mathrm{L}.$$

The function H is equal to the total energy when the kinetic energy T is a homogeneous quadratic function of q_k, that is, when V is time independent (conservative system). The function H defined by (10′) is known as Hamilton's function.

‡ Thus in studying the motion of projectiles, the classical problem involves the motion to a known point along a trajectory starting from some origin with known initial velocity (and therefore known energy).

Knowledge of the function $S(x_r, E)$ allows us to get the differential equations of the path, and thus the time of flight, by:

(14) $$m\dot{x}_r = \frac{\partial S}{\partial x_r}; \qquad t = \frac{\partial S}{\partial E}.$$

The function S is known as the Maupertuis action function. Equation (12) shows that for $t = t_0$ with $W = 0$, we must have $S_0 = Et_0$. Because of this, it is preferable to replace Et by $E(t - t_0)$ in (12). To simplify the writing, we retain (12) and take the origin in time as the instant that the object leaves M_0.

It is necessary to note that the independent variables here are x_r and E, and that by contrast, t represents a function of these variables defined as follows: we consider the trajectory of known energy E passing through the coordinates x_r, which are well defined (in most cases we select them), while t, representing the time at which the other point is reached, is calculated from the integral on the trajectory:

(15) $$t = \int_{M_0}^{M_1} \frac{1}{v}\, ds = \int_{M_0}^{M} \left[\frac{m}{2(E - V)}\right]^{1/2} ds.$$

Properties of the Maupertuis Action Function. This function has properties analogous to those of W. Like W, it can be expressed as an integral along the trajectory, as we can see immediately from the definition (12) which, from the definition of W, gives (the index 1 for the end-point is introduced for clarity):

(16) $$S = \int_{M_0}^{M_1} L\,dt + Et = \int_{M_0}^{M_1} (L + E)\,dt = \int_{M_0}^{M_1} 2T\,dt.$$

The last integral takes the form of a line integral by expressing the kinetic energy $2T = mv^2$, and by introducing the line element of the trajectory $ds = v\,dt$, so that:

(17) $$S = \int_{M_0}^{M_1} mv\, ds.$$

At each point the speed is fixed for a given total energy so that the integral can be rewritten:

(17′) $$S = \int_{M_0}^{M_1} \sqrt{2m(E - V)}\, ds.$$

This expression is time independent and the integral may be evaluated on any curve joining M_0 and M_1 and not just on the real trajectory. However, it is possible to state a theorem which fixes the complete parallel between the Maupertuis and the Jacobi actions: The real motion of a particle in a field of force derived from a potential $V(x_r)$ corresponds to the extremum of the action S for any trajectory with the same end-points and the same energy as the real trajectory.

We note that we are dealing here with a variation at constant energy, and not constant time interval as in the case of the Jacobi action

principle. This may be demonstrated without recourse to the variational calculation. Consider the following trajectories (fig. 79):
(a) The real trajectory (C) of energy E, starting from M_0 at $t = 0$ and arriving at M' at t' (calculated using equation 15).
(b) A varied trajectory (C′) with the same end-points and the same energy as the real trajectory. The integral (15) calculated on the new path leads to a different time of transit $t_1 + \delta t_1$.
(c) The real trajectory $(C)_0'$ corresponding to a time of transit $t_1 + \delta t_1$, gotten between the same end-points M_0 and M_1.

We now consider the action integral W calculated on these three curves between the same end-points. Hamilton's principle is applicable during

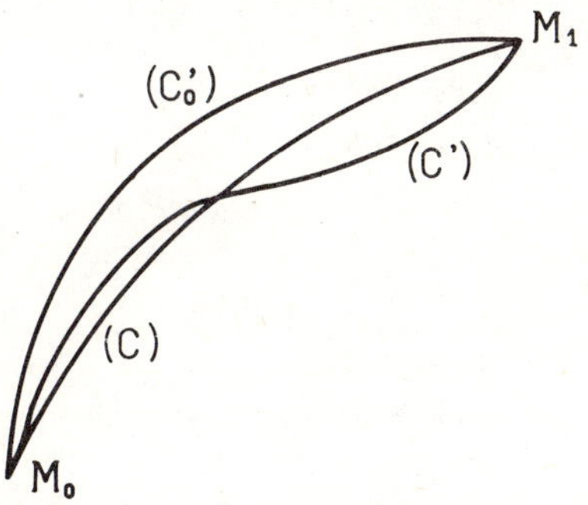

Figure 79. Diagram of the trajectories used to pass from the Jacobi action to the Maupertuis action: (C), the real trajectory of energy E and transit time t_1; (C′), the varied trajectory of the same energy and transit time $t_1 + \delta t_1$; (C_0'), the real trajectory corresponding to the transit time $t_1 + \delta t_1$.

passage to the real trajectory (C_0') from the varied trajectory (C′) since they correspond by hypothesis to the same time interval $t_1 + \delta t_1$, where:

$$\delta_1 W = W(C') - W(C_0') = 0.$$

On the other hand, equation (10) is applicable during passage from the real trajectory (C) to the real trajectory (C_0'). These two trajectories have common end-points and the relationship reduces to:

$$\delta_2 W = W(C_0') - W(C) = -E\delta t_1.$$

Combining these two we see that:

$$\delta W = \delta_1 W + \delta_2 W = W(C') - W(C) = -E\delta t_1.$$

Now the variation takes place at constant energy, and differentiation of (12) for S yields:

$$\delta S = \delta W + E\delta t_1 = 0$$

which demonstrates the required property that S is invariant at constant energy.

The Basis for the Analogy between Mechanics and Optics

We begin by stating that the title for this section should probably read: analogies between the laws of particle dynamics and wave propagation. Optics is referred to only because it is a particularly familiar example of wave propagation.

First method. We begin with the fact that the laws of particle mechanics, as well as those of geometric optics, can be expressed by means of a variational principle. For optics, the Descartes' laws of refraction are equivalent to Fermat's principle (see *Exercise 2*) which states: the trajectory of a ray in a locally isotropic medium corresponds to the extremum of the optical length of the curve joining the end-points M_0 and M_1 of this ray. The optical length is defined by:

$$L = \int_{M_0}^{M_1} n(x_r)ds$$

$n(x_r)$ representing the index of refraction which varies from point to point in a non-homogeneous medium. The fact that the index is well defined at each point implies that the ray is monochromatic.

Time does not appear in this integral, which suggests a comparison with the Maupertuis action defined by (17′) as:

$$S = \int_{M_0}^{M_1} \sqrt{2m(E - V)}\, ds.$$

We now arrive at a rigorous statement, which simply says that two different physical phenomena have the same mathematical formalism: the trajectories of a particle of mass m and energy E, in a field derived from a time-independent potential $V(x_r)$ are identical to monochromatic light waves in a medium where the index is given at each point by:

$$n(x_r) = k\sqrt{2m(E - V)} \tag{18}$$

where k is a proportionality constant for the system.

An equivalent form of this correspondence is obtained by introducing variables characteristic of each point, for the particle the velocity v, and for the ray the velocity of propagation u or the wavelength λ, related to the index by:

$$u = c/n\,; \qquad \lambda = \lambda_0/n,$$

c and λ_0 represent the velocity and wavelength in vacuum. The two variational principles are written by introducing these variables:

Mechanics: $$\delta\int_{M_0}^{M_1} mv\,ds = 0.$$

Optics: $$\delta\int_{M_0}^{M_1} \frac{1}{u}\,ds = 0 \quad \text{or} \quad \delta\int_{M_0}^{M_1} \frac{1}{\lambda}\,ds = 0.$$

These two equations have the same form when the wavelength λ at each point is inversely proportional to the momentum $p = mv$ of the particle. The proportionality factor has action units, and we are therefore lead to the de Broglie relation:

(19) $$\lambda = \frac{h}{p}.$$

The preceding arguments are rigorous and suggest a common formalism for mechanics and optics. Louis de Broglie has suggested that this correspondence is not fortuitous but reflects a real physical situation, namely, that a system of waves can be associated with a particle. The difficult problem is to fix the precise nature of the correspondence, the basic idea of relating the phase velocity u and the speed of the particle v appearing impossible. De Broglie has solved this problem by introducing the following particularly fruitful idea: the correspondence between the wave propagation and the motion of the particle implies a special localization of the wave and thus a "wave packet," which leads us to examine not the phase velocity u, but rather, the group velocity v_g, derived from the Rayleigh relation (1-28), namely:

$$v_g = \frac{d\nu}{d(1/\lambda)}.$$

The use of this equation requires introduction of the frequency ν of the wave, which does not arise when we limit ourselves to the correspondence with the principles of Fermat and Maupertuis. The reconciliation of monochromatic rays of frequency ν, and trajectories of constant energy E, implies the existence of a correspondence equation between these two quantities. It is possible to derive this relationship from de Broglie's concept, by assuming that the group velocity v_g of the associated wave is equal to the velocity v of the particle. In this case, the Rayleigh equation is used to derive the wavelength expression (19) of de Broglie:

$$v = \frac{hd\nu}{mdv} \qquad \text{so that} \qquad mv\frac{\partial v}{\partial \nu} = h,$$

the partial derivatives are used here to emphasize that the Rayleigh relationship, generally written for a homogeneous substance, is also applicable in the more general case where the group velocity depends not only on the frequency but also on the position. Upon integration this gives:

$$h\nu = \tfrac{1}{2}mv^2 + f(x_r).$$

with an arbitrary function of the coordinates. This equation can be compared to the energy expression:

$$\mathrm{E} = \frac{1}{2}mv^2 + \mathrm{V}(x_r).$$

Here we have a precise correspondence between ν and E by comparing these two equations, which leads to the expression:

(20) $$\mathrm{E} = h\nu$$

that is, the Planck relation. Additionally, $f(x_r)$, determined by the integration, involves an arbitrary constant which refers to the definition of ν, and which the Planck condition arbitrarily associates with the definition of energy.

Second method. We now introduce a more systematic argument, starting from the form of the Jacobi action function, which we write explicitly as:

$$W(x_r, t) = S(x_r, E) - Et$$

t being the time at which the particle of energy E reaches the point x_r. The variables E and t are not independent in this equation. To generalize and to move toward the wave interpretation it is profitable to consider the same equation, but with E and t in this case being independent variables for which we have a function $W(x_r, t, E)$ rather than $W(x_r, t)$. Moreover, t loses its former significance and no longer designates the time at which the particle with energy E reaches the point x_r. Also, to eliminate confusion, and to show that the only invariant function in the two equations is the Maupertuis action function, we now write the relation in the form:

$$\Phi(x_r, \tau) = S(x_r, E) - E\tau$$

where Φ depends on the parameter E. For a fixed value of E, we examine the system of surfaces $S(x_r, E) = \text{const.}$, where this constant takes the increasing values $S_0, S_1, \ldots, S_r$. At arbitrary time τ_0, the function $\Phi(x_r, \tau_0)$, as defined, has a constant value at each point of the surface $S(x_r, E) = S_0$, so that:

$$\Phi(x_r, \tau_0) = S_0 - E\tau_0.$$

At τ_1 we find:

$$\Phi(x_r, \tau_1) = S(x_r, E) - E\tau_1,$$

and the value of the function Φ will again be the same if we have:

$$S(x_r, E) = S_0 + E(\tau_1 - \tau_0) = S_1.$$

Thus, in the course of time, the surface $\Phi(x_r, \tau) = \text{const.}$ undergoes displacement, and will occupy successively the different surfaces of the system $S(x_r, E) = S_0, S_1, \ldots, S_k$. This is exactly the characteristic of a wave variable. It is necessary to note carefully that in this representation the different points of $S = \text{const.}$ are reached by the wave for the same value of the variable, even if the particle of energy E does not arrive at the same time, t, at different points on this surface.

We now see the possibility of associating a wave representation with a system of surfaces of constant action introduced via mechanics. An important point in this correspondence is found by examining the differential expression (13), which for constant E leads to a displacement on $S = \text{const.}$ of:

$$dS = \sum_r m\dot{x}_r dx_r = \sum_r p_r dx_r = 0$$

This shows that the vector with components p_r tangent to the trajectory is normal to the surface of constant action. The surfaces S = constant are therefore orthogonal to the constant energy trajectories, which is a fundamental property of wave surfaces relative to rays, which here correspond exactly to mechanical trajectories (fig. 80).

The definitions that we will give, will allow the determination of the phase velocity at each point by examining the propagation for a differential interval $d\tau$, which yields:

$$d\Phi = dS - E d\tau = 0.$$

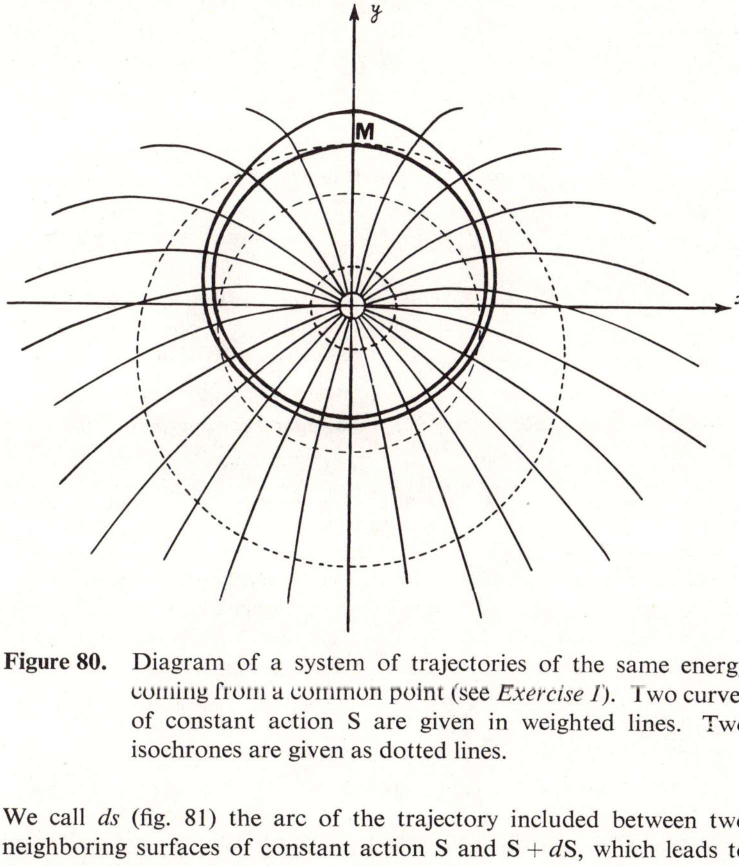

Figure 80. Diagram of a system of trajectories of the same energy coming from a common point (see *Exercise 1*). Two curves of constant action S are given in weighted lines. Two isochrones are given as dotted lines.

We call ds (fig. 81) the arc of the trajectory included between two neighboring surfaces of constant action S and S + dS, which leads to the desired phase velocity:

$$u = \frac{ds}{d\tau}$$

or with:

$$dS = \overrightarrow{\text{grad}}\, S . \overrightarrow{ds} = |\, \text{grad } S \,|\, ds$$

(21) $$u = \frac{E}{|\text{grad } S|} = \frac{E}{mv}.$$

Thus it is just a matter of another interpretation of a system of equations of mechanics. The essential postulate is that this interpretation corresponds to a physically real situation in which we will identify the value of the group velocity calculated from the Rayleigh relation with the speed of the particle. We have here $1/\lambda = \nu/u = mv\nu/E$, with the result that the Rayleigh relation can be written as:

$$v\frac{\partial(1/\lambda)}{\partial\nu} = v\frac{\partial}{\partial\nu}\left(\frac{mv\nu}{E}\right) = 1.$$

To integrate, we note that the desired correspondence requires a relation $E(\nu)$ so that we can express $\partial v/\partial\nu$ from the energy $E = \frac{1}{2}mv^2 + V$ which yields:

$$mv\frac{\partial v}{\partial\nu} = \frac{dE}{d\nu}.$$

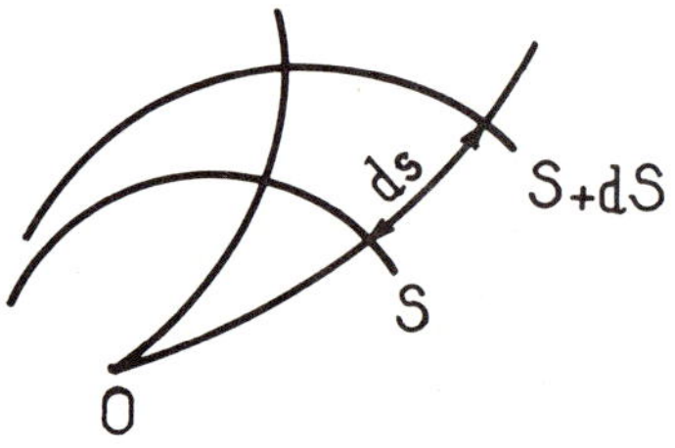

Figure 81. The relative disposition of two curves of constant action S and S + dS on two trajectories coming from a common point.

On expansion of the Rayleigh relation, and substituting from above:

$$(E - mv^2)\left(1 - \frac{\nu}{E}\frac{dE}{d\nu}\right) = 0$$

which implies $dE/E = d\nu/\nu$, thus a relationship of the form $E = h\nu$ which with (21) leads to the construction of the de Broglie relation. We can note that an analogous calculation can be developed by taking with (21) the relativistic form of the energy $E = mc^2$, which leads to the relation $u = c^2/v$ used by de Broglie. The Rayleigh relation is then used as above.

This result can be gotten in yet another way by noting that the differential of the function Φ, namely:

$$d\Phi = \vec{p}.\vec{ds} - E d\tau$$

has the same structure as the phase variation $d\varphi$ of a wave during variations with time and space so that:

$$\frac{d\varphi}{2\pi} = \vec{k}\cdot\vec{ds} - \nu\, d\tau$$

where $\vec{k}$ is the propagation vector. The comparison requires the introduction of a coefficient having the dimensions of action and again gives the two relations:

(22) $$\vec{p} = h\vec{k}; \qquad E = h\nu.$$

Exercises:

(*1*) Consider the two-dimensional motion of a point particle, mass *m*, subject to a constant force directed along the line of descent.

Find the expressions for x and y as a function of time by taking the particle initially at the origin. We consider trajectories of the same energy and thus the same initial velocity. By taking as a parameter the angle θ of the initial velocity with the horizontal axis x, we will have, with the conventional values of the variables defining the system:

$$x = t \cos \theta ; \qquad y = t \sin \theta - \frac{1}{20} t^2.$$

Graph a group of trajectories by indicating the isochronic curves. Calculate the action S as a function of t and θ, which will allow the tracing of the curves of constant action. Note that these curves are the orthogonal trajectories of the preceding group of trajectories and that they distinguish the isochrones (see fig. 80).

(2) Fermat's principle. The laws of propagation of a light ray in a medium of index n, which varies from point to point in the medium, can be gotten from the extremum of the optical length of the path, so that:

$$\mathscr{L} = \int_{M_0}^{M_1} n ds = \int_0^{t_1} n(x, y, z)\sqrt{\dot{x}^2 + \dot{y}^2 + \dot{z}^2}\, dt,$$

the latter expression being derived by introducing a parametric representation of the trajectory. We get the extremum conditions by using the Lagrange equations (9) and can show that the result can be written $d(n\vec{k}) = \text{grad}\, n \cdot \vec{ds}$, where $\vec{k}$ is a unit tangent to the ray. Verify that this relation gives the Descartes laws of the ray path in a zone defined by equi-index surfaces n and $n + dn$, with $dn = \text{grad}\, n \cdot \vec{ds}$.

(3) Obtain the dynamical equations starting from the extremum conditions of the Maupertuis action. The method of calculation will follow the preceding exercise. The only difference is in the use of the results, since we are interested not only in the trajectory, but also in the motion. The differential equations of the motion can be found by projecting on the trajectory and on the principal normal to the curve.

(4) The motion of a particle with charge e and mass m in an electric field $\vec{E}$ and magnetic field $\vec{B}$ can be studied by introducing a suitable Lagrange function. The calculation will be done in cartesian coordinates.

(*a*) We will start with the dynamical equations in the classical form:

$$\vec{F} = e(\vec{E} + \vec{v} \wedge \vec{B}) = m\vec{\gamma}$$

and will express the equations for each of the three coordinate axes. The result can be put into the form of the Lagrange equations (9) under the condition that the function L, containing the components $\dot{x}$, $\dot{y}$, $\dot{z}$ of the velocity, is a first-order term. Test the form:

$$L = T + e(-V + a\dot{x} + b\dot{y} + c\dot{z}),$$

where V is the electrostatic potential ($\vec{E} = -\text{grad}\,V$) and a, b, and c are functions of x, y, and z. Show that we can get the dynamical equations when the functions a, b, and c, are the components of a vector $\vec{A}$ such that $\vec{B} = \text{curl}\,\vec{A}$, which is, therefore, a vector potential.
(*b*) Calculate the momentum components (one finds $p = mv + eA$). Calculate the function H as a function of p. We find:

$$H = \sum p_k \dot{q}_k - L = \frac{1}{2m} | \vec{p} - e\vec{A} |^2 + eV.$$

(*c*) Form the Schrödinger equation for the system by replacing p_x by the operator $-ih(\partial/\partial x)$ and similarly for the other momentum components. This gives:

$$-\frac{\hbar^2}{2m}\Delta\varphi - \frac{e\hbar}{2im}(\text{div}\,\vec{A}\varphi + \vec{A}.\overrightarrow{\text{grad}}\,\varphi) + \left(\frac{e^2}{2m}A^2 + eV\right)\varphi = E_0\varphi.$$

Examine the particular case of a uniform magnetic field along the direction taken as the z axis. We can take:

$$A_x = -\frac{1}{2}By, \quad A_y = \frac{1}{2}Bx, \quad A_z = 0.$$

(5) Calculate the path of a light ray in a medium of continuously variable (with position) index of refraction n, by using the fact that the trajectory of the ray is identical to that of a particle of mass m in a field of force derived from a potential V, such that $n = [2m(E - V)]^{1/2}$. This requires only representation in optical terms of the classical mechanical expression $F_N = mv^2/R$.

APPENDIX II

ORTHOGONAL FUNCTIONS

The idea of orthogonal functions arises in the study of square summable functions, that is in functions such that we can define the integral:

$$\text{(1)} \qquad L^2 = \int_{-\infty}^{+\infty} \psi^\star \psi dx,$$

limited here to the case of functions ψ of a single real variable x defined on the domain $-\infty$ to $+\infty$. L^2 is a positive quantity and $L^2 = 0$ only when ψ is identically zero. We note the analogy between this definition and that of the length of a vector which can only be equal to zero when all the vector components are equal to zero.

We next consider a function ψ, the sum of two square summable functions ψ_1 and ψ_2. Definition (1) leads us to consider the integral:

$$L^2 = \int_{-\infty}^{+\infty} \psi^\star \psi dx = L_1^2 + L_2^2 + \int_{-\infty}^{+\infty} \psi_1^\star \psi_2 dx + \int_{-\infty}^{+\infty} \psi_2^\star \psi_1 dx,$$

L_1^2 and L_2^2 being determined by the two functions under consideration. This relation can be compared to that which gives the squared modulus of a vector sum of two vectors, and which requires their scalar product —the scalar product being zero—when the vectors are orthogonal. By analogy, the functions ψ_1 and ψ_2 are called orthogonal if they satisfy the condition:

$$\text{(2)} \qquad \int_{-\infty}^{+\infty} \psi_1^\star \psi_2 dx = \int_{-\infty}^{+\infty} \psi_2^\star \psi_1 dx = 0.$$

A set of normalized functions ψ_r, that is, functions such that in each case the integral (1) is equal to 1, and such that they are orthogonal to each other, constitutes an orthonormal set of functions.

Given such an orthonormal set, we can try to represent a function $f(x)$ by means of the series:

$$\text{(3)} \qquad f = c_1\psi_1 + c_2\psi_2 + \dots .$$

The coefficients c_r can be found immediately on the basis of the orthonormality, which we usually write with the Kronecker δ_{rs} as:

$$\int_{-\infty}^{+\infty} \psi_r^\star \psi_s dx = \delta_{rs}, \quad \text{with} \quad \begin{aligned} \delta_{rs} &= 1, \ \text{if} \ r = s \\ \delta_{rs} &= 0, \ \text{if} \ r \neq s. \end{aligned}$$

Thus we get:

$$c_r = \int_{-\infty}^{+\infty} \psi_r^\star f dx.$$

There is no *a priori* assurance that a function $f(x)$ can be represented by a series of the form (3).

In order for this to be true, it is necessary that the series $\sum_r c_r \psi_r$ converge, and that its sum be equal to f, and thus certain conditions on f and on the set ψ_r must be met. The function f should be square summable which implies:

$$\int_{-\infty}^{+\infty} f^\star f dx = \sum_r c_r^\star c_r .$$

In addition, it must satisfy certain conditions at the limit, as the following examples will demonstrate. When the set of functions allows the expansion of any f satisfying the above requirements, it is called a complete orthonormal set, and this property for the set can only be established through a detailed study of the set.

Finally, the functions of an orthonormal system are of necessity linearly independent, since a relation of the form:

$$c_1\psi_1 + c_2\psi_2 + \ldots + c_r\psi_r = 0$$

followed by multiplication by $\psi_1^\star$ and integration gives:

$$c_1 \int_{-\infty}^{+\infty} \psi_1^\star \psi_1 dx + c_2 \int_{-\infty}^{+\infty} \psi_1^\star \psi_2 dx + \ldots = c_1 = 0$$

and similarly for the other coefficients.

The eigenfunctions of a hermitian operator constitute an orthonormal set, and this shows their importance in quantum theory. We cite the following examples:

(*a*) The eigenfunctions of the energy in the case of the one-dimensional well (§9). These functions written for an infinitely deep well give:

Symmetric function:

$$(\psi_k)_s = \mathrm{C} \cos \frac{(2k+1)\pi x}{\mathrm{L}} .$$

Antisymmetric function:

$$(\psi_k)_a = \mathrm{C} \sin \frac{2k\pi x}{\mathrm{L}}$$

with

$$\mathrm{C} = \sqrt{\frac{2}{\mathrm{L}}}; \quad -\mathrm{L}/2 \leqslant x \leqslant +\mathrm{L}/2.$$

The system is composed of two families of functions, so that it seems better to take the origin not at the center of the well but at one of the edges. This is accomplished through the change in variables:

$$\xi = x + \frac{\mathrm{L}}{2}; \ (0 \leqslant \xi \leqslant \mathrm{L}).$$

Two types of function appear, ignoring a factor eventually seen to be -1:

$$(\varphi_k)_s = \mathrm{C} \sin \frac{(2k+1)\pi\xi}{\mathrm{L}}; \qquad (\varphi_k)_a = \mathrm{C} \sin \frac{2k\pi\xi}{\mathrm{L}}$$

these take the form:

$$\varphi_k = C \sin \frac{k\pi\xi}{L},$$

where k is any positive integer.

For a function f to be represented in this domain by a series of the form (3), it is not only necessary that it be square summable but that it disappear at the edges of the well. This is directly analogous to the case of the vibrating string with fixed ends.

(*b*) The eigenfunctions of the planar rotor (§15). The functions of the angle χ are defined in the interval 0–2π and are given by:

$$\psi_k = Ce^{ik\chi},$$

with k a positive or negative integer or zero and:

$$C = \sqrt{\frac{1}{2\pi}}.$$

This is the same as the expansion on the series of positive integers k, if we take for each of these functions either $\cos k\chi$ or $\sin k\chi$, which is the basis of the Fourier series, and which allows the expansion of a periodic function f which is square summable.

(*c*) The spherical functions $Y_l^m(\theta, \chi)$ which are the eigenfunctions of the rigid rotor (§16), and which form an orthonormal set on the unit sphere. As a result, it is possible to represent a function defined at each point on this sphere.

APPENDIX III
LAGRANGE MULTIPLIERS

In a number of problems in quantum mechanics we are lead to deal with the extremum of the energy E along with the normalization condition. We begin with the two corresponding expressions:

(1) $$E = F(c_1, c_2, \ldots, c_n)$$

(2) $$f(c_1, c_2, \ldots, c_n) = 1.$$

$c_1, c_2, \ldots, c_n$ designate the parameters on which the state function of the system depends.

The extremum condition on the energy is given by the following expression, which we write using the abbreviation $\partial F/\partial c_r = F_r$:

(3) $$dE = F_1 dc_1 + \ldots + F_n dc_n = 0.$$

If the variables c_r are independent, the variations dc_r can be arbitrary, with the result that the preceding condition implies $F_1 = F_2 \ldots = F_n = 0$. This condition is impossible, since equation (2) must be continuously satisfied, which implies the following relationship between the differentials:

(4) $$f_1 dc_1 + \ldots + f_n dc_n = 0.$$

We can replace the two simultaneous conditions (3) and (4) by two equivalent conditions, one of which is (4), and the other of which is taken as a linear combination of these two relations, so that:

(5) $$(F_1 - \lambda f_1)dc_1 + \ldots + (F_n - \lambda f_n)dc_n = 0$$

λ designates an arbitrary function of c_r. One can select λ by imposing the condition:

$$F_1 - \lambda f_1 = 0$$

with the result that condition (5) is written:

$$(F_2 - \lambda f_2)dc_2 + \ldots + (F_n - \lambda f_n)dc_n = 0,$$

and only makes use of $(n - 1)$ differentials, which may, as a result, be chosen arbitrarily. Thus the extremum condition suggests that the c_r's should be such that we can take:

$$F_2 - \lambda f_2 = \ldots = F_n - \lambda f_n = 0.$$

In conclusion, the extremum of F can be found by using a quantity λ, called the Lagrange multiplier, such that the n relations:

(7) $$F_r = \lambda f_r \quad \text{or} \quad \frac{\partial F}{\partial c_r} = \lambda \frac{\partial f}{\partial c_r}$$

are simultaneously satisfied.

This method is frequently used in molecular orbital studies (Chapter 8). The functions F and f are two real, homogeneous, quadratic functions of the coefficients c_r of the orbital. Euler's relation allows us to write:

$$2F = 2E = \sum_r c_r F_r\,; \qquad 2f = 2 = \sum_r c_r f_r$$

with the result that equation (7) implies:

$$E = \lambda$$

which gives a direct significance to the Lagrange multiplier λ, which, in this case, represents the energy E of the system under study. This parameter is calculated by using the fact that the n conditions (7) take the form of linear homogeneous equations in c_r, which can only give solutions not all zero in c_r if their determinant is zero. This gives the desired equation in λ where the degree is equal to the number n of parameters c_r.

APPENDIX IV
SYSTEMS OF ORTHOGONAL CURVILINEAR COORDINATES

Definition 1

We will examine a change in coordinates which will allow us to express the cartesian coordinates x, y, and z of a point M as a function of three coordinates q_1, q_2, and q_3. An infinitely small displacement $\overrightarrow{MM'} = d\vec{r}$ characterized by dq_1, dq_2, and dq_3, has as its components:

$$dx = \frac{\partial x}{\partial q_1} dq_1 + \frac{\partial x}{\partial q_2} dq_2 + \frac{\partial x}{\partial q_3} dq_3$$

(1) $$dy = \frac{\partial y}{\partial q_1} dq_1 + \frac{\partial y}{\partial q_2} dq_2 + \frac{\partial y}{\partial q_3} dq_3$$

$$dz = \frac{\partial z}{\partial q_1} dq_1 + \frac{\partial z}{\partial q_2} dq_2 + \frac{\partial z}{\partial q_3} dq_3.$$

The coordinates q_r constitute an orthogonal system if the three vectors $d\vec{r}_1$, $d\vec{r}_2$, and $d\vec{r}_3$, corresponding respectively to (dq_1 with $dq_2 = dq_3 = 0$),

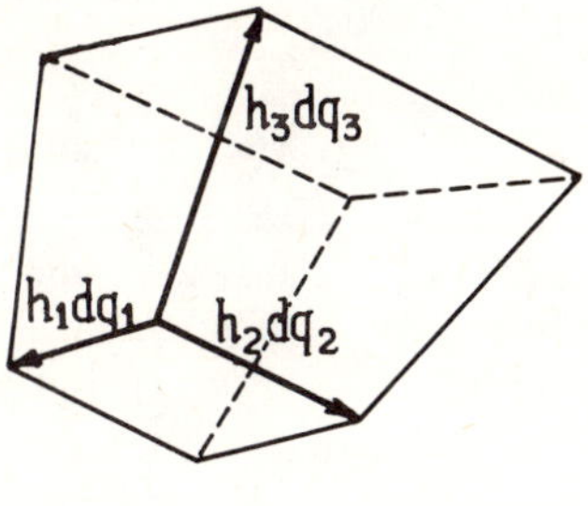

Figure 82. The volume element in orthogonal curvilinear coordinates; the volume tends toward a parallelepiped for infinitesimal dq_1, dq_2, and dq_3.

(dq_2 with $dq_1 = dq_3 = 0$), and (dq_3 with $dq_1 = dq_2 = 0$), are orthogonal to each other. This, in turn, implies (see fig. 82):

(2) $$\frac{\partial x}{\partial q_r} \cdot \frac{\partial x}{\partial q_s} + \frac{\partial y}{\partial q_r} \cdot \frac{\partial y}{\partial q_s} + \frac{\partial z}{\partial q_r} \cdot \frac{\partial z}{\partial q_s} = 0, \quad \text{if} \quad r \neq s.$$

The three vectors thus defined have as their respective lengths:

(3) $$h_1dq_1\,; \quad h_2dq_2\,; \quad h_3dq_3$$

with

$$h_r = \sqrt{\left(\frac{\partial x}{\partial q_r}\right)^2 + \left(\frac{\partial y}{\partial q_r}\right)^2 + \left(\frac{\partial z}{\partial q_r}\right)^2}.$$

The volume of the parallelepiped formed by these three infinitesimal vectors represents the volume element in the new reference system, and this is given by (fig. 82):

$$d\tau = h_1\, h_2\, h_3\, dq_1\, dq_2\, dq_3. \tag{4}$$

Another expression for this volume is given by the absolute value of the determinant:

$$\begin{vmatrix} dx_1 & dy_1 & dz_1 \\ dx_2 & dy_2 & dz_2 \\ dx_3 & dy_3 & dz_3 \end{vmatrix} = \begin{vmatrix} \frac{\partial x}{\partial q_1} & \frac{\partial y}{\partial q_1} & \frac{\partial z}{\partial q_1} \\ \frac{\partial x}{\partial q_2} & \frac{\partial y}{\partial q_2} & \frac{\partial z}{\partial q_2} \\ \frac{\partial x}{\partial q_3} & \frac{\partial y}{\partial q_3} & \frac{\partial z}{\partial q_3} \end{vmatrix} dq_1\, dq_2\, dq_3 \tag{5}$$

$$= \frac{D(x,\, y,\, z)}{D(q_1,\, q_2,\, q_3)}\, dq_1\, dq_2\, dq_3.$$

This determinant, often written $D(x,y,z)/D(q_1,q_2,q_3)$, is known as the Jacobian of the transformation.

2 Differential Operators in Orthogonal Coordinates

Gradient of a function f. This is defined by the variation of the function under a displacement $\vec{dr}$:

$$df = \overrightarrow{\text{grad}}\, f . \vec{dr} = \frac{\partial f}{\partial q_1}\, dq_1 + \frac{\partial f}{\partial q_2}\, dq_2 + \frac{\partial f}{\partial q_3}\, dq_3. \tag{6}$$

We demonstrated above that $\vec{dr}$ defined by the dq_r can be broken down into three components on the orthogonal axes given by (3). The differential df thus takes the form of a scalar product, which allows us to evaluate the three corresponding components of grad f, so that:

$$\overrightarrow{\text{grad}}\, f : \frac{1}{h_1}\frac{\partial f}{\partial q_1}; \qquad \frac{1}{h_2}\frac{\partial f}{\partial q_2}; \qquad \frac{1}{h_3}\frac{\partial f}{\partial q_3}. \tag{7}$$

Divergence of a Vector Field $\vec{u}$. This idea can be introduced by means of Gauss's theorem, which allows transformation of the integral giving the flux of $\vec{u}$ crossing a closed surface S into a volume integral:

$$\int_S \vec{u} . \vec{n}\, d\sigma = \int \text{div}\, \vec{u}\, d\tau, \tag{8}$$

$\vec{n}$ represents the unit outgoing normal to the surface S, and the volume integral is taken over the volume enclosed by S. When the volume V becomes infinitely small the equation goes to the limit:

$$\text{div}\, \vec{u} = \lim_{V \to 0} \frac{1}{V} \int \vec{u} . \vec{n}\, d\sigma. \tag{9}$$

We will evaluate div $\vec{u}$ in orthogonal coordinates by applying equation (9) to a volume element defined by the dq_r. The calculation can be made immediately when the vector $\vec{u}$ is defined at each point by its components u_r on the three axes defined respectively by $\vec{dr}_1$, $\vec{dr}_2$, and $\vec{dr}_3$, by treating the two infinitely small parallel faces defined by dq_1 and dq_2 (fig. 83). The contribution of these two surfaces to the flux is written:

$$d\Phi = (u_3\, h_1\, h_2\, dq_1\, dq_2)_{q_3 + dq_3} - (u_3\, h_1\, h_2\, dq_1\, dq_2)_{q_3}$$

so that:

$$d\Phi = \frac{\partial}{\partial q_3}(u_3\, h_1\, h_2)\, dq_1\, dq_2\, dq_3.$$

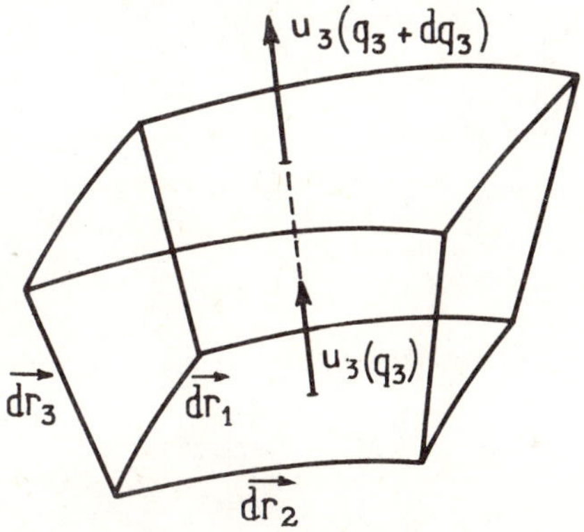

Figure 83. Calculation of the contributions of two faces of the volume element separated by distance $\vec{dr}_3$ to the flux of the vector $\vec{u}$ crossing the surface of the element.

By grouping the contribution of the other two pairs of surfaces of the volume element we get:

$$\text{(10)}\quad \text{div}\,\vec{u} = \frac{1}{h_1\, h_2\, h_3}\left[\frac{\partial}{\partial q_3}(h_1\, h_2\, u_3) + \frac{\partial}{\partial q_1}(h_2\, h_3\, u_1) + \frac{\partial}{\partial q_2}(h_3\, h_1\, u_2)\right].$$

Laplacian of a Function f. The laplacian is expressed by the formula:

$$\text{(11)}\qquad \Delta f = \text{div}\,\overrightarrow{\text{grad}}\, f$$

which leads immediately to the classical form in cartesian coordinates. The equations (7) and (10) give the general form:

$$\text{(12)}\quad \Delta f = \frac{1}{h_1\, h_2\, h_3}\left[\frac{\partial}{\partial q_1}\left(\frac{h_2\, h_3}{h_1}\frac{\partial f}{\partial q_1}\right) + \frac{\partial}{\partial q_2}\left(\frac{h_3\, h_1}{h_2}\frac{\partial f}{\partial q_2}\right) + \frac{\partial}{\partial q_3}\left(\frac{h_1\, h_2}{h_3}\frac{\partial f}{\partial q_3}\right)\right].$$

Applications 3

We will now consider two systems frequently used in quantum mechanics.

Spherical Coordinates (fig. 84) r, θ, χ. The equations of transformation from cartesian coordinates are written:

$$(13) \qquad x = r \sin \theta \cos \chi \,; \qquad y = r \sin \theta \sin \chi \,; \qquad z = r \cos \theta .$$

The coefficients h_r relative to these three coordinates are found immediately using (3):

$$(14) \qquad h_r = 1 \,; \qquad h_\theta = r \,; \qquad h_\chi = r \sin \theta$$

with the volume element being $d\tau = r^2 \sin\theta \, dr \, d\theta \, d\chi$.

Equation (12) becomes:

$$(15) \qquad \Delta f = \frac{1}{r^2} \frac{\partial}{\partial r} \left(r^2 \frac{\partial f}{\partial r} \right) + \frac{1}{r^2 \sin \theta} \frac{\partial}{\partial \theta} \left(\sin \theta \frac{\partial f}{\partial \theta} \right) + \frac{1}{r^2 \sin^2 \theta} \frac{\partial^2 f}{\partial \chi^2} .$$

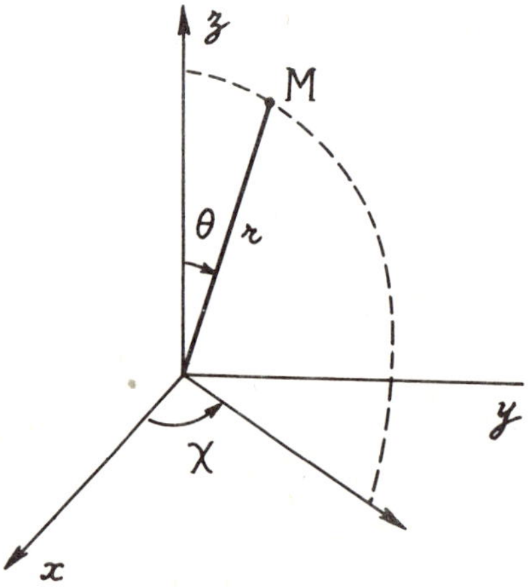

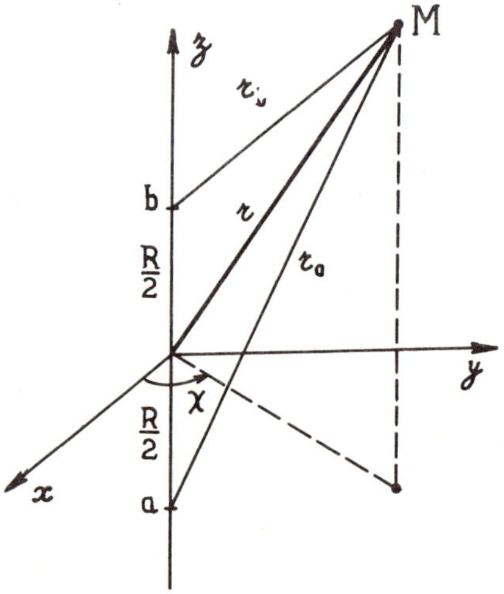

Figure 84. Definition of the spherical coordinates.
Figure 85. Definition of the elliptical coordinates.

This equation is often written in the abbreviated form:

$$(16) \qquad \Delta f = \frac{1}{r^2} \frac{\partial}{\partial r} \left(r^2 \frac{\partial f}{\partial r} \right) + \frac{1}{r^2} \Lambda f$$

with

$$\Lambda f = \frac{1}{\sin \theta} \frac{\partial}{\partial \theta} \left(\sin \theta \frac{\partial f}{\partial \theta} \right) + \frac{1}{\sin^2 \theta} \frac{\partial^2 f}{\partial \chi^2} .$$

Elliptical Coordinates. These are defined by the distances to point M from the fixed points a and b (fig. 85) using:

$$\xi = \frac{r_a + r_b}{\mathrm{R}} \,; \qquad 1 \leqslant \xi \leqslant +\infty$$

$$\eta = \frac{r_a - r_b}{\mathrm{R}} \,; \qquad -1 \leqslant \eta \leqslant +1 .$$

The third coordinate is the meridian angle of M with the reference xz plane and varies from 0 to 2π.

The transformation of elliptical coordinates to spherical coordinates is accomplished starting with the two equations:

$$r_b^2 = \left(z - \frac{R}{2}\right)^2 + x^2 + y^2 ; \qquad r_a^2 = \left(z + \frac{R}{2}\right)^2 + x^2 + y^2$$

which yield:

$$r_a^2 + r_b^2 = 2r^2 + \frac{R^2}{2} ; \qquad r_a^2 - r_b^2 = 2rR \cos \theta$$

from which we get:

$$(17) \qquad r = \frac{R}{2}\sqrt{\xi^2 + \eta^2 - 1} ; \qquad \cos \theta = \frac{\xi\eta}{\sqrt{\xi^2 + \eta^2 - 1}} .$$

The volume element can be gotten quickly by using a classical property of the Jacobian resulting from the formula for determinant multiplication:

$$\frac{D(x, y, z)}{D(\xi, \eta, \chi)} = \frac{D(x, y, z)}{D(r, \theta, \chi)} \cdot \frac{D(r, \theta, \chi)}{D(\xi, \eta, \chi)} .$$

The first Jacobian appears in the expression for the volume element in spherical coordinates and its value is thus $r^2 \sin \theta$. The second reduces to:

$$\frac{D(r, \theta, \chi)}{D(\xi, \eta, \chi)} = \begin{vmatrix} \frac{\partial r}{\partial \xi} & \frac{\partial r}{\partial \eta} & 0 \\ \frac{\partial \theta}{\partial \xi} & \frac{\partial \theta}{\partial \eta} & 0 \\ 0 & 0 & 1 \end{vmatrix} = \left(\frac{\partial r}{\partial \xi} \cdot \frac{\partial \theta}{\partial \eta} - \frac{\partial r}{\partial \eta} \cdot \frac{\partial \theta}{\partial \xi}\right).$$

The absolute value of this Jacobian is:

$$\left|\frac{D(r, \theta, \chi)}{D(\xi, \eta, \chi)}\right| = \frac{1}{\sin \theta} \cdot \frac{R(\xi^2 - \eta^2)}{2(\xi^2 + \eta^2 - 1)} .$$

Finally, the volume element in elliptical coordinates is given by:

$$(18) \qquad d\tau = \frac{R^3}{8}(\xi^2 - \eta^2) d\xi \, d\eta \, d\chi .$$

APPENDIX V
THE SPHERICAL FUNCTIONS

Laplace's Equation 1

This equation is written in spherical coordinates as:

(1) $$\Delta u = \frac{\partial^2 u}{\partial x^2} + \frac{\partial^2 u}{\partial y^2} + \frac{\partial^2 u}{\partial z^2} = 0.$$

One important property of this equation is that, if $u(x, y, z)$ is a solution of (1), so are all functions generated from u by any rotation about an axis passing through the origin. This property explains our interest in a choice of variables which allows a particularly simple expression for the rotations about an axis, taken here as the z-axis, so that:

(2) $$\xi = x + iy\,; \qquad \eta = x - iy\,; \qquad z.$$

The usual differentiation formulas give as a first derivative:

$$\frac{\partial u}{\partial x} = \frac{\partial u}{\partial \xi}\frac{\partial \xi}{\partial x} + \frac{\partial u}{\partial \eta}\frac{\partial \eta}{\partial x} = \frac{\partial u}{\partial \xi} + \frac{\partial u}{\partial \eta}$$

likewise:

$$\frac{\partial u}{\partial y} = i\left(\frac{\partial u}{\partial \xi} - \frac{\partial u}{\partial \eta}\right).$$

The second derivatives are gotten in the same way, and thus we get Laplace's equation in ξ, η, z:

(3) $$\Delta u = 4\frac{\partial^2 u}{\partial \xi\, \partial \eta} + \frac{\partial^2 u}{\partial z^2} = 0.$$

Polynomial Solutions of Laplace's Equation 2

Laplace's equation is homogeneous with constant coefficients and it is natural to look for solutions in the form of a power series which, as we shall see, reduces to a polynomial. The most general homogeneous polynomial of degree l written in terms of the variables introduced in (2) is:

(4) $$u_l = \sum_{\alpha, \beta, \gamma} c(\alpha, \beta, \gamma)\xi^{\alpha}\eta^{\beta}z^{\gamma}, \quad \text{with} \quad \alpha + \beta + \gamma = l.$$

We assume that u_l is a solution of Laplace's equation, and by substituting in (3) and taking the coefficient of $\xi^{\alpha}\eta^{\beta}z^{\gamma-2}$ identically zero, we get:

(5) $$4(\alpha + 1)(\beta + 1)\,c(\alpha + 1, \beta + 1, \gamma - 2) + \gamma(\gamma - 1)\,c(\alpha, \beta, \gamma) = 0.$$

When one coefficient $c(\alpha, \beta, \gamma)$ is given, this recurrence relation allows us to calculate one-by-one $c(\alpha+1, \beta+1, \gamma-2)$, $c(\alpha+2, \beta+2, \gamma-4)$, ..., $c(\alpha', \beta', \gamma'$, with $\alpha' = \alpha + r$, $\beta' = \beta + r$, and $\gamma' = \gamma - 2r$. All of these terms correspond to the same value of the difference $\alpha' - \beta'$, and we take:

$$\alpha' - \beta' = \alpha - \beta = m.$$

The integer m is positive or negative depending on whether $\alpha > \beta$ or $\alpha < \beta$, and its absolute value cannot exceed the degree l of u_l.

If we take as zero all of the coefficients $c(\alpha, \beta, \gamma)$ corresponding to other combinations of indices, we get a solution of Laplace's equation in the form (4) characterized by m, and all of whose coefficients can be gotten from the first by means of the recurrence relation (5). We envision two possible cases:

$m > 0$. We begin from $c(m,0,l-m)$, and equation (5) allows us to calculate successively $c(m+1, 1, l-m-2), \ldots, c(m+r, r, l-m-2r), \ldots$. The use of (5) remains feasible as long as we have $\gamma(\gamma - 1) \neq 0$. The values of $\gamma = l - m - 2r$ form a decreasing series of ratio 2, with the result that we will necessarily have $\gamma = 0$ or $\gamma = 1$, depending on the parity of $l - m$. The additional coefficients are thus all zero, and this demonstrates the possibility of getting a solution in polynomial form defined to within a constant factor, which we call u_l^m.

$m < 0$. The same considerations apply here but the initial coefficient used in (5) should be $c(0, -m, l+m)$.

The condition $|m| \leqslant l$ leads to $2l+1$ possible values of m from l to $-l$, and thus to $2l+1$ solutions of the Laplace equation, each defined to within a constant factor. The most general polynomial of degree l satisfying Laplace's equation is thus of the form:

$$u_l = \sum_{m=-l}^{+l} c_m u_l^m \tag{6}$$

and depends on $2l+1$ arbitrary coefficients.

Each of the solutions u_l^m can be written in abbreviated form by taking $c(m+\beta, \beta, l-m-2\beta) = c_\beta$:

$$u_l^m = \sum_\beta c_\beta \, \xi^{m+\beta} \eta^\beta z^{l-m-2\beta}.$$

Generally we shall use u_l^m in polar coordinates r, θ, χ, so that:

$$\xi = r \sin\theta . e^{i\chi}; \qquad \eta = r \sin\theta . e^{-i\chi}; \qquad z = r\cos\theta. \tag{7}$$

Substitution in u_l^m gives:

$$u_l^m = r^l e^{im\chi} \sum_\beta c_\beta \, (\sin\theta)^{m+2\beta} (\cos\theta)^{l-m-2\beta} = r^l Y_l^m(\theta, \chi) \tag{8}$$

introducing the spherical function $Y_l^m(\theta, \chi)$, which we express in a series, and which, like u_l^m is defined to within an arbitrary factor. It must be noted that it is the product $r^l Y_l^m$, which is the solution of Laplace's equation, and not the spherical function itself. In this notation, the most general polynomial of degree l in the form (6) satisfying this equation can be written:

$$u_l = r^l Y_l(\theta, \chi), \quad \text{with} \quad Y_l = \sum_m c_m Y_l^m.$$

We now return to the structure of the spherical functions defined by (8). If in this sum we introduce a factor $(\sin\theta)^{-m}$, we apparently make a sum f_l^m which is a polynomial in $t = \cos\theta$. We can now introduce the following notation for writing the spherical functions:

$$Y_l^m = \frac{1}{\sqrt{2\pi}} e^{im\chi} P_l^m(\theta), \tag{9a}$$

by using the normalization factor $(1/\sqrt{2\pi})$ relative to the coordinate χ. The function P_l^m is usually known as the *associated Legendre function* and is written with a normalization factor N_l^m to be determined later†:

$$\begin{aligned} P_l^m &= N_l^m (\sin\theta)^{-m} f_l^m (\cos\theta) \\ &= N_l^m (1 - t^2)^{-\frac{m}{2}} f_l^m(t), \quad \text{with} \quad t = \cos\theta. \end{aligned} \tag{9b}$$

The associated Legendre function reduces to a polynomial in t for even values of m.

At the outset we indicated that use of the coordinates ξ, η, and z allows us to express the rotations about the z-axis simply. The functions u_l^m and Y_l^m have a particularly simple behavior under such rotations, given by $r' = r$, $\theta' = \theta$, and $\chi' = \chi + \chi_0$. These functions, which all include the factor $e^{im\chi}$, are obtained through multiplication by the factor $e^{im\chi_0}$.

The Equations of the Functions: Y_l^m, P_l^m, and f_l^m. 3

The definition of the spherical functions by (8) from the solution of Laplace's equation, allows us to determine their form by substitution. We use spherical coordinates in which Laplace's equation can be written in the abbreviated form (Appendix IV-16):

$$\Delta u = \frac{1}{r^2}\frac{\partial}{\partial r}\left(r^2 \frac{\partial u}{\partial r}\right) + \frac{1}{r^2}\Lambda u.$$

† The notation P_l^m is nearly universal, at least to within the normalization coefficient. There does not seem to be a general notation for the polynomial given here by $f_l^m(t)$, its definition varying from author to author, since we get a polynomial in t both through the introduction of the factors $(\sin\theta)^m$ and $(\sin\theta)^{-m}$ in P_l^m.

We compute the first term of the equation taking u as in (8):

$$\frac{1}{r^2}\frac{\partial}{\partial r}\left(r^2\frac{\partial u_l^m}{\partial r}\right) = \frac{1}{r^2}\frac{d}{dr}\left(r^2\frac{dr^l}{dr}\right).\mathrm{Y}_l^m = l(l+1)r^{l-2}\,\mathrm{Y}_l^m.$$

Laplace's equation then becomes:

$$\Lambda \mathrm{Y}_l^m + l(l+1)\,\mathrm{Y}_l^m = 0. \tag{10}$$

The operator Λ (Appendix IV-16) is given by:

$$\Lambda = \frac{1}{\sin\theta}\frac{\partial}{\partial\theta}\left(\sin\theta.\frac{\partial}{\partial\theta}\right) + \frac{1}{\sin^2\theta}\cdot\frac{\partial^2}{\partial\chi^2}.$$

Equation (10) takes the following form when we replace Y_l^m by (9*a*) and divide through by the common factor $(1/\sqrt{2\pi})e^{im\chi}$:

$$\frac{1}{\sin\theta}\frac{d}{d\theta}\left(\sin\theta\frac{d\mathrm{P}_l^m}{d\theta}\right) + \left[l(l+1) - \frac{m^2}{\sin^2\theta}\right]\mathrm{P}_l^m = 0 \tag{11}$$

which is the differential equation for the associated Legendre function. By replacing these functions by their equivalent (9*b*) a long but easy differentiation gives the differential equation of the polynomials $f_l^m(t)$ where $t = \cos\theta$:

$$(1-t^2)\frac{d^2f_l^m}{dt^2} + 2(m-1)t\frac{df_l^m}{dt} + [l(l+1) - m(m-1)]f_l^m = 0. \tag{12}$$

This equation allows us to study the polynomials f_l^m. A basic relationship is gotten by differentiating (12) and regrouping the terms:

$$(1-t^2)\frac{d^2}{dt^2}\left(\frac{df_l^m}{dt}\right) + 2(m-2)t\frac{d}{dt}\left(\frac{df_l^m}{dt}\right)$$
$$+ [l(l+1) - (m-1)(m-2)]\frac{df_l^m}{dt} = 0.$$

The polynomials df_l^m/dt and f_l^{m-1} which satisfy the same equation should be identical to within a constant factor.† Since these polynomials are only defined to within a constant factor, we can eliminate this factor and take:

$$f_l^{m-1} = \frac{df_l^m}{dt}. \tag{13}$$

Thus it seems possible to calculate the set of polynomials f_l^m starting from f_l^l by means of successive differentiations:

$$f_l^m = \frac{d}{dt}f_l^{m+1} = \frac{d^2}{dt^2}f_l^{m+2} = \ldots = \frac{d^r}{dt^r}f_l^{m+r} = \ldots = \frac{d^{l-m}}{dt^{l-m}}f_l^l. \tag{14}$$

---

† This result is only valid because we limited the solutions of (12) to the form of a polynomial, since the general solution depends upon two arbitrary constants. A rigorous derivation is given later (Appendix VI).

The polynomial f_l^l is gotten immediately from the defining equations (9*a*) and (9*b*), since the series (8) which gives u_l^l reduces to a single term ($m = l, \beta = 0$). Thus we have:

$$u_l^l = c_0 r^l e^{il\chi} (\sin\theta)^l = r^l Y_l^l$$

and as a result:

$$Y_l^l = c_0 e^{il\chi} (\sin\theta)^l$$

$$\frac{1}{\sqrt{2\pi}} P_l^l = c_0 (\sin\theta)^l = \frac{1}{\sqrt{2\pi}} N_l^l (\sin\theta)^{-l} f_l^l$$

$$f_l^l = \frac{\sqrt{2\pi}\, c_0}{N_l^l} (\sin\theta)^{2l} = \frac{\sqrt{2\pi}\, c_0}{N_l^l} (1 - t^2)^l.$$

We have introduced two arbitrary constants c_0 and N_l^l. The polynomials are usually written by taking:

$$\frac{\sqrt{2\pi}\, c_0}{N_l^l} = [2^l (l\,!)]^{-1}.$$

In addition, we introduce the factor $(-1)^l$, since this leads to positive values in the calculation of the matrix elements of the coordinate z. From (14) we get:

(15) $$f_l^m = \frac{d^{l-m}}{dt^{l-m}} f_l^l = \frac{1}{2^l (l\,!)} \cdot \frac{d^{l-m}}{dt^{l-m}} (t^2 - 1)^l.$$

The function P_l^m defined by (9*b*) is now satisfactorily determined, with the exception of fixing the normalization constant N_l^m. We usually adopt the normalization condition:

(16) $$\int | Y_l^m(\theta, \chi) |^2 \sin\theta \, d\theta \, d\chi = 1,$$

the integration being on a sphere of unit radius. Integration relative to χ leads to the factor 2π already mentioned. The rest of the calculation, although laborious, is not difficult, and leads to:

(17) $$| N_l^m | = \left[\frac{(2l + 1).(l + m)\,!}{2.(l - m)\,!}\right]^{1/2}.$$

In conclusion, we summarize the fundamental relations of the spherical functions that we have established:

(18)

$$Y_l^m = \frac{1}{\sqrt{2\pi}} e^{im\chi} P_l^m(\theta)$$

$$P_l^m = N_l^m (\sin\theta)^{-m} f_l^m (\cos\theta),$$

with $$| N_l^m | = \left[\frac{(2l + 1).(l + m)\,!}{2.(l - m)\,!}\right]^{1/2}$$

$$f_l^m(t) = \frac{1}{2^l (l\,!)} \frac{d^{l-m}}{dt^{l-m}} (t^2 - 1)^l.$$

Note: Y_l^m is the solution of (10) with real coefficients, so that the complex conjugate function is also a solution of the same equation. The form (9*a*) of the function Y_l^m shows that its conjugate should be identical to Y_l^{-m} to within a factor λ, so that:

$$(Y_l^m)^* = \lambda Y_l^{-m}.$$

Passing from Y_l^m to its complex conjugate does not affect the normalization factor, with the result that, the functions θ being real, we must have $|\lambda| = 1$. Following simplification of the preceding equation this is written:

$$N_l^m (\sin\theta)^{-m} f_l^m = \lambda N_l^{-m} (\sin\theta)^{+m} f_l^{-m}$$

where:

$$N_l^m \frac{d^{l-m}}{dt^{l-m}} (t^2 - 1)^l = \lambda N_l^{-m} (1 - t^2)^m \frac{d^{l+m}}{dt^{l+m}} (t^2 - 1)^l.$$

For the determination of λ it is sufficient to consider the terms of higher order in these two expressions, namely, in t^{l+m}, and we see that $\lambda = (-1)^m$. We can take the absolute value of the numerical coefficient and can then get the value of the ratio N_l^m/N_l^{-m}, the two normalization constants as defined in (17).

This result allows us to represent Y_l^m and Y_l^{-m} by a single equation, which we will not bother to write. In addition, in numerous applications (one-electron atom §17) only the modulus of the spherical function is used, with the result that it becomes useless to distinguish between these two functions.

4 Recurrence Relations between the Spherical Functions

These relations allow us to calculate the matrix elements of the position coordinates which arise in the problems dealing with transitions in the rigid rotor and in the atom. We will have two relations, both obtained by the same method.

(*a*) We start with expression (18) for the polynomials f_l^m, which we write for order $l + 1$ as:

$$f_{l+1}^m = \frac{1}{2^{l+1}.(l+1)!} \cdot \frac{d^{l+1-m}}{dt^{l+1-m}} (t^2 - 1)^{l+1}$$

$$= \frac{1}{2^{l+1}.(l+1)!} \cdot \frac{d^{l-m}}{dt^{l-m}} \cdot \frac{d}{dt} (t^2 - 1)^{l+1}$$

and by taking:

$$f_{l+1}^m = \frac{1}{2^l.l!} \cdot \frac{d^{l-m}}{dt^{l-m}} \left[t(t^2 - 1)^l\right].$$

The derivative of the product obeys the recurrence theorem:

$$\frac{d^r}{dt^r}(tu) = t\frac{d^r u}{dt^r} + r\frac{d^{r-1}u}{dt^{r-1}}$$

from which we get:

$$f_{l+1}^m = t f_l^m + (l - m) f_l^{m+1}.$$

We now pass to the associated Legendre function by multiplying by $(\sin\theta)^{-m}$ and by the normalization factor. To simplify the writing we first eliminate N_l^m by taking:

$$\Theta_l^m = \frac{1}{N_l^m} P_l^m = (\sin\theta)^{-m} f_l^m.$$

and thus we get:

$$\Theta_{l+1}^m = t\Theta_l^m + (l - m)\sin\theta . \Theta_l^{m+1}. \tag{19}$$

(*b*) The second relation is gotten by applying this process twice. We get:

$$f_{l+2}^m = \frac{1}{2^{l+2}.(l+2)!} \cdot \frac{d^{l+2-m}}{dt^{l+2-m}} (t^2 - 1)^{l+2}$$

$$= \frac{1}{2^{l+2}.(l+2)!} \cdot \frac{d^{l-m}}{dt^{l-m}} \cdot \frac{d^2(t^2 - 1)^{l+2}}{dt^2}$$

so that, by expanding and regrouping the terms:

$$f_{l+2}^m = (2l + 3) f_{l+1}^{m+1} + f_l^m.$$

We turn again to the function Θ_l^m and write the result, replacing l by $l - 1$ in such a way that the function multiplied by $\sin\theta$ is the same as in (19).

$$\Theta_{l+1}^m = (2l + 1)\sin\theta . \Theta_l^{m+1} + \Theta_{l-1}^m. \tag{20}$$

With these two relations established, elimination of the term in $\sin\theta$ leads to:

$$(2l + 1)t\Theta_l^m = (l + m + 1)\Theta_{l+1}^m + (l - m)\Theta_{l-1}^m.$$

We get the Legendre functions back by returning the normalization coefficient N_l^m. The relation keeps the same form as the spherical functions when multiplied by $(1/\sqrt{2\pi})e^{im\chi}$. Thus we get:

$$\cos\theta . Y_l^m = \left[\frac{(l + 1 + m)(l + 1 - m)}{(2l + 3)(2l + 1)}\right]^{1/2} Y_{l+1}^m + \left[\frac{(l + m)(l - m)}{(2l + 1)(2l - 1)}\right]^{1/2} Y_{l-1}^m \tag{21}$$

and the expansion of zY_l^m by multiplying by r.

Equation (20) permits, in addition, the calculation of ξY_l^m and ηY_l^m. For this purpose we use it to find $\sin\theta \cdot \Theta_l^m$, so that:

$$(2l + 1)\sin\theta . \Theta_l^m = \Theta_{l+1}^{m-1} - \Theta_{l-1}^{m-1}.$$

We get the spherical functions Y_l^m by introducing the normalization coefficients N_l^m, as well as the factor $(1/\sqrt{2\pi})e^{i(m-1)\chi}$.

$$\sin\theta . e^{-i\chi}\, Y_l^m = \left[\frac{(l+2-m)\,(l+1-m)}{(2l+3)\,(2l+1)}\right]^{1/2} Y_{l+1}^{m-1}$$

(22)

$$- \left[\frac{(l+m)\,(l-1+m)}{(2l+1)\,(2l-1)}\right]^{1/2} Y_{l-1}^{m-1}.$$

We thus have, to within a factor r, the expansion of ηY_l^m. That for ξY_l^m can likewise be gotten by taking the complex conjugate of (22); the preceding note suggests replacement of $(Y_l^m)^\star$ by $(-1)_m Y_l^{-m}$. Thus, replacing m by $-m$:

$$\sin\theta . e^{i\chi}\, Y_l^m = - \left[\frac{(l+2+m)(l+1+m)}{(2l+3)(2l+1)}\right]^{1/2} Y_{l+1}^{m+1}$$

(23)

$$+ \left[\frac{(l-m)(l-1-m)}{(2l+1)(2l-1)}\right]^{1/2} Y_{l-1}^{m+1}.$$

APPENDIX VI
ANGULAR MOMENTUM OPERATORS

(a) Definition

The angular momentum $\vec{L}$ plays an important role in the study of the rigid rotor and therefore in the study of the atom. From the classical point of view this vector is defined by $\vec{L} = \vec{r} \times \vec{p}$, and has as cartesian components $L_x = yp_z - zp_y$, $L_y = zp_x - xp_z$, and $L_z = xp_y - yp_x$. The products which appear in these expressions combine only compatible variables, with the result that construction of the associated operators is not difficult. It is only necessary to replace the variables with their corresponding operators, which leads to the following expressions:

$$\mathbf{L}_x = y\mathbf{P}_z - z\mathbf{P}_y = \frac{\hbar}{i}\left(y\frac{\partial}{\partial z} - z\frac{\partial}{\partial y}\right)$$

(1) $$\mathbf{L}_y = z\mathbf{P}_x - x\mathbf{P}_z = \frac{\hbar}{i}\left(z\frac{\partial}{\partial x} - x\frac{\partial}{\partial z}\right)$$

$$\mathbf{L}_z = x\mathbf{P}_y - y\mathbf{P}_x = \frac{\hbar}{i}\left(x\frac{\partial}{\partial y} - y\frac{\partial}{\partial x}\right).$$

These operators do not commute and it is easy to show that:

$$\mathbf{L}_x\mathbf{L}_y - \mathbf{L}_y\mathbf{L}_x = i\hbar\mathbf{L}_z$$

(2) $$\mathbf{L}_y\mathbf{L}_z - \mathbf{L}_z\mathbf{L}_y = i\hbar\mathbf{L}_x$$

$$\mathbf{L}_z\mathbf{L}_x - \mathbf{L}_x\mathbf{L}_z = i\hbar\mathbf{L}_y.$$

The variables L_x, L_y, and L_z are not compatible. However, it is possible to find a variable compatible with these three variables, and hence, an operator which commutes with $\mathbf{L}_x$, $\mathbf{L}_y$, and $\mathbf{L}_z$. This variable is the square of the angular momentum defined by:

$$L^2 = L_x^2 + L_y^2 + L_z^2.$$

The associated operator is written:

(3) $$\mathbf{L}^2 = \mathbf{L}_x^2 + \mathbf{L}_y^2 + \mathbf{L}_z^2.$$

Equation (2) permits us to verify immediately the commutation of $\mathbf{L}^2$, and any one of the three operators, $\mathbf{L}_x$, $\mathbf{L}_y$ or $\mathbf{L}_z$. We take $\mathbf{L}_z$ for example. Multiply the second equation in (2) both from the left and from the right by $\mathbf{L}_y$:

$$\mathbf{L}_y\mathbf{L}_z\mathbf{L}_y - \mathbf{L}_z\mathbf{L}_y^2 = i\hbar\mathbf{L}_x\mathbf{L}_y$$

$$\mathbf{L}_y^2\mathbf{L}_z - \mathbf{L}_y\mathbf{L}_z\mathbf{L}_y = i\hbar\mathbf{L}_y\mathbf{L}_x$$

which gives, adding these together:

$$\mathbf{L}_y^2\,\mathbf{L}_z - \mathbf{L}_z\,\mathbf{L}_y^2 = i\hbar(\mathbf{L}_x\,\mathbf{L}_y + \mathbf{L}_y\,\mathbf{L}_x).$$

We can do the same with the last equation in (2), multiplying this time by $\mathbf{L}_x$:

$$\mathbf{L}_x^2\,\mathbf{L}_z - \mathbf{L}_z\,\mathbf{L}_x^2 = -\,i\hbar(\mathbf{L}_x\,\mathbf{L}_y + \mathbf{L}_y\,\mathbf{L}_x).$$

We add the obvious identity:

$$\mathbf{L}_z^2\,\mathbf{L}_z - \mathbf{L}_z\,\mathbf{L}_z^2 = 0.$$

Addition of these three equations gives:

$$\mathbf{L}^2\,\mathbf{L}_z - \mathbf{L}_z\,\mathbf{L}^2 = 0$$

which establishes the commutability of the operators $\mathbf{L}^2$ and $\mathbf{L}_z$, and this has been accomplished without reference to the nature of the functions to which these operators are applied. This will be found useful for many generalizations about these operators.

(b) Connection to the study of Spherical Functions

As we shall see, the commuting operators $\mathbf{L}^2$ and $\mathbf{L}_z$ can have a common system of eigenfunctions. We will show initially that a particular solution to this problem is given by the spherical functions which we studied previously (Appendix V). For this, we must explicitly write the operators $\mathbf{L}^2$ and $\mathbf{L}_z$. We begin with $\mathbf{L}^2$, which requires the calculation of the operators $\mathbf{L}_x^2$, $\mathbf{L}_y^2$, and $\mathbf{L}_z^2$. For example:

$$\mathbf{L}_z^2 = -\,\hbar^2\left(x\frac{\partial}{\partial y} - y\frac{\partial}{\partial x}\right)\left(x\frac{\partial}{\partial y} - y\frac{\partial}{\partial x}\right)$$

so that by adding the expressions gotten by circular permutation of the indices:

$$\mathbf{L}^2 = -\,\hbar^2\left\{(x^2+y^2+z^2)\Delta - \left[x^2\frac{\partial^2}{\partial x^2} + y^2\frac{\partial^2}{\partial y^2} + z^2\frac{\partial^2}{\partial z^2} + 2\left(x\frac{\partial}{\partial x} + y\frac{\partial}{\partial y} + z\frac{\partial}{\partial z}\right) + 2xy\frac{\partial^2}{\partial x\,\partial y} + 2yz\frac{\partial^2}{\partial y\,\partial z} + 2zx\frac{\partial^2}{\partial z\,\partial x}\right]\right\}.$$

This expression is simplified by noting that one has:

$$\left(x\frac{\partial}{\partial x} + y\frac{\partial}{\partial y} + z\frac{\partial}{\partial z}\right) = r\frac{\partial}{\partial r}$$

$$\left(x\frac{\partial}{\partial x} + y\frac{\partial}{\partial y} + z\frac{\partial}{\partial z}\right)^2 = x^2\frac{\partial^2}{\partial x^2} + y^2\frac{\partial^2}{\partial y^2} + z^2\frac{\partial^2}{\partial z^2} + x\frac{\partial}{\partial x} + y\frac{\partial}{\partial y} + z\frac{\partial}{\partial z} + 2xy\frac{\partial^2}{\partial x\,\partial y} + 2yz\frac{\partial^2}{\partial y\,\partial z} + 2zx\frac{\partial^2}{\partial z\,\partial x}.$$

And therefore:

$$\mathbf{L}^2 = -\,\hbar^2\left[r^2\Delta - \left(r\frac{\partial}{\partial r}\right)^2 - \left(r\frac{\partial}{\partial r}\right)\right].$$

We have already derived (Appendix IV) the expression for the laplacian in spherical coordinates and introducing the operator Λ allows us to write:

$$r^2\Delta = \frac{\partial}{\partial r}\left(r^2 \frac{\partial}{\partial r}\right) + \Lambda = \left(r \frac{\partial}{\partial r}\right)^2 + r\frac{\partial}{\partial r} + \Lambda$$

and as a result:

$$\mathbf{L}^2 = -\hbar^2\Lambda. \tag{4}$$

We now turn to the determination of $\mathbf{L}_z$ in spherical coordinates, which is easily done. Beginning from:

$$x = r \sin\theta . \cos\chi\,; \qquad y = r \sin\theta . \sin\chi,$$

we get:

$$\frac{\partial x}{\partial \chi} = -\, r \sin\theta . \sin\chi = -\, y\,; \qquad \frac{\partial y}{\partial \chi} = r \sin\theta . \cos\chi = x$$

where:

$$\frac{\partial}{\partial \chi} = \frac{\partial}{\partial x} \cdot \frac{\partial x}{\partial \chi} + \frac{\partial}{\partial y} \cdot \frac{\partial y}{\partial \chi} = -\, y\frac{\partial}{\partial x} + x\frac{\partial}{\partial y}$$

so that to within a factor, expression (1) for $\mathbf{L}_z$ is:

$$\mathbf{L}_z = \frac{\hbar}{i}\left(x\frac{\partial}{\partial y} - y\frac{\partial}{\partial x}\right) = \frac{\hbar}{i}\frac{\partial}{\partial \chi}. \tag{5}$$

We have shown that the spherical functions Y_l^m satisfy a partial differential equation (Appendix V-10), which, using (4), can be written as:

$$\mathbf{L}^2\, Y_l^m = -\,\hbar^2\Lambda Y_l^m = l(l+1)\hbar^2 Y_l^m. \tag{6a}$$

On the other hand, the form of the spherical functions (Appendix V 9*a*) allows us to see immediately that we have:

$$\mathbf{L}_z\, Y_l^m = \frac{\hbar}{i}\frac{\partial}{\partial \chi} Y_l^m = m\hbar\, Y_l^m. \tag{6b}$$

The spherical functions are thus the common eigenfunctions of the operators $\mathbf{L}^2$ and $\mathbf{L}_z$, and it is this property that explains their importance in the treatment of the rigid rotor and the atom.

There appears to be another way of studying the spherical functions, based on relation (2) between the operators $\mathbf{L}_x$, $\mathbf{L}_y$, and $\mathbf{L}_z$, without having to write the operators explicitly. As we shall see, this presents advantages and leads to the most general results.

(c) The Common Spectrum of the Operators L^2 and L_z

The commutability of these operators assures the existence of common eigenfunctions. We repeat the proof of this principle established earlier (Chapter 3). To do this we consider an eigenvalue of $\mathbf{L}_z$, which we denote by $m\hbar$, where m is a dimensionless number. We call ψ_m the corresponding eigenfunction, without necessarily specifying the variables on which it depends, since the subsequent study is based only on the system of equations (2) between operators.

Thus, by hypothesis:

(7) $$\mathbf{L}_z\psi_m = m\hbar\psi_m.$$

The commutability of $\mathbf{L}^2$ and $\mathbf{L}_z$ allows us to write:

$$\mathbf{L}_z(\mathbf{L}^2\psi_m) = \mathbf{L}^2(\mathbf{L}_z\psi_m) = m\hbar(\mathbf{L}^2\psi_m),$$

which shows that $\mathbf{L}^2\psi_m$ is also an eigenfunction of $\mathbf{L}_z$, with the same eigenvalue $m\hbar$ as ψ_m. In the case of the spherical functions, these eigenvalues are non-degenerate and, for the sake of convenience we will assume that this is always true. We can conclude that $\mathbf{L}^2\psi_m$ should equal ψ_m to within a constant factor, which we take in the form $\lambda\hbar^2$ so that λ is dimensionless.

Thus we can write:

(8) $$\mathbf{L}^2\psi_m = \lambda\hbar^2\psi_m$$

which establishes the existence of common eigenfunctions for the two operators $\mathbf{L}^2$ and $\mathbf{L}_z$.

We pursue the study of these eigenfunctions by studying their transformations by the operators $\mathbf{L}_x$ and $\mathbf{L}_y$. We start with the last two relations (2) which we have combined so as to form the operators $\mathbf{A} = \mathbf{L}_x - i\mathbf{L}_y$ and $\mathbf{B} = \mathbf{L}_x + i\mathbf{L}_y$. We see that:

(9) $$\begin{aligned} \mathbf{AL}_z - \mathbf{L}_z\mathbf{A} &= \hbar\mathbf{A} \\ \mathbf{BL}_z - \mathbf{L}_z\mathbf{B} &= -\hbar\mathbf{B}. \end{aligned}$$

Applying the first of these operator equations to the eigenfunctions ψ_m of $\mathbf{L}_z$ verifies (7). We get:

$$\mathbf{L}_z\mathbf{A}\psi_m = \mathbf{AL}_z\psi_m - \hbar\mathbf{A}\psi_m = (m-1)\hbar\mathbf{A}\psi_m.$$

The function $\mathbf{A}\psi_m$ is thus an eigenfunction of $\mathbf{L}_z$, and has as an eigenvalue $m-1$, with the result that it is, to within a constant factor, the function ψ_{m-1}. Thus we can show:

(10) $$\mathbf{A}\psi_m = \hbar a_{m-1,m}\psi_{m-1}.$$

By the same procedure with $\mathbf{B}$, we get:

(11) $$\mathbf{B}\psi_m = \hbar b_{m+1,m}\psi_{m+1}.$$

In addition, the operators $\mathbf{L}_x$ and $\mathbf{L}_y$ commute with $\mathbf{L}^2$, and so do their linear combinations $\mathbf{A}$ and $\mathbf{B}$ (as the above argument for $\mathbf{L}_z$ and $\mathbf{L}^2$ proved), since ψ_m belongs to a certain eigenvalue $\lambda\hbar^2$ of $\mathbf{L}^2$ the same should be true for $\mathbf{A}\psi_m$ and $\mathbf{B}\psi_m$. We thus see the interest in $\mathbf{A}$ and $\mathbf{B}$, as they lead to the construction of a system of eigenfunctions of $\mathbf{L}^2$ and

$\mathbf{L}_z$ from any one of the functions. We will now derive the simultaneous eigenvalues of these functions.

We note first of all that the successive action of **A** and **B** leaves the function ψ_m invariant to within a constant factor. We must therefore expect that the products **AB** and **BA** can be expressed completely in terms of $\mathbf{L}^2$ and $\mathbf{L}_z$. After working them out using (2), we get:

$$\begin{aligned} \mathbf{AB} &= (\mathbf{L}_x - i\mathbf{L}_y)(\mathbf{L}_x + i\mathbf{L}_y) = \mathbf{L}^2 - \mathbf{L}_z^2 - \hbar\mathbf{L}_z \\ \mathbf{BA} &= (\mathbf{L}_x + i\mathbf{L}_y)(\mathbf{L}_x - i\mathbf{L}_y) = \mathbf{L}^2 - \mathbf{L}_z^2 + \hbar\mathbf{L}_z. \end{aligned} \tag{12}$$

We now consider the function $\psi_{m'}$ of the maximal eigenvalue $m'\hbar$ of $\mathbf{L}_z$. There should be no eigenfunction for a higher eigenvalue which implies $\mathbf{B}\psi_{m'} = 0$, or, after applying **A**, $\mathbf{AB}\psi_{m'} = 0$. By using the first relation of (12) we can write:

$$\mathbf{AB}\psi_{m'} = (\mathbf{L}^2 - \mathbf{L}_z^2 - \hbar\mathbf{L}_z)\psi_{m'} = (\lambda - m'^2 - m')\hbar^2\psi_{m'} = 0$$

where:

$$\lambda = m'^2 + m'.$$

By considering only the function $\psi_{m''}$ belonging to the minimum eigenvalue of $\mathbf{L}_z$ we must have $\mathbf{BA}\psi_{m''} = 0$, so that explicitly:

$$\mathbf{BA}\psi_{m''} = (\mathbf{L}^2 - \mathbf{L}_z^2 + \hbar\mathbf{L}_z)\psi_{m''} = (\lambda - m''^2 + m'')\hbar^2\psi_{m''} = 0$$

where:

$$\lambda = m''^2 - m''.$$

Combining these two relations gives:

$$(m' + m'')(m' - m'' + 1) = 0.$$

By hypothesis, $m' \geqslant m''$, with the result that we necessarily have:

$$m' + m'' = 0, \qquad \lambda = m'(m' + 1).$$

In addition, the process of generation of eigenfunctions by means of the operators **A** and **B** implies that $m' - m''$ must be a non-negative integer k, which, taking into account the previous equation, suggests:

$$m' = \frac{k}{2} = l; \qquad \lambda = l(l + 1). \tag{13}$$

If the number k is even, l becomes the series 0, 1, 2, ..., and we get the system of eigenvalues of $\mathbf{L}^2$ and $\mathbf{L}_z$ which arise in the study of spherical functions. But there is another family of solutions in this study corresponding to odd values of k, and l then belongs to the series $\frac{1}{2}$, $\frac{3}{2}$, ...
We encounter such solutions in the study of spin.

(d) Calculation of the Coefficients: $a_{m-1,\,m}$, $b_{m+1,\,m}$

This calculation requires the introduction of a normalization condition for the functions ψ_m, which we write:

$$\int \psi_m^* \psi_m d\tau = 1 \tag{14}$$

without specifying the variables appearing in $d\tau$. The relations (10) and (11) allow us to obtain the desired coefficients by the relations:

$$\hbar a_{m-1,m} = \int \psi^{\star}_{m-1} \mathbf{A} \psi_m d\tau\,; \qquad \hbar b_{m+1,m} = \int \psi^{\star}_{m+1} \mathbf{B} \psi_m d\tau.$$

The hermitian character of the operators $\mathbf{L}_x$ and $\mathbf{L}_y$ involves the relations between the two sets of coefficients, since we can write:

$$\int \psi^{\star}_{m-1} \mathbf{A} \psi_m d\tau = \int \psi^{\star}_{m-1} (\mathbf{L}_x - i\mathbf{L}_y) \psi_m d\tau$$

$$= \int [(\mathbf{L}_x + i\mathbf{L}_y) \psi_{m-1}]^{\star} \psi_m d\tau = \int (\mathbf{B}\psi_{m-1})^{\star} \psi_m d\tau$$

where:

$$a_{m-1,m} = b^{\star}_{m,m-1}.$$

Thus it is possible to calculate the modulus of the coefficients by applying one of the relations (12) between the operators on ψ_m. For example:

$$\mathbf{AB}\psi_m = \hbar \mathbf{A} b_{m+1,m} \psi_{m+1} = \hbar^2 a_{m,m+1} \cdot b_{m+1,m} \psi_m = (\mathbf{L}^2 - \mathbf{L}_z^2 - \hbar \mathbf{L}_z)\psi_m$$

where:

(15)
$$|a_{m,m+1}| = |b_{m+1,m}| = \sqrt{(l-m)(l+m+1)}$$
$$|a_{m-1,m}| = |b_{m.m-1}| = \sqrt{(l+m)(l-m+1)}.$$

The form of these expressions shows that we cannot have an eigenvalue of $\mathbf{L}_z$ greater than l or less than $-l$ since $b_{l+1,\,l} = a_{-(l+1),\,-l} = 0$.

(e) Application to the Study of Spherical Functions

The relations we have established constitute one way of studying the spherical functions, namely, by considering the case of integral values of l. For this purpose we give the operators $\mathbf{A}$ and $\mathbf{B}$ in spherical coordinates. The change of variables from cartesian coordinates is accomplished in two steps. First we get them in the coordinates ξ, η, and z (Appendix IV-2) so that:

$$\mathbf{A} = \mathbf{L}_x - i\mathbf{L}_y = \hbar\left(\eta \frac{\partial}{\partial z} - 2z \frac{\partial}{\partial \xi}\right)$$

$$\mathbf{B} = \mathbf{L}_x + i\mathbf{L}_y = \hbar\left(-\xi \frac{\partial}{\partial z} + 2z \frac{\partial}{\partial \eta}\right).$$

Change to spherical coordinates requires the partial derivatives calculated from the equations relating the coordinates:

$$r^2 = \xi\eta + z^2\,; \qquad \operatorname{tg}\theta = \frac{\sqrt{\xi\eta}}{z}\,; \qquad e^{2i\chi} = \frac{\xi}{\eta}.$$

For functions of θ and χ independent of r, the usual laws of differentiation lead to:

$$\frac{\partial}{\partial \xi} = \frac{1}{2r} e^{-i\chi}\left(\cos\theta \cdot \frac{\partial}{\partial \theta} - \frac{i}{\sin\theta} \cdot \frac{\partial}{\partial \chi}\right)$$

$$\frac{\partial}{\partial \eta} = \frac{1}{2r} e^{i\chi}\left(\cos\theta \cdot \frac{\partial}{\partial \theta} + \frac{i}{\sin\theta} \cdot \frac{\partial}{\partial \chi}\right)$$

$$\frac{\partial}{\partial z} = -\frac{\sin\theta}{r} \cdot \frac{\partial}{\partial \theta}$$

which immediately yields:

(16)
$$\mathbf{A} = \hbar e^{-i\chi}\left(-\frac{\partial}{\partial \theta} + \frac{i\cos\theta}{\sin\theta} \cdot \frac{\partial}{\partial \chi}\right)$$
$$\mathbf{B} = \hbar e^{i\chi}\left(\frac{\partial}{\partial \theta} + \frac{i\cos\theta}{\sin\theta} \cdot \frac{\partial}{\partial \chi}\right).$$

To these equations we add equation (5):

$$\mathbf{L}_z = -i\hbar \frac{\partial}{\partial \chi}.$$

We apply the operator **A** in the form (16) to the spherical function defined by (Appendix V-9*a*) so that:

$$Y_l^m(\theta, \chi) = \frac{1}{\sqrt{2\pi}} e^{im\chi} P_l^m(\theta).$$

Thus we get:

$$\hbar a_{m-1,m} Y_l^{m-1} = \mathbf{A} Y_l^m = \frac{1}{\sqrt{2\pi}} \hbar e^{i(m-1)\chi}\left(-\frac{dP_l^m}{d\theta} - \frac{m\cos\theta}{\sin\theta} P_l^m\right).$$

The factor $(1/\sqrt{2\pi})e^{i(m-1)\chi}$ appears on the right as required to obtain Y_l^{m-1}. Division by this common factor gives:

$$a_{m-1,m} P_l^{m-1} = -\frac{dP_l^m}{d\theta} - \frac{m\cos\theta}{\sin\theta} P_l^m.$$

We now introduce expression (Appendix V-9*b*) for P_l^m so that:

$$P_l^m = N_l^m (\sin\theta)^{-m} f_l^m(t), \quad \text{with} \quad t = \cos\theta$$

and we get:

$$a_{m-1,m} N_l^{m-1} f_l^{m-1} = N_l^m \frac{df_l^m}{dt}.$$

To find the normalization condition (Appendix V-13) of the polynomials, which consists of showing:

$$f_l^{m-1} = \frac{df_l^m}{dt}$$

it is necessary to have the relation:

$$N_l^m = N_l^{m-1} . a_{m-1,m}.$$

The coefficients N_l^m can thus be calculated one by one from N_l^l so that:

$$N_l^m = N_l^l [a_{m,m+1} \, a_{m+1,m+2} \, \dots \, a_{l-1,l}]^{-1}$$

and by taking (15) for $a_{m,\, m+1}$:

(17) $$N_l^m = N_l^l \left[\frac{(l+m)!}{(2l)!\,(l-m)!}\right]^{1/2}.$$

The first coefficient N_l^l is calculated directly from the normalization condition. The corresponding spherical function is written:

$$Y_l^l = c_0 e^{il\chi} (\sin \theta)^l.$$

c_0 is determined by (Appendix V-16):

$$\int |Y_l^l|^2 d\sigma = |c_0^2| \int (\sin \theta)^{2l} \sin \theta \, d\theta \, d\chi = 1.$$

The integral above can be calculated quickly by recurrence and we take:

$$J_l = \int_0^\pi (\sin \theta)^{2l} \sin \theta \, d\theta = \int_0^\pi (\sin \theta)^{2l} \, d(-\cos \theta).$$

Integration by parts gives:

$$J_l = \int_0^\pi \cos \theta . d \, (\sin \theta)^{2l} = 2l \int_0^\pi (\sin \theta)^{2l-1} \cos^2 \theta \, d\theta$$

hence the relation:

$$J_l = 2l(J_{l-1} - J_l), \quad \text{or} \quad (2l+1)J_l = 2lJ_{l-1}.$$

Thus we see:

$$J_l = J_0 \frac{2l.2(l-1) \dots 2}{(2l+1)(2l-1) \dots 1}$$

with

$$J_0 = \int_0^\pi \sin \theta \, d\theta = 2.$$

We get the definitive result:

$$J_l = 2 \frac{2^{2l}.(l!)^2}{(2l+1)!}; \qquad c_0 = \frac{1}{2^l.l!} \left[\frac{(2l+1)!}{4\pi}\right]^{\frac{1}{2}}$$

We have adopted the following convention (Appendix V) between c_0 and N:

$$\frac{\sqrt{2\pi}\, c_0}{N_l^l} = [2^l.l!]^{-1}.$$

The value obtained above for c_0 allows us to get the normalization coefficient N_l^l:

$$N_l^l = \left[\frac{(2l+1)!}{2}\right]^{\frac{1}{2}}$$

from which, using (17):

$$N_l^m = \left[\frac{2l+1}{2} \cdot \frac{(l+m)!}{(l-m)!}\right]^{\frac{1}{2}}.$$

and we get the previously indicated expression (Appendix V-17).

APPENDIX VII
WAVE FUNCTIONS BY THE BWK METHOD

Wave Functions of the One-Dimensional Well 1

The treatment of this problem (§25) by the BWK method has lead to an approximate solution whose analytical form differs according to the region under consideration (fig. 86). We reproduce here the expressions obtained in section 25 (equations 38, 39, and 40) so that:

$$(1)\quad \text{Log}\ \psi_1 = \pm\frac{1}{\hbar}\int_{x_a}^{x}\sqrt{2m(\mathrm{V}-\mathrm{E})}\,dx - \frac{1}{4}\text{Log}\,[2m(\mathrm{V}-\mathrm{E})] + \gamma_1$$

$$(2)\quad \text{Log}\ \psi_2 = \pm\frac{i}{\hbar}\int_{x_a}^{x}\sqrt{2m(\mathrm{E}-\mathrm{V})}\,dx - \frac{1}{4}\text{Log}\,[2m(\mathrm{E}-\mathrm{V})] + \gamma_2$$

$$(3)\quad \text{Log}\ \psi_3 = \pm\frac{1}{\hbar}\int_{x_b}^{x}\sqrt{2m(\mathrm{V}-\mathrm{E})}\,dx - \frac{1}{4}\text{Log}\,[2m(\mathrm{V}-\mathrm{E})] + \gamma_3.$$

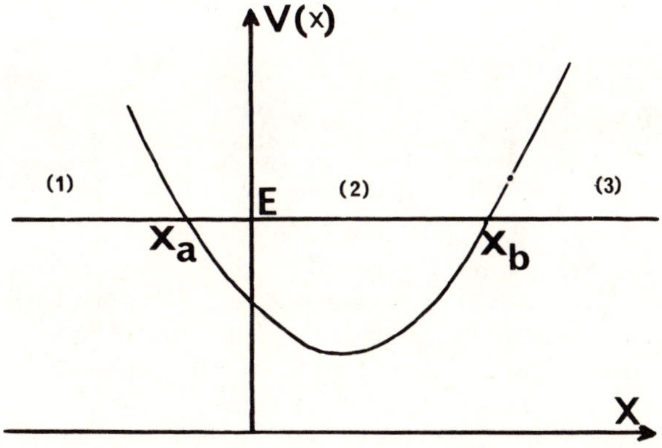

Figure 86. Diagram of a potential well indicating the three regions used in the BWK method for an energy E.

We have left open the choice of the sign of the integral and thus that of the integration constant. Our first step is to note that the wave function must be normalized and must disappear at infinity and, in addition, that the wave functions being real, must give real solutions, since we are dealing with non-degenerate energy levels. This last condition requires that we take a combination of the two solutions (2) corresponding to the two possible signs, with the complex conjugate constants of the form $\gamma_2' + i\gamma_2''$ and $\gamma_2' - i\gamma_2''$ respectively. Our sum thus leads to the three following expressions for the wave function, according to the region:

$$(4) \qquad \psi_1 = \frac{e^{\Upsilon_1}}{[2m(V - E)]^{1/4}} \exp \left[\frac{1}{\hbar} \int_{x_a}^{x} \sqrt{2m(V - E)}\, dx\right]$$

$$(5) \qquad \psi_2 = \frac{2e^{\Upsilon_2'}}{[2m(E - V)]^{1/4}} \cos \left[\frac{1}{\hbar} \int_{x_a}^{x} \sqrt{2m(E - V)}\, dx + \Upsilon_2''\right]$$

$$(6) \qquad \psi_3 = \frac{e^{\Upsilon_3}}{[2m(V - E)]^{1/4}} \exp \left[-\frac{1}{\hbar} \int_{x_b}^{x} \sqrt{2m(V - E)}\, dx\right].$$

What remains is to establish the relationship between the four real constants:

$$\Upsilon_1, \Upsilon_2', \Upsilon_2'', \Upsilon_3.$$

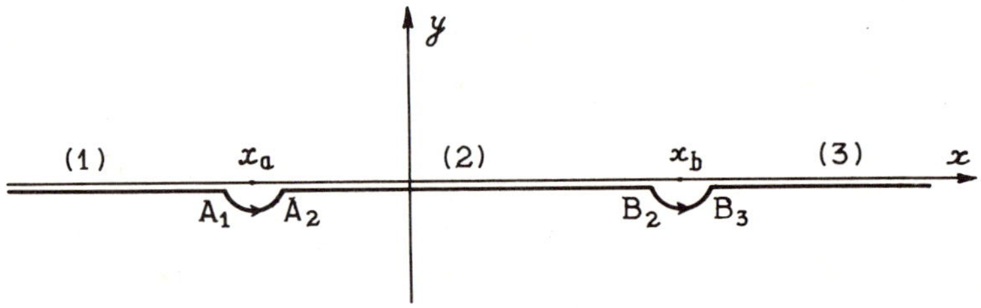

Figure 87. Complex contour taken about the junction points x_a and x_b.

We have shown in section 25 that the normal procedure of writing the boundary conditions for $x = x_a$ and $x = x_b$ is impractical. We overcome this difficulty by suggesting that the solution must be an analytic function that must be extended on a path in the complex plane (fig. 87). Writing the boundary conditions for this becomes possible by considering the critical points in the solution. We call φ_1, φ_2, and φ_3, the three functions derived respectively from ψ_1, ψ_2, and ψ_3, over the path. We can take $\varphi_1 = \psi_1$, and $\varphi_3 = \psi_3$. On the other hand, the two partial complex conjugates forming ψ_2 behave differently. For a point below the real axis, we can, in effect, write:

$$\int_{x_a}^{z = x - iy} \sqrt{2m(E - V)}\, dx = \int_{x_a}^{x} \sqrt{2m(E - V)}\, dx + \int_{x}^{x - iy} \sqrt{2m(E - V)}\, dx$$

so that:

$$\frac{1}{\hbar} \int_{x_a}^{z} \sqrt{2m(E - V)}\, dx \simeq \frac{1}{\hbar} \int_{x_a}^{x} \sqrt{2m(E - V)}\, dx - i \frac{y}{\hbar} \sqrt{2m(E - V)}.$$

For $y > 0$, which corresponds to an expected path, one of the two exponentials in the expansion of $\cos[(i/\hbar)\int_{x_a}^{x}\sqrt{2m(E-V)}\,dx + \gamma_2'']$ become negligible, with the result that we must take, with $\gamma_2 = \gamma_2' + i\gamma_2''$:

$$\varphi_2 = \frac{e^{\gamma_2'}}{[2m(E-V)]^{1/4}}\exp\left[\frac{i}{\hbar}\int_{x_a}^{x}\sqrt{2m(E-V)}\,dx + i\gamma_2''\right]$$

or its logarithmic form:

$$\text{Log}\ \varphi_2 = \frac{i}{\hbar}\int_{x_a}^{x}\sqrt{2m(E-V)}\,dx - \frac{1}{4}\text{Log}\,[2m(E-V)] + \gamma_2.$$

We will now write the boundary conditions of the three solutions φ_1, φ_2, and φ_3, respectively, in the neighborhood of x_a and x_b.

Boundary of φ_1 and φ_2. The integral disappears for these two functions, with the result that only the logarithmic term is taken. Near x_a where $E = V(x_a)$ we can write, by taking $(dV/dx)_{x_a} = V_a'$:

$$2m(V(z) - E) \simeq 2mV_a'(z - x_a).$$

To study the behavior along the arc about x_a, we take:

$$z = x_a + \rho e^{i\theta},$$

and taking account of the negative sign of V_a', we write:

$$\text{Log}\,[2m(V(z) - E)] \simeq \text{Log}\,(-2mV_a'\rho) + i(\theta - \pi).$$

We obtain at A_2 $(\theta = 2\pi)$:

$$(\text{Log}\ \varphi_1)_{x_a+\rho} = -\frac{1}{4}\text{Log}\,(-2mV_a'\rho) - \frac{i\pi}{4} + \gamma_1$$

and at the same point we have:

$$(\text{Log}\ \varphi_2)_{x_a+\rho} = -\frac{1}{4}\text{Log}\,(-2mV_a'\rho) + \gamma_2$$

hence the boundary condition:

$$\gamma_2 = \gamma_1 - \frac{i\pi}{4}.$$

Boundary of φ_2 and φ_3. The method of calculation is the same, but it is necessary to note that the integral in φ_2 does not vanish at x_b. We thus have at B_3:

$$(\text{Log}\ \varphi_2)_{x_b+\rho} \simeq \frac{i}{\hbar}\int_{x_a}^{x_b}\sqrt{2m(E-V)}\,dx - \frac{1}{4}\text{Log}\,(2mV_b'\rho) - \frac{i\pi}{4} + \gamma_2$$

$$(\text{Log}\ \varphi_3)_{x_b+\rho} = \frac{1}{4}\text{Log}\,(2mV_b'\rho) + \gamma_3$$

hence the boundary condition:

$$\gamma_3 = \gamma_2 - \frac{i\pi}{4} + \frac{i}{\hbar}\int_{x_a}^{x_b}\sqrt{2m(E-V)}\,dx$$

$$= \gamma_1 - \frac{i\pi}{2} + \frac{i}{\hbar}\int_{x_a}^{x_b}\sqrt{2m(E-V)}\,dx$$

γ_3 and γ_1 are real, so that the preceding condition implies that their real parts must be the same; their imaginary parts can differ by $r\pi i$ with r integral. Thus we get the condition:

$$(7) \qquad \gamma_3 = \gamma_1 + r\pi i, \quad \text{with} \quad \left(r + \frac{1}{2}\right)h = 2\int_{x_a}^{x_b} \sqrt{2m(\mathrm{E} - \mathrm{V})}\, dx.$$

This is the quantum condition indicated without proof (41) in section 25. Finally we give the form of the wave functions (4), (5) and (6) in terms of the single parameter $\mathrm{C} = e^{\gamma_1}$ which is fixed by normalization:

$$(8) \qquad \psi_1 = \frac{\mathrm{C}}{[2m(\mathrm{V} - \mathrm{E})]^{1/4}} \exp\left[\frac{1}{\hbar}\int_{x_a}^{x_b} \sqrt{2m(\mathrm{V} - \mathrm{E})}\, dx\right]$$

$$(9) \qquad \psi_2 = \frac{2\mathrm{C}}{[2m(\mathrm{E} - \mathrm{V})]^{1/4}} \cos\left[\frac{1}{\hbar}\int_{x_a}^{x_b} \sqrt{2m(\mathrm{E} - \mathrm{V})}\, dx - \frac{\pi}{4}\right]$$

$$(10) \qquad \psi_3 = \frac{(-1)^r\mathrm{C}}{[2m(\mathrm{V} - \mathrm{E})]^{1/4}} \exp\left[-\frac{1}{\hbar}\int_{x_a}^{x_b} \sqrt{2m(\mathrm{V} - \mathrm{E})}\, dx\right]$$

with the quantum condition†:

$$\left(r + \frac{1}{2}\right)h = 2\int_{x_a}^{x_b} \sqrt{2m(\mathrm{E} - \mathrm{V})}\, dx.$$

2 The Potential Barrier

The preceding method can be applied to the study of the crossing of a potential barrier and to calculation of the transmission coefficient (§8). The BWK approximation again permits the determination of the approximate form of the wave function in the three regions as defined in figure 88, so that:

$$(11) \qquad \psi_1 = \frac{1}{[2m(\mathrm{E} - \mathrm{V})]^{1/4}}\left[\mathrm{C}_1 \exp\left(\frac{i}{\hbar}\int_{x_a}^{x} \sqrt{2m(\mathrm{E} - \mathrm{V})}\, dx\right) + \mathrm{C}_1' \exp\left(-\frac{i}{\hbar}\int_{x_a}^{x} \sqrt{2m(\mathrm{E} - \mathrm{V})}\, dx\right)\right]$$

$$(12) \qquad \psi_2 = \frac{1}{[2m(\mathrm{V} - \mathrm{E})]^{1/4}}\left[\mathrm{C}_2 \exp\left(\frac{1}{\hbar}\int_{x_a}^{x} \sqrt{2m(\mathrm{V} - \mathrm{E})}\, dx\right) + \mathrm{C}_2' \exp\left(-\frac{1}{\hbar}\int_{x_a}^{x} \sqrt{2m(\mathrm{V} - \mathrm{E})}\, dx\right)\right]$$

$$(13) \qquad \psi_3 = \frac{1}{[2m(\mathrm{E} - \mathrm{V})]^{1/4}}\left[\mathrm{C}_3 \exp\left(\frac{i}{\hbar}\int_{x_b}^{x} \sqrt{2m(\mathrm{E} - \mathrm{V})}\, dx\right) + \mathrm{C}_3' \exp\left(-\frac{i}{\hbar}\int_{x_b}^{x} \sqrt{2m(\mathrm{E} - \mathrm{V})}\, dx\right)\right]$$

† This quantum condition can be obtained directly by using the preceding integral on a closed contour in the complex plane, taking one side or the other of the segment $x_a x_b$, and joining the ends (R. Mazet, *J. Mathem.*, XXIII, 4 (1944)).

We assume that the particles approach the barrier only from the left, which makes $C_3' = 0$.

We are again going to find solutions φ_1, φ_2, and φ_3, gotten by projection on a convenient complex path.

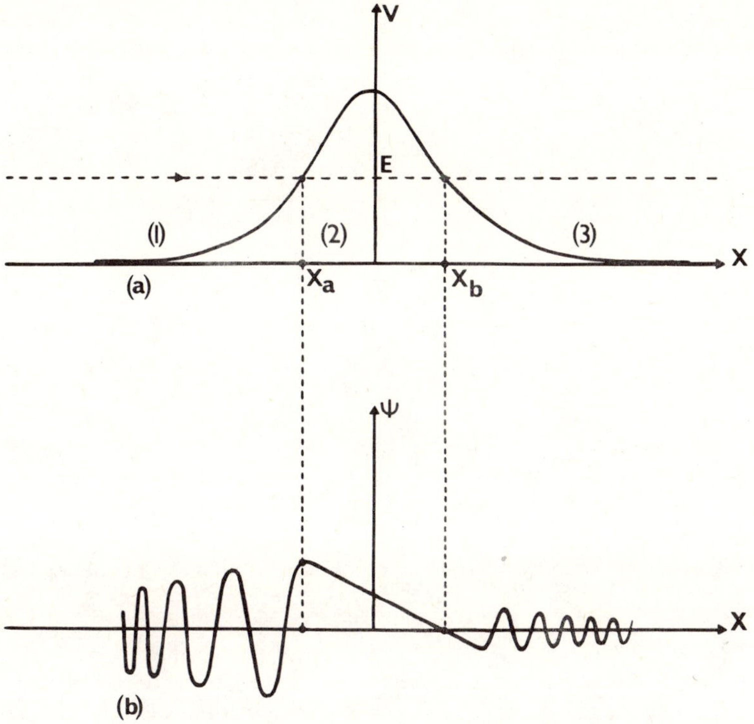

Figure 88. (*a*) Energy diagram for the crossing of a potential barrier. (*b*) Behavior of the real part of the continuous solution ψ which beyond the boundary points is given by the wave functions ψ_1, ψ_2, and ψ_3.

It is preferable to begin the argument with ψ_3, which depends on only one parameter, so that we can take $\varphi_3 = \psi_3$ in its logarithmic form with $C_3 = e^{\gamma_3}$:

$$\text{(14)} \qquad \operatorname{Log} \varphi_3 = \frac{i}{\hbar} \int_{x_b}^{x} \sqrt{2m(E - V)}\, dx - \frac{1}{4} \operatorname{Log} [2m(E - V)] + \gamma_3.$$

The correspondence between φ_r and ψ_r permits approximations which do not arise in the case of the potential well, so that we are committed to slightly modifying the line of reasoning by first studying the boundary condition at x_b, which will permit us to anticipate the form of φ_2.

Boundary at x_b. This point can be passed by a semicircle in the complex plane either above or below the real axis, but the results obtained are not the same in both cases. We get an acceptable solution† from the first path (fig. 89). The expansion limit of $(i/\hbar)\sqrt{2m(E-V)}$ near x_b is written $(i/\hbar)\sqrt{-2mV'_b\rho}\,e^{i\theta/2+r\pi i}$. We can take $r=0$ to determine the coefficient of i for $\theta=0$ (B_3). If we take $\theta=\pi$, we get at B_2:

$$-\frac{1}{\hbar}\sqrt{-2mV'_b\rho},$$

so that the exponential in ψ_2 affects the coefficient C'_2 and corresponds to a wave damped in the positive sense of x. By assuming that we can neglect the other exponential (an approximation) we are lead to take:

$$\text{(15)}\qquad \text{Log}\,\varphi_2 = -\frac{1}{\hbar}\int_{x_a}^{x}\sqrt{2m(V-E)}\,dx - \frac{1}{4}\,\text{Log}\,[2m(V-E)] + \gamma'_2$$

with

$$C'_2 = e^{\gamma'_2}.$$

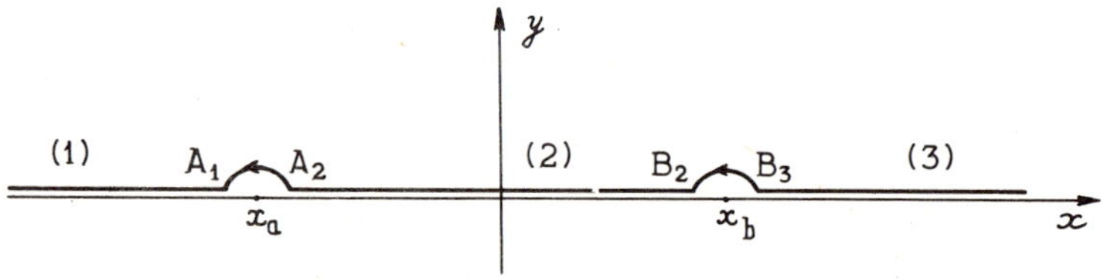

Figure 89. Complex contour taken about the junction points x_a and x_b.

The boundary condition at x_b, between φ_3 and φ_2, is written by explicitly following the reasoning from the case of the potential well:

$$\text{(16)}\qquad \gamma_3 - \frac{i\pi}{4} = -\frac{1}{\hbar}\int_{x_a}^{x_b}\sqrt{2m(V-E)}\,dx + \gamma'_2.$$

Boundary at x_a. Near this point, the expansion of $-(1/\hbar)\sqrt{2m(V-E)}$ is written:

$$-\frac{1}{\hbar}\sqrt{2mV'_a\rho}\,e^{\frac{i\theta}{2}},$$

which gives the desired value at A_2 ($\theta=0$). We pass to point A_1 ($\theta=\pi$) and we get $-(i/\hbar)\sqrt{2mV'_a\rho}$ corresponding to the reflected wave of ψ. Note that, if we depart from the real axis in the direction $y>0$, it is exactly

† The solution obtained from the case corresponding to particles striking the potential barrier coming from $-\infty$. The symmetric path situated in the complex plane below xx' corresponds to the arrival of particles on the other side of the barrier and, as a result, to a wave damped in the inverse sense in region 2.

this exponential which becomes predominant, so that we are justified in showing that the extension of ψ_1 is φ_1, such that:

$$\text{(17)} \qquad \text{Log } \varphi_1 = -\frac{i}{\hbar}\int_{x_a}^{x} \sqrt{2m(\mathrm{E}-\mathrm{V})}\, dx - \frac{1}{4}\text{Log }[2m(\mathrm{E}-\mathrm{V})] + \gamma_1'$$

with

$$\mathrm{C}_1' = e^{\gamma_1'}.$$

The boundary condition of φ_2 and φ_1 at A_1 is written:

$$\text{(18)} \qquad \gamma_2' - \frac{i\pi}{4} = \gamma_1'.$$

Equations (16) and (18) lead by definition to:

$$\gamma_3 = \gamma_1' + \frac{i\pi}{2} - \frac{1}{\hbar}\int_{x_a}^{x_b} \sqrt{2m(\mathrm{V}-\mathrm{E})}\, dx.$$

We get as the ratio of the wave amplitudes:

$$\left|\frac{\mathrm{C}_3}{\mathrm{C}_1'}\right| = \left|e^{\gamma_3-\gamma_1'}\right| = \exp\left[-\frac{1}{\hbar}\int_{x_a}^{x_b} \sqrt{2m(\mathrm{V}-\mathrm{E})}\, dx\right] = k.$$

The conservation of the number of particles in the incident (C_1), reflected (C_1'), and transmitted (C_3) beams is written:

$$\left|\mathrm{C}_1^2\right| = \left|\mathrm{C}_1'^2\right| + \left|\mathrm{C}_3^2\right|$$

and we get the transmission coefficient:

$$\text{(19)} \qquad \tau = \left|\frac{\mathrm{C}_3^2}{\mathrm{C}_1^2}\right| = \frac{k^2}{1+k^2} \simeq \exp\left[-\frac{2}{\hbar}\int_{x_a}^{x_b} \sqrt{2m(\mathrm{V}-\mathrm{E})}\, dx\right].$$

It is possible to make a simplification in the tunnel effect problem by neglecting one of the damped waves inside the potential barrier. Also, in the case of a rectangular barrier, we do not get precisely equation (19) of section 8, but the coefficient which is introduced there, $16\mathrm{E}(\mathrm{V}_0 - \mathrm{E})/\mathrm{V}_0^2$, is of the order of magnitude of unity, and the predominant term is the exponential factor whose magnitude is exactly that obtained in the more complete treatment.

GENERAL BIBLIOGRAPHY

Coulson, C. A. *Valence.* New York: Oxford University Press, 1961.

Daudel, R., Lefevre, R. and Moser, C. *Quantum Chemistry.* New York: Interscience Publishers, 1959.

Eyring, H., Walter, J. and Kimball, G. E. *Quantum Chemistry.* New York: John Wiley, 1947.

Herzberg, G. *Spectra of Diatomic Molecules.* New York: D. Van Nostrand, 1951.

Kemble, E. C. *The Fundamental Principles of Quantum Mechanics.* New York: McGraw-Hill, 1937.

Messiah, A. *Quantum Mechanics.* New York: John Wiley, 1962.

Pauling, L., and Wilson, E. B. *Introduction to Quantum Mechanics,* New York: McGraw-Hill, 1939.

Pitzer, K. S. *Quantum Chemistry.* New York: Prentice Hall, 1953.

Schiff, L. I. *Quantum Mechanics.* 3rd ed. New York: McGraw-Hill, 1968.

Slater, J. C. *Quantum Theory of Atomic Structure.* Vols. 1 and 2. New York: McGraw-Hill, 1960.

INDEX